Valuations, Orderings, and Milnor K-Theory

Mathematical
Surveys
and
Monographs

Volume 124

Valuations, Orderings, and Milnor *K*-Theory

Ido Efrat

American Mathematical Society

2000 *Mathematics Subject Classification*. Primary 12J10, 12J15;
Secondary 12E30, 12J20, 19F99.

For additional information and updates on this book, visit
www.ams.org/bookpages/surv-124

Library of Congress Cataloging-in-Publication Data

Efrat, Ido, 1963–
Valuations, orderings, and Milnor K-theory / Ido Efrat.
p. cm. — (Mathematical surveys and monographs, ISSN 0076-5376 ; v. 124)
Includes bibliographical references and index.
ISBN 0-8218-4041-X (alk. paper)
1. Valuation theory. 2. Ordered fields. 3. K-theory I. Title. II. Mathematical surveys and monographs ; no. 124.

QA247.E3835 2006
515′.78–dc22 2005057091

∞ The paper used in this book is acid-free and falls within the guidelines
established to ensure permanence and durability.
Visit the AMS home page at http://www.ams.org/

10 9 8 7 6 5 4 3 2 1   11 10 09 08 07 06

Contents

Introduction

The *fundamental theorem of arithmetic* describes the structure of the multiplicative group $\mathbb{Q}^\times$ of the field $\mathbb{Q}$ of rational numbers as a direct sum

$$\mathbb{Q}^\times \cong (\mathbb{Z}/2) \oplus \bigoplus_{p \text{ prime}} \mathbb{Z}.$$

Namely, a non-zero rational number a has a unique decomposition $a = \pm \prod_p p^{v_p(a)}$, where the exponents $v_p(a)$ are integers and are zero for all but finitely many primes p. This very basic fact brings together the three main objects studied in this book: *multiplicative groups of fields*, *valuations*, and *orderings*. In fact, as we shall see later on, the maps v_p are all non-trivial valuations on $\mathbb{Q}$, and the $\pm$ sign corresponds to its unique ordering.

The attempts to generalize the fundamental theorem of arithmetic to arbitrary number fields F led to the creation of algebraic number theory. Of course, to make such a generalization possible, one had to modify the mathematical language used. The right generalization of both the notion of a prime number as well as of the $\pm$ sign turned out to be that of an *absolute value*: a map $|\cdot|$ from F to the non-negative real numbers such that $|x| = 0$ if and only if $x = 0$, and such that

$$|x \cdot y| = |x| \cdot |y| \quad \text{and} \quad |x + y| \leq |x| + |y|$$

for all x, y in F. For instance, on $\mathbb{Q}$ the usual ordering gives an absolute value $|\cdot|_\infty$ in the standard way, and each map v_p as above gives the *p-adic* absolute value $|x|_p = 1/p^{v_p(x)}$. For the p-adic absolute value $|\cdot| = |\cdot|_p$ the triangle inequality can be strengthened to the so-called *ultrametric inequality*

$$|x + y| \leq \max\{|x|, |y|\}.$$

Absolute values having this stronger property are called *non-Archimedean*, the rest being referred to as *Archimedean*. Using these concepts it was possible to develop one of the most beautiful branches of algebraic number theory: the so-called *ramification theory*, which describes the behavior of absolute values under field extensions, and especially their reflection in Galois groups.

At this point, it was natural to ask for a generalization of this theory to arbitrary fields F. Unfortunately, the notion of an absolute value, which was satisfactory in the number field case, is inadequate in general, so better concepts had to be found. The right substitute for the notion of an Archimedean absolute value has been systematically developed by E. Artin and O. Schreier in the late 1920s ([**Ar**], [**AS1**], [**AS2**]), following an earlier work by Hilbert: this is the notion of an *ordering* on F, i.e., an additively closed subgroup P of the multiplicative group $F^\times$ of F (standing for the set of "positive" elements) such that $F^\times = P \cup -P$.

The proper definition in the non-Archimedean case is more subtle, and was introduced by W. Krull in his landmark 1931 paper [**Kru2**]. Roughly speaking, instead of looking at the absolute value $|\cdot|$ itself, Krull focused on the group homomorphism $v = -\log|\cdot|: F^\times \to \mathbb{R}$. Of course, this minor modification cannot change much, and is still insufficient for general fields. However, Krull's conceptual breakthrough was to replace the additive group $\mathbb{R}$ by an arbitrary *ordered abelian group* $(\Gamma, \leq)$. Thus what we now call a *Krull valuation* on the field F is a group homomorphism $v: F^\times \to \Gamma$, where $(\Gamma, \leq)$ is an ordered abelian group, which satisfies the following variant of the ultrametric inequality:

$$v(x+y) \geq \min\{v(x), v(y)\}$$

for $x \neq -y$.

Krull's seminal work [**Kru2**] paved the way to modern valuation theory. Starting from this definition, he introduced some of the other key ingredients of the theory: valuation rings, the analysis of their ideals, the convex subgroups of $(\Gamma, \leq)$, and the connections between all these objects and coarsenings of valuations. He adapted for his general setting the (already existent) notions of decomposition, inertia, and ramification subgroups of Galois groups over F. Furthermore, he studied maximality properties of valued fields with respect to field extensions. In a somewhat more implicit way he also studied a notion which will later on become central in valuation theory, namely, *Henselian* valued fields (although he does not give it a name). This notion turned out to be the right algebraic substitute in the setup of Krull valuations for the topological property of completeness. It is analogous to the notion of a *real closed field* introduced by Artin and Schreier in the context of ordered fields. The term "Henselian" is in honor of K. Hensel, who discovered the field $\mathbb{Q}_p$ of p-adic numbers, and proved (of course, under a different terminology) that its canonical valuation is Henselian [**He**]. We refer to [**Ro**] for a comprehensive study of the early (pre-Krull) history of valuation theory.

The classical theory of valuations from the point of view of Krull and his followers is well presented in the already classical books by O. Endler [**En**], P. Ribenboim [**Ri1**], and O.F.G. Schilling [**Schi**]. Yet, over the decades that elapsed since the publication of these books, valuation theory went through several conceptual developments, which we have tried to present in this monograph.

First, the different definitions in the Archimedean and non-Archimedean cases caused a split of the unified theory into two separate branches of field arithmetic: the theory of ordered fields on one hand, and valuation theory on the other hand. While Krull still keeps in [**Kru2**] a relatively unified approach (at least to the extent possible), later expositions on general valuation theory have somewhat abandoned the connections with orderings. Fortunately, the intensive work done starting in the 1970s on ordered fields and quadratic forms (which later evolved into real algebraic geometry) revived the interest in this connection, and led to a reintegration of these two sub-theories. T.Y. Lam's book [**Lam2**] beautifully describes this interplay between orderings and valuations from the more restrictive viewpoint of the reduced theory of quadratic forms, i.e., quadratic forms modulo a preordering (see also [**Lam1**] and [**Jr**]). In the present book we adopt this approach in general, and whenever possible study orderings and valuations jointly, under the common name *localities*.

Second, starting already from Krull's paper [**Kru2**], the emphasis in valuation theory has been on its Galois-theoretic aspects. These will be discussed in detail in

Part III of the book. However, by their mere definitions, valuations and orderings are primarily related to the multiplicative group $F^\times$ of the field F, and much can be said when studying them in this context. This approach has become dominant in the ordered field case (as in [**Lam2**]). However, it is our feeling that in the valuation case this viewpoint has been somewhat neglected in favor of the Galois-theoretic one. Therefore, in addition to presenting the classical theory of Galois groups of valued field extensions, we devote several sections (in Parts II and IV of the book) to developing the theory with emphasis on subgroups S of $F^\times$. In particular, we focus on valuations satisfying a natural condition called *S-compatibility*, which is the analog of Henselity in the multiplicative group context.

Part IV takes this approach one step further, and studies the Milnor K-theory of valued and ordered fields F. We recall that the Milnor K-group of F of degree r is just the tensor product $F^\times \otimes_{\mathbb{Z}} \cdots \otimes_{\mathbb{Z}} F^\times$ (r times) modulo the simple relations $a_1 \otimes \cdots \otimes a_r = 0$ whenever $a_i + a_j = 1$ for some $i < j$. Several important results (or conjectures) in arithmetic geometry indicate that there should be some kind of parallelism between Milnor's K-theory and Galois theory of fields. For instance, the *Bloch-Kato conjecture* predicts a canonical isomorphism between $K_r^M(F)/n$ and the Galois cohomology group $H^r(F, \mu_n^{\otimes r})$ (where $r \geq 0$ and $n \geq 1$ are integers with $\operatorname{char} F \nmid n$, and the cohomology is with respect to the r-times twisted cyclotomic action); this has been proved in several important cases by A.S. Merkurjev, A.A. Suslin, M. Rost, V. Voevodsky, and others (see §24.3). It is therefore not surprising that large parts of the Galois theory of valued and ordered fields have analogs in this natural framework of Milnor's K-theory. These analogs will be presented in Part IV. In some sense, this shift of viewpoint resembles the introduction of the K-theoretic approach to higher class field theory, complementing the earlier Galois-theoretic approach (see [**FV**, Appendix B] and [**FK**]).

Finally, there has been much interest lately in construction of non-trivial valuations on fields. Such constructions emerged in the context of ordered fields (in particular, L. Bröcker's *"trivialization of fans"* theorem [**Br1**]), and later in an elementary and explicit way by B. Jacob, R. Ware, J.K. Arason, R. Elman, and Y.S. Hwang ([**J1**], [**War2**], [**AEJ**], [**HwJ**]). Such constructions became especially important in recent years in connection with the so-called *birational anabelian geometry*. This line of research originated from ideas of A. Grothendieck ([**G1**], [**G2**]) as well as from works of J. Neukirch ([**N1**], [**N2**]). Here one wants to recover the arithmetic structure of a field (if possible, up to an isomorphism) from its various canonical Galois groups. The point is that usually the first step is to recover enough valuations from their cohomological (or K-theoretic) "footprints"; see, e.g., [**BoT**], [**Ef1**], [**Ef7**], [**EfF**], [**NSW**, Ch. XII], [**P1**], [**P2**], [**P3**], [**Sp**], [**Sz**] for more details. In §11 we give a new presentation of the above-mentioned line of elementary constructions, based on the coarsening relation among valuations. While these constructions were considered for some time to be somewhat mysterious, they fit very naturally into the multiplicative group approach as discussed above, especially when one uses the K-theoretic language. In §26 we use this language to prove the main criterion for the existence of "optimal" valuations, as is required in the applications to the birational anabelian geometry. This is further related to the notion of *fans* in the theory of ordered fields, thus closing this fruitful circle of ideas that began with [**Br1**].

The prerequisites of this book are quite minimal. We assume a good algebraic knowledge at a beginning graduate level, including of course familiarity with general field theory and Galois theory. The generalization of finite Galois theory to infinite normal extensions is reviewed for the reader's convenience in §13. Likewise we develop the basic facts and formalisms of Milnor's K-theory in §§23–24 in order not to assume any prior knowledge in this area. On the other hand, we do assume familiarity with the language of homological algebra (exact sequences, commutative diagrams, direct and inverse limits, etc.). The presentation is mostly self-contained, and only very few facts are mentioned without proofs: the "snake lemma" and some basic properties of flatness in §1.1, the structure theory of finitely generated modules over a principal ideal domain and the Nakayama lemma in §17.4, short cohomological discussions in §22.2, §24.3 and Remark 25.1.7, and some facts from local class field theory in §27.1.

Unlike most existing texts on valuation theory, we chose not to develop the theory using commutative algebra machinery, but rather to use the machinery of abelian groups. This simplifies the presentation in many respects. The required results about abelian groups (and in particular ordered abelian groups) are developed in Part I of the book.

Needless to say, we have not pretended to fully describe here the vast research work done on valued and ordered fields throughout the twentieth century and which still goes on today. The choice of material reflects only the author's personal taste (and even more so, his limitations). More material can be found in the texts by Ax [**Ax**], Bourbaki [**Bou1**], Endler [**En**], Jarden [**Jr**], Ribenboim ([**Ri1**], [**Ri3**]), Schilling [**Schi**], and Zariski and Samuel [**ZS**] on valuation theory, as well as those by Knebusch and Scheiderer [**KnS**], Lam ([**Lam1**], [**Lam2**]), Prestel [**Pr**] and Scharlau [**Sch2**] on ordered fields. Likewise, the reference list at the end of this monograph surely covers only a small portion of the possible bibliography. Other and more comprehensive lists of references on valuation theory can be found in [**FV**], [**Ro**], and at the Valuation Theory internet site at http://math.usask.ca/fvk/Valth.html. A comprehensive bibliography on the work done until 1980 on ordered fields is given in [**Lam1**].

I thank Eli Shamovich as well as the anonymous referees for their very valuable comments on previous versions of this manuscript.

This book was typeset using $\mathcal{AMS}$-TeX, the TeX macro system of the American Mathematical Society.

Be'er-Sheva 2005 *I.E.*

Conventions

The image, kernel, and cokernel of a group homomorphism $f\colon A \to B$ will be denoted as usual by $\mathrm{Im}(f)$, $\mathrm{Ker}(f)$, $\mathrm{Coker}(f)$, respectively. Thus $\mathrm{Coker}(f) = B/\mathrm{Im}(f)$. Given an abelian group A and a positive integer n, we denote the image, kernel, and cokernel of the homomorphism $A \to A$ of multiplication by n by nA, ${}_nA$, and A/n, respectively.

For a prime number p we set $\mathbb{Z}_p = \varprojlim \mathbb{Z}/p^i$. Likewise, we set $\hat{\mathbb{Z}} = \varprojlim \mathbb{Z}/n$, where n ranges over all positive integers, and the inverse limit is with respect to the divisibility relation.

Unless explicitly stated otherwise, all rings will be tacitly assumed to be commutative with 1, and all modules two-sided (an important exception will be the κ-structures, discussed in Part IV, which are *anti*-commutative rings). The group of invertible elements in a ring R will be denoted by $R^\times$. In particular, the multiplicative group $F \setminus \{0\}$ of a field F will be denoted by $F^\times$. A grading on a ring will always be by the nonnegative integers.

Given a subset A of a group, we denote the subgroup it generates by $\langle A\rangle$. The notation $B \leq A$ will mean that B is a subgroup of the group A.

Given a subsets A, B of a field F and an element c of F we set

$$A \pm B = \{a \pm b \mid a \in A,\ b \in B\}, \quad AB = \{ab \mid a \in A,\ b \in B\}$$

$$-A = \{-a \mid a \in A\}, \quad cA = \{ca \mid a \in A\},$$

etc.

We denote the fixed field of a group G of automorphisms of a field E by E^G. If a is an element of some field extension of E and is algebraic over E, then we denote its irreducible polynomial over E by $\mathrm{irr}(a, E)$. An extension $F \subseteq E$ of fields will be written as E/F, and its transcendence degree will be denoted by $\mathrm{tr.deg}(E/F)$.

Part I

ABELIAN GROUPS

By their definition, valuations are homomorphisms from the multiplicative group of a field into ordered abelian groups (subject to the ultrametric inequality). It is therefore not surprising that many results about valuations reduce to general facts about the category of ordered abelian groups, or even that of torsion-free abelian groups. Of course, the latter categories have a simpler structure than that of valued fields, and therefore allow simpler proofs. For this reason we develop in this part of the book several themes from the theory of abelian groups which will ultimately be applied in field-theoretic considerations, but do so in a purely group-theoretic manner. For the general theory of abelian groups see [**Fu**] or [**Kap2**].

CHAPTER 1

Preliminaries on Abelian Groups

1.1 General facts

An abelian group Γ is called **divisible** (resp., **uniquely divisible**, **torsion-free**) if for every positive integer n the homomorphism $\Gamma \to \Gamma$, $\alpha \mapsto n\alpha$, is surjective (resp., bijective, injective).

Let Γ_{tor} be the torsion subgroup of the abelian group Γ, i.e., the subgroup of all elements of Γ of finite order. Note that $\Gamma/\Gamma_{\mathrm{tor}}$ is torsion-free. One says that Γ is a **torsion group** if $\Gamma = \Gamma_{\mathrm{tor}}$. The **$p$-primary subgroup** Γ_p of Γ, where p is a prime number, is the subgroup of Γ consisting of all elements of p-power order. One says that Γ is **p-primary** if $\Gamma = \Gamma_p$. The fundamental theorem of arithmetic gives the following **primary decomposition** for the torsion part of an abelian group:

LEMMA 1.1.1. *For an abelian group Γ one has $\Gamma_{\mathrm{tor}} = \bigoplus_p \Gamma_p$.*

A homomorphism $\Gamma_1 \to \Gamma_2$ of abelian groups Γ_1, Γ_2, both containing the abelian group Γ, will be called a **Γ-homomorphism** if it is the identity map on Γ.

LEMMA 1.1.2. *Let $\Gamma, \Gamma_1, \Gamma_2$ be abelian groups such that $\Gamma \le \Gamma_1, \Gamma_2$ and let $\varphi\colon \Gamma_1 \to \Gamma_2$ be a Γ-homomorphism.*

(a) *If Γ_1 is torsion-free and Γ_1/Γ is a torsion group, then φ is injective.*

(b) *If Γ_1 is divisible, Γ_2 is torsion-free, and Γ_2/Γ is a torsion group, then φ is surjective.*

(c) *If Γ_1/Γ is a torsion group, Γ_2 is torsion-free, and $\Gamma_1 \le \Gamma_2$, then $\varphi = \mathrm{Id}_{\Gamma_1}$.*

PROOF. (a) Suppose that $\alpha \in \mathrm{Ker}(\varphi)$. Since Γ_1/Γ is a torsion group, there exists a positive integer n such that $n\alpha \in \Gamma$. Then

$$n\alpha = \varphi(n\alpha) = n\varphi(\alpha) = 0 \quad .$$

Since Γ_1 is torsion-free, $\alpha = 0$.

(b) Take $\beta \in \Gamma_2$. Since Γ_2/Γ is a torsion group there exists $n > 0$ such that $n\beta \in \Gamma \le \Gamma_1$. The divisibility of Γ_1 yields $\alpha \in \Gamma_1$ with $n\alpha = n\beta$. Then

$$n\varphi(\alpha) = \varphi(n\alpha) = \varphi(n\beta) = n\beta \quad .$$

Since Γ_2 is torsion-free, $\varphi(\alpha) = \beta$.

(c) Given $\alpha \in \Gamma_1$, we take a positive integer n such that $n\alpha \in \Gamma$. Then $n\alpha = \varphi(n\alpha) = n\varphi(\alpha)$. Since α and $\varphi(\alpha)$ are elements of the torsion-free group Γ_2, this implies that $\alpha = \varphi(\alpha)$. □

We recall the following useful homological fact. For the easy, but rather tedious, proof, see, e.g., [**Bou1**, Ch. I, §1.4, Prop. 2(iii)].

LEMMA 1.1.3 ("The Snake Lemma"). *A commutative diagram of abelian groups with exact rows*

$$\begin{array}{ccccccccc} 0 & \longrightarrow & A & \longrightarrow & B & \longrightarrow & C & \longrightarrow & 0 \\ & & \downarrow{\scriptstyle f} & & \downarrow{\scriptstyle g} & & \downarrow{\scriptstyle h} & & \\ 0 & \longrightarrow & A' & \longrightarrow & B' & \longrightarrow & C' & \longrightarrow & 0 \end{array}$$

induces an exact sequence

$$0 \to \mathrm{Ker}(f) \to \mathrm{Ker}(g) \to \mathrm{Ker}(h) \to \mathrm{Coker}(f) \to \mathrm{Coker}(g) \to \mathrm{Coker}(h) \to 0.$$

LEMMA 1.1.4. *Let Γ be a subgroup of an abelian group Γ'. Let m be a positive integer.*

(a) *There is a canonical exact sequence*

$$0 \to {}_m\Gamma \to {}_m\Gamma' \to {}_m(\Gamma'/\Gamma) \to \Gamma/m \to \Gamma'/m \to (\Gamma'/\Gamma)/m \to 0.$$

(b) *If ${}_m\Gamma' = 0$ and $\Gamma = m\Gamma$, then ${}_m(\Gamma'/\Gamma) = 0$.*

(c) *If the order of every element of Γ'/Γ is finite and prime to m, then the canonical map $\Gamma/m \to \Gamma'/m$ is an isomorphism.*

(d) *If $(\Gamma' : \Gamma) < \infty$ and ${}_m\Gamma' = 0$, then*

$$(\Gamma : m\Gamma) = (\Gamma' : m\Gamma') \quad .$$

PROOF. (a) Apply the snake lemma with respect to the following commutative diagram of multiplications by m:

$$\begin{array}{ccccccccc} 0 & \longrightarrow & \Gamma & \longrightarrow & \Gamma' & \longrightarrow & \Gamma'/\Gamma & \longrightarrow & 0 \\ & & \downarrow{\scriptstyle m} & & \downarrow{\scriptstyle m} & & \downarrow{\scriptstyle m} & & \\ 0 & \longrightarrow & \Gamma & \longrightarrow & \Gamma' & \longrightarrow & \Gamma'/\Gamma & \longrightarrow & 0. \end{array}$$

(b) This is immediate from (a).

(c) Here multiplication by m is an automorphism of Γ'/Γ. Now use (a).

(d) For a finite cyclic group Δ of order n, both ${}_m\Delta$ and Δ/m are cyclic of order $\gcd(n, m)$. Since the functors $\Delta \mapsto {}_m\Delta$ and $\Delta \mapsto \Delta/m$ commute with direct sums of abelian groups, we have $|{}_m\Delta| = |\Delta/m|$ for every finite abelian group Δ, and in particular for $\Delta = \Gamma'/\Gamma$. Hence, by (a), the kernel and cokernel of the map $\Gamma/m \to \Gamma'/m$ have the same finite size. Therefore Γ/m and Γ'/m have the same cardinality. □

Given a commutative ring R, an R-module M is **flat** if the functor $- \otimes_R M$ on R-modules is exact, i.e., whenever

$$N \to N' \to N''$$

is an exact sequence of R-modules, so is the induced sequence

$$N \otimes_R M \to N' \otimes_R M \to N'' \otimes_R M.$$

If $\varphi: R \to S$ is a ring homomorphism and M is an R-module, then $S \otimes_R M$ is an S-module in the obvious way. Furthermore, every S-module N can be considered as an R-module via φ, and one has

$$N \otimes_S (S \otimes_R M) \cong N \otimes_R M$$

canonically. It follows that if M is a flat R-module, then $S \otimes_R M$ is a flat S-module.

1.1.5 EXAMPLES.

(1) If R is a principal ideal domain and M is an R-module with trivial R-torsion, then M is flat [**Bou1**, Ch. I, §2.4].
(2) In particular, any torsion-free abelian group is a flat $\mathbb{Z}$-module.
(3) A free R-module is flat.
(4) Let R be a local ring with a nilpotent maximal ideal $\mathfrak{m}$, and let M be a flat R-module. Then M is free; furthermore, in this case any collection of elements of M which lifts an $R/\mathfrak{m}$-linear basis of $(R/\mathfrak{m}) \otimes_R M$ is an R-basis of M [**Mat**, 3.G].

LEMMA 1.1.6. *Let Γ be a torsion-free abelian group and let l be a prime number. There exist elements α_i, $i \in I$, of Γ such that:*

(i) $|I| = \dim_{\mathbb{F}_l}(\Gamma/l)$;
(ii) *for every positive integer r, the $\mathbb{Z}/l^r$-module Γ/l^r is free with basis $\alpha_i + l^r\Gamma$, $i \in I$.*

PROOF. Choose $\alpha_i \in \Gamma$, $i \in I$, such that the cosets $\alpha_i + l\Gamma$, $i \in I$, form an $\mathbb{F}_l$-linear basis of Γ/l. Then (i) is clear.

To see (ii) take $r \geq 1$. By Example 1.1.5(2), Γ is flat as a $\mathbb{Z}$-module. Therefore $\Gamma/l^r = (\mathbb{Z}/l^r) \otimes_{\mathbb{Z}} \Gamma$ is a flat $\mathbb{Z}/l^r$-module. Now $\mathbb{Z}/l^r$ is a local ring with a nilpotent maximal ideal $l\mathbb{Z}/l^r\mathbb{Z}$, so we may use Example 1.1.5(4). □

1.2 Divisible hulls

Let Γ be an abelian group and consider it as a $\mathbb{Z}$-module. We define the **divisible hull** of Γ to be

$$\Gamma_{\text{div}} = \Gamma \otimes_{\mathbb{Z}} \mathbb{Q} \quad .$$

The map $\Gamma \mapsto \Gamma_{\text{div}}$ is a covariant functor on the category of abelian groups. Since $\mathbb{Q}$ is a flat $\mathbb{Z}$-module (Example 1.1.5(2)), the functor is exact. By the general properties of the tensor product, this functor commutes with direct sums and direct limits. We obtain:

EXAMPLE 1.2.1. $(\mathbb{Z}^r)_{\text{div}} \cong \mathbb{Q}^r$ for $r \geq 0$.

By definition, every element of $\alpha \in \Gamma_{\text{div}}$ can be expressed as $\sum_{i=1}^k \beta_i \otimes r_i$ with $\beta_i \in \Gamma$ and $r_i \in \mathbb{Q}$. We may write $r_i = n_i/m$, where $n_1, \dots, n_k$ and $m > 0$ are integers, to obtain that $\alpha = (\sum_{i=1}^k n_i\beta_i) \otimes (1/m)$. Thus every element of Γ_{div} can be written as $\beta \otimes (1/m)$ for some $\beta \in \Gamma$ and some $m > 0$.

PROPOSITION 1.2.2. *Let Γ be an abelian group.*

(a) *If Γ is a torsion group, then $\Gamma_{\text{div}} = \{0\}$.*
(b) *$\Gamma_{\text{div}} \cong (\Gamma/\Gamma_{\text{tor}})_{\text{div}}$ canonically.*
(c) *Γ_{div} is uniquely divisible.*
(d) *Γ_{div} is torsion-free.*

PROOF. (a) Let $\alpha \in \Gamma_{\mathrm{div}}$ and express it as $\alpha = \beta \otimes (1/m)$ with $\beta \in \Gamma$ and $m > 0$. By assumption, there exists a positive integer n such that $n\beta = 0$. It follows that $\alpha = n\beta \otimes (1/nm) = 0$.

(b) Consider the exact sequence

$$0 \to \Gamma_{\mathrm{tor}} \to \Gamma \to \Gamma/\Gamma_{\mathrm{tor}} \to 0.$$

By (a), $(\Gamma_{\mathrm{tor}})_{\mathrm{div}} = \{0\}$. Now apply the exactness of the divisible hull functor.

(c) Let n be a positive integer and consider the exact sequences

$$0 \to {}_n\Gamma \to \Gamma \xrightarrow{n} n\Gamma \to 0 \quad , \quad 0 \to n\Gamma \hookrightarrow \Gamma \to \Gamma/n \to 0.$$

Since ${}_n\Gamma$ and Γ/n are torsion groups, they have trivial divisible hulls by (a). The exactness of the divisible hull functor therefore yields an isomorphism

$$\Gamma_{\mathrm{div}} \xrightarrow{n} (n\Gamma)_{\mathrm{div}} = \Gamma_{\mathrm{div}}.$$

(d) This is contained in (c). □

COROLLARY 1.2.3. *Let Γ be a subgroup of an abelian group Γ' such that Γ'/Γ is a torsion group. Then the functorial map $\Gamma_{\mathrm{div}} \to \Gamma'_{\mathrm{div}}$ is an isomorphism.*

PROOF. By Proposition 1.2.2(a), $(\Gamma'/\Gamma)_{\mathrm{div}} = \{0\}$. Now apply the exactness of the divisible hull functor with respect to the exact sequence

$$0 \to \Gamma \to \Gamma' \to \Gamma'/\Gamma \to 0.$$ □

For an abelian group Γ we define a group homomorphism

$$\iota_\Gamma \colon \Gamma \to \Gamma_{\mathrm{div}}, \qquad \alpha \mapsto \alpha \otimes 1.$$

Any homomorphism $\Gamma \to \Gamma'$ of abelian groups induces a commutative square

$$\begin{array}{ccc} \Gamma & \longrightarrow & \Gamma' \\ {\scriptstyle \iota_\Gamma}\downarrow & & \downarrow{\scriptstyle \iota_{\Gamma'}} \\ \Gamma_{\mathrm{div}} & \longrightarrow & \Gamma'_{\mathrm{div}}. \end{array} \tag{1.2.1}$$

PROPOSITION 1.2.4. *Let Γ be an abelian group.*
(a) *If Γ is torsion-free, then ι_Γ is injective.*
(b) *If Γ is divisible, then ι_Γ is surjective.*

PROOF. (a) When Γ is finitely generated, it is isomorphic to $\mathbb{Z}^r$ for some non-negative integer r. Then $\Gamma_{\mathrm{div}} \cong \mathbb{Q}^r$ and ι_Γ coincides with the embedding $\mathbb{Z}^r \hookrightarrow \mathbb{Q}^r$ (Example 1.2.1).

In the general case we express Γ as a direct limit $\varinjlim \Gamma_i$ of finitely generated subgroups Γ_i. The assertion then follows from the finitely generated case using the compatibility of the functor $\Gamma \mapsto \Gamma_{\mathrm{div}}$ with direct limits.

(b) Let $\alpha \in \Gamma_{\mathrm{div}}$. Write $\alpha = \beta \otimes (1/m)$ for some $\beta \in \Gamma$ and some $m > 0$. Choose $\gamma \in \Gamma$ such that $\beta = m\gamma$. Then

$$\alpha = m\gamma \otimes (1/m) = \gamma \otimes 1 = \iota_\Gamma(\gamma).$$ □

When Γ is torsion-free we will often use Proposition 1.2.4(a) to tacitly identify Γ as a subgroup of Γ_{div} via the monomorphism ι_Γ.

LEMMA 1.2.5. *Let Γ be an abelian group. Then:*
(a) $\mathrm{Ker}(\iota_\Gamma) = \Gamma_{\mathrm{tor}}$*;*
(b) $\mathrm{Coker}(\iota_\Gamma) = \Gamma_{\mathrm{div}}/\iota_\Gamma(\Gamma)$ *is a torsion group.*

PROOF. (a) Set $\bar\Gamma = \Gamma/\Gamma_{\mathrm{tor}}$. By Proposition 1.2.2(b), there is a commutative diagram with exact rows:

$$\begin{array}{ccccccccc}
0 & \longrightarrow & \Gamma_{\mathrm{tor}} & \longrightarrow & \Gamma & \longrightarrow & \bar\Gamma & \longrightarrow & 0 \\
 & & \big\downarrow & & \big\downarrow{\scriptstyle \iota_\Gamma} & & \big\downarrow{\scriptstyle \iota_{\bar\Gamma}} & & \\
0 & \longrightarrow & 0 & \longrightarrow & \Gamma_{\mathrm{div}} & \longrightarrow & \bar\Gamma_{\mathrm{div}} & \longrightarrow & 0.
\end{array}$$

Since $\bar\Gamma$ is torsion-free, $\iota_{\bar\Gamma}$ is injective (Proposition 1.2.4(a)). The assertion therefore follows from the snake lemma (Lemma 1.1.3).

(b) Take $\alpha \in \Gamma_{\mathrm{div}}$ and express it as before as $\alpha = \beta \otimes (1/m)$ with $\beta \in \Gamma$ and $m > 0$. Then $m\alpha = \beta \otimes 1 = \iota_\Gamma(\beta)$, as required. □

Our next theorem characterizes the divisible hull Γ_{div} of a torsion-free abelian group Γ by means of two universal properties. Roughly speaking, among the torsion-free abelian groups Γ' containing Γ, and up to a Γ-isomorphism, Γ_{div} is the maximal group for which Γ'/Γ is torsion. Also, it is the minimal such group which is divisible. This second universal property justifies the name "divisible hull" for Γ_{div}.

THEOREM 1.2.6. *Let Γ be a torsion-free abelian group and let Γ' be an abelian group containing Γ. Identify Γ as a subgroup of Γ_{div} via ι_Γ. The following conditions are equivalent:*

(a) Γ' *is Γ-isomorphic to* Γ_{div}.
(b)(b1) Γ' *is torsion-free and* Γ'/Γ *is a torsion group; and*
(b2) *for every torsion-free abelian group Γ'' containing Γ such that Γ''/Γ is a torsion group there exists a Γ-monomorphism $\Gamma'' \to \Gamma'$.*
(c)(c1) Γ' *is uniquely divisible; and*
(c2) *for every uniquely divisible abelian group Γ'' containing Γ there exists a Γ-monomorphism $\Gamma' \to \Gamma''$.*
(d) Γ' *is uniquely divisible and* Γ'/Γ *is a torsion group.*

PROOF. (a)⇒(b): Suppose that Γ' is Γ-isomorphic to Γ_{div}. Then (b1) follows from Proposition 1.2.2(d) and from Lemma 1.2.5(b).

For Γ'' as in (b2), we have the commutative square

$$\begin{array}{ccc}
\Gamma & \longrightarrow & \Gamma'' \\
{\scriptstyle \iota_\Gamma}\big\downarrow & & \big\downarrow{\scriptstyle \iota_{\Gamma''}} \\
\Gamma_{\mathrm{div}} & \longrightarrow & \Gamma''_{\mathrm{div}}.
\end{array} \tag{1.2.2}$$

By Proposition 1.2.4(a), $\iota_{\Gamma''}$ is injective. By Corollary 1.2.3, the lower horizontal map is an isomorphism. By composition, we obtain the desired Γ-monomorphism $\Gamma'' \to \Gamma_{\mathrm{div}} \cong \Gamma'$.

(b)$\Rightarrow$(a): As just observed, the assumptions of (b2) hold for $\Gamma''=\Gamma_{\rm div}$, so we get a Γ-monomorphism $\Gamma_{\rm div}\to\Gamma'$. By Lemma 1.1.2(b) and by (b1), it is surjective, whence a Γ-isomorphism.

(a)$\Rightarrow$(c): Suppose that Γ' is Γ-isomorphic to $\Gamma_{\rm div}$. Then (c1) was proved in Proposition 1.2.2(c).

For (c2) suppose that Γ'' is a uniquely divisible abelian group containing Γ, and consider again the commutative square (1.2.2). By Proposition 1.2.4, $\iota_{\Gamma''}\colon\Gamma''\to\Gamma''_{\rm div}$ is a Γ-isomorphism. Moreover, $\Gamma_{\rm div}$ is torsion-free (Proposition 1.2.2(d)) and $\Gamma_{\rm div}/\Gamma$ is a torsion group (Lemma 1.2.5(b)). The functorial map $\Gamma_{\rm div}\to\Gamma''_{\rm div}$ is a Γ-homomorphism. By Lemma 1.1.2(a) it is injective. Composing it with $\iota_{\Gamma''}^{-1}$ we obtain a Γ-monomorphism $\Gamma'\cong\Gamma_{\rm div}\to\Gamma''$ as required in (c2).

(c)$\Rightarrow$(a): Since $\Gamma_{\rm div}$ is uniquely divisible (Proposition 1.2.2(c)), we may take in (c2) $\Gamma''=\Gamma_{\rm div}$ to obtain a Γ-monomorphism $\Gamma'\to\Gamma_{\rm div}$. Hence Γ'/Γ embeds in $\Gamma_{\rm div}/\Gamma$. Since the latter group is torsion (Lemma 1.2.5(b)), so is the former. We conclude from Corollary 1.2.3 that the lower map in the square (1.2.1) is an isomorphism. By Proposition 1.2.4 and (c1), so is the right vertical map of (1.2.1). This gives the desired Γ-isomorphism $\Gamma'\cong\Gamma_{\rm div}$.

(a)$\Rightarrow$(d): This was shown in Proposition 1.2.2(c) and Lemma 1.2.5(b).

(d)$\Rightarrow$(a): We have already seen that $\Gamma_{\rm div}$ has the universal property (b). Since Γ' is torsion-free and Γ'/Γ is torsion, we therefore obtain a Γ-monomorphism $\Gamma'\to\Gamma_{\rm div}$. The divisibility of Γ' and Lemma 1.1.2(b) imply that this monomorphism is also surjective. □

Now let Γ be a torsion-free abelian group and identify it again as a subgroup of $\Gamma_{\rm div}$ via ι_Γ. For a positive integer n we write $\frac1n\Gamma$ for the preimage of ${}_n(\Gamma_{\rm div}/\Gamma)$ under the projection $\Gamma_{\rm div}\to\Gamma_{\rm div}/\Gamma$. In other words, $\frac1n\Gamma$ is the subgroup of $\Gamma_{\rm div}$ consisting of all elements α such that $n\alpha\in\Gamma$.

LEMMA 1.2.7. *For a torsion-free abelian group Γ and for a positive integer n one has $(\frac1n\Gamma)/\Gamma\cong\Gamma/n$ via multiplication by n.*

PROOF. There is a commutative diagram of multiplications by n:

$$\begin{array}{ccccccccc} 0 & \longrightarrow & \Gamma & \longrightarrow & \frac1n\Gamma & \longrightarrow & (\frac1n\Gamma)/\Gamma & \longrightarrow & 0\\ & & \downarrow{\scriptstyle n} & & \downarrow{\scriptstyle n} & & \downarrow{\scriptstyle n} & & \\ 0 & \longrightarrow & n\Gamma & \longrightarrow & \Gamma & \longrightarrow & \Gamma/n & \longrightarrow & 0. \end{array}$$

The left vertical map here is bijective. The middle vertical map is injective, since $\Gamma_{\rm div}$ is also torsion-free, and is surjective by the definition of $\frac1n\Gamma$. It follows from the snake lemma (Lemma 1.1.3) that the right vertical map is an isomorphism. □

For a prime number l we set

$$\tfrac{1}{l^\infty}\Gamma=\bigcup_{i=0}^{\infty}\tfrac{1}{l^i}\Gamma.$$

Observe that $(\Gamma_{\rm div}/\Gamma)_l=(\frac{1}{l^\infty}\Gamma)/\Gamma$. Hence we have the primary decomposition

$$\Gamma_{\rm div}/\Gamma=\bigoplus_l\bigl[(\tfrac{1}{l^\infty}\Gamma)/\Gamma\bigr],$$

with l ranging over all prime numbers (Lemma 1.1.1). We will need the following "local" form of the universal property (b) of Theorem 1.2.6.

LEMMA 1.2.8. *Let l be a prime number and let Γ be a subgroup of a torsion-free abelian group Γ'. The following conditions are equivalent:*
(a) *Γ' is isomorphic to $\frac{1}{l^\infty}\Gamma$ over Γ;*
(b) *Γ'/Γ is an l-primary torsion group and $\Gamma' = l\Gamma'$.*

PROOF. (a)$\Rightarrow$(b): Immediate.

(b)$\Rightarrow$(a): By the universal property of the divisible hull, we may assume that $\Gamma' \leq \Gamma_{\mathrm{div}}$. Hence

$$\Gamma'/\Gamma = (\Gamma'/\Gamma)_l \leq (\Gamma_{\mathrm{div}}/\Gamma)_l = \tfrac{1}{l^\infty}\Gamma/\Gamma \quad ,$$

and therefore $\Gamma' \leq \frac{1}{l^\infty}\Gamma$.

Conversely, by Lemma 1.2.7, $(\frac{1}{l^i}\Gamma')/\Gamma' \cong \Gamma'/l^i = 0$ for every $i \geq 1$. Hence $\frac{1}{l^\infty}\Gamma' = \Gamma'$. Since $\Gamma \leq \Gamma'$, this implies that $\frac{1}{l^\infty}\Gamma \leq \frac{1}{l^\infty}\Gamma' = \Gamma'$. $\square$

1.3 Rational ranks

While an abelian group Γ is only a $\mathbb{Z}$-module, its divisible hull Γ_{div} carries the structure of a $\mathbb{Q}$-linear space. We may therefore define the **rational rank** of Γ to be

$$\mathrm{rr}(\Gamma) = \dim_{\mathbb{Q}}(\Gamma_{\mathrm{div}})$$

considered as a cardinal number.

1.3.1 REMARKS.
(1) By Proposition 1.2.2(b),

$$\mathrm{rr}(\Gamma) = \mathrm{rr}(\Gamma/\Gamma_{\mathrm{tor}}) \quad .$$

(2) When Γ is torsion-free, it embeds in Γ_{div} (Proposition 1.2.4(a)). Hence in this case $\mathrm{rr}(\Gamma) = 0$ if and only if $\Gamma = \{0\}$.
(3) For an arbitrary abelian group Γ we deduce from remarks (1) and (2) that $\mathrm{rr}(\Gamma) = 0$ if and only if Γ is a torsion group.
(4) Let Γ be a subgroup of an abelian group Γ'. The exactness of the divisible hull functor implies that

$$\mathrm{rr}(\Gamma') = \mathrm{rr}(\Gamma) + \mathrm{rr}(\Gamma'/\Gamma) \quad .$$

LEMMA 1.3.2. *Let Γ be a torsion-free abelian group and let l be a prime number. Then* $\dim_{\mathbb{F}_l}(\Gamma/l) \leq \mathrm{rr}(\Gamma)$.

PROOF. Take $\gamma_1, \ldots, \gamma_n \in \Gamma$ and assume that $\iota_\Gamma(\gamma_1), \ldots, \iota_\Gamma(\gamma_n)$ are $\mathbb{Q}$-linearly dependent in Γ_{div}. Since Γ_{div} is torsion-free (Proposition 1.2.2(d)), there exist integers $m_1, \ldots, m_n$ such that $\sum_{i=1}^n m_i \iota_\Gamma(\gamma_i) = 0$ and $l \not| m_i$ for some i. The injectivity of ι_Γ (Proposition 1.2.4(a)) implies that $\sum_{i=1}^n m_i\gamma_i = 0$. Hence

$$\gamma_1 + l\Gamma, \ldots, \gamma_n + l\Gamma$$

are $\mathbb{F}_l$-linearly dependent in Γ/l. The assertion follows. $\square$

REMARK 1.3.3. There is a useful analogy between the notions discussed in the last two sections and the main notions of the theory of field extensions. Namely, torsion-free abelian groups Γ' containing a group Γ with Γ'/Γ torsion correspond to algebraic extensions of fields, while uniquely divisible abelian groups are analogous to algebraically closed fields. Then the divisible hull of a torsion-free abelian group is analogous to the algebraic closure of a field. Finally, the rational rank corresponds to the absolute transcendence degree. We will pursue this analogy further in Part III of this book.

1.4 Characters

Fix a subgroup $\bar{\mu}$ of $\mathbb{Q}/\mathbb{Z}$. Given an abelian group Γ let

$$\chi_{\bar{\mu}}(\Gamma) = \mathrm{Hom}(\Gamma, \bar{\mu})$$

be its group of characters into $\bar{\mu}$. The map $\Gamma \mapsto \chi_{\bar{\mu}}(\Gamma)$ is a contravariant functor on the category of abelian groups. From the general properties of Hom we obtain:

REMARK 1.4.1.
(a) The functor $\chi_{\bar{\mu}}$ is left-exact, i.e., if

$$\Gamma_1 \to \Gamma_2 \to \Gamma_3 \to 0$$

is an exact sequence of abelian groups, then the induced sequence

$$0 \to \chi_{\bar{\mu}}(\Gamma_3) \to \chi_{\bar{\mu}}(\Gamma_2) \to \chi_{\bar{\mu}}(\Gamma_1)$$

is also exact.
(b) If $\Gamma = \varinjlim \Gamma_i$ is a direct limit of abelian groups, then $\chi_{\bar{\mu}}(\Gamma) \cong \varprojlim \chi_{\bar{\mu}}(\Gamma_i)$ canonically.
(c) For abelian groups Γ_i, $i \in I$, one has

$$\chi_{\bar{\mu}}\Big(\bigoplus_{i\in I}\Gamma_i\Big) \cong \prod_{i\in I}\chi_{\bar{\mu}}(\Gamma_i)$$

canonically.

From the primary decomposition of torsion abelian groups (Lemma 1.1.1) and from Remark 1.4.1(c) we deduce:

LEMMA 1.4.2. *Let Γ be an abelian group and for each prime number p let Γ_p be its p-primary subgroup. Then*

$$\chi_{\bar{\mu}}(\Gamma_{\mathrm{tor}}) \cong \prod_p \chi_{\bar{\mu}}(\Gamma_p)$$

canonically.

LEMMA 1.4.3. *Let Γ be a torsion-free abelian group and let l be a prime number. Let $m = \dim_{\mathbb{F}_l}(\Gamma/l)$, considered as a cardinal number. Then*

$$\chi_{\bar{\mu}}\Big(\big(\tfrac{1}{l^\infty}\Gamma\big)/\Gamma\Big) \cong (\varprojlim{}_{l^r}\bar{\mu})^m,$$

where the inverse limit is with respect to multiplications by l.

PROOF. Lemma 1.1.6 yields α_i, $i \in I$, in Γ such that $|I| = m$ and for every positive integer r the $\mathbb{Z}/l^r$-module Γ/l^r is free with basis $\alpha_i + l^r\Gamma$, $i \in I$. By Lemma 1.2.7, $(\frac{1}{l^r}\Gamma)/\Gamma \cong \Gamma/l^r$ via multiplication by l^r. Hence $(\frac{1}{l^r}\Gamma)/\Gamma$ is a free $\mathbb{Z}/l^r$-module with basis $\frac{1}{l^r}\alpha_i + \Gamma$, $i \in I$. It follows that

$$\chi_{\bar\mu}((\tfrac{1}{l^r}\Gamma)/\Gamma) \cong ({}_{l^r}\bar\mu)^m$$

via the evaluation map

$$\psi \mapsto (\psi(\tfrac{1}{l^r}\alpha_i + \Gamma))_{i\in I}.$$

These evaluation maps are compatible in the sense that the following square commutes:

$$\begin{array}{ccc} \chi_{\bar\mu}((\frac{1}{l^{r+1}}\Gamma)/\Gamma) & \longrightarrow & \chi_{\bar\mu}((\frac{1}{l^r}\Gamma)/\Gamma) \\ \cong\downarrow & & \downarrow\cong \\ ({}_{l^{r+1}}\bar\mu)^m & \xrightarrow{\;l\;} & ({}_{l^r}\bar\mu)^m. \end{array}$$

Consequently,

$$\chi_{\bar\mu}\Big((\tfrac{1}{l^\infty}\Gamma)/\Gamma\Big) \cong \chi_{\bar\mu}(\varinjlim((\tfrac{1}{l^r}\Gamma)/\Gamma)) \cong \varprojlim \chi_{\bar\mu}((\tfrac{1}{l^r}\Gamma)/\Gamma) \cong (\varprojlim {}_{l^r}\bar\mu)^m. \qquad \square$$

1.4.4 EXAMPLES.

(1) For a prime number l Lemma 1.4.3 gives:

$$\chi_{\mathbb{Q}/\mathbb{Z}}(\tfrac{1}{l^\infty}\mathbb{Z}/\mathbb{Z}) \cong \varprojlim {}_{l^i}(\mathbb{Q}/\mathbb{Z}) \cong \varprojlim(\tfrac{1}{l^i}\mathbb{Z})/\mathbb{Z} \cong \varprojlim \mathbb{Z}/l^i\mathbb{Z} = \mathbb{Z}_l,$$

where $\mathbb{Z}_l$ is the pro-l completion of $\mathbb{Z}$.

(2) From Lemma 1.4.2 and from (1) we obtain:

$$\chi_{\mathbb{Q}/\mathbb{Z}}(\mathbb{Q}/\mathbb{Z}) \cong \chi_{\mathbb{Q}/\mathbb{Z}}(\bigoplus_l(\tfrac{1}{l^\infty}\mathbb{Z}/\mathbb{Z})) \cong \prod_l \chi_{\mathbb{Q}/\mathbb{Z}}(\tfrac{1}{l^\infty}\mathbb{Z}/\mathbb{Z}) \cong \prod_l \mathbb{Z}_l \cong \hat{\mathbb{Z}}$$

(of course, this can also be computed directly).

Still in the special case $\bar\mu = \mathbb{Q}/\mathbb{Z}$, we observe:

LEMMA 1.4.5. *For Γ finite and abelian, $\Gamma \cong \chi_{\mathbb{Q}/\mathbb{Z}}(\Gamma)$ (non-canonically).*

PROOF. This is straightforward when Γ is cyclic. In the general case we write Γ as a direct sum of finite cyclic groups and use Remark 1.4.1(c). $\square$

COROLLARY 1.4.6. *Let Γ be a finite abelian group. Then $|\chi_{\bar\mu}(\Gamma)| \le |\Gamma|$.*

PROOF. By assumption, $\bar\mu$ embeds in $\mathbb{Q}/\mathbb{Z}$, so $\chi_{\bar\mu}(\Gamma)$ embeds in $\chi_{\mathbb{Q}/\mathbb{Z}}(\Gamma)$, and we apply Lemma 1.4.5. $\square$

If Γ is a torsion group, then we write it as a direct limit $\Gamma = \varinjlim \Gamma_i$ of finite abelian groups Γ_i to obtain $\chi_{\bar\mu}(\Gamma) = \varprojlim \chi_{\bar\mu}(\Gamma_i)$ (Remark 1.4.1(b)). By Corollary 1.4.6, the groups $\chi_{\bar\mu}(\Gamma_i)$ are finite. We see that $\chi_{\bar\mu}(\Gamma)$ carries a natural structure of a profinite abelian group.

LEMMA 1.4.7. *Suppose that p is a prime number such that $\bar{\mu}$ does not contain elements of order p. Let Γ be an abelian group. Then:*
(a) *$\chi_{\bar{\mu}}(\Gamma)$ has no elements of order p;*
(b) *when Γ is a p-primary torsion group, $\chi_{\bar{\mu}}(\Gamma) = 0$;*
(c) *when $|\chi_{\bar{\mu}}(\Gamma)| = |\Gamma| < \infty$ we have $\Gamma_p = 0$.*

PROOF. (a) Suppose that $\psi \in \chi_{\bar{\mu}}(\Gamma)$ has order p. Thus $p(\psi(\alpha)) = (p\psi)(\alpha) = 0$ for every $\alpha \in \Gamma$. Since $\bar{\mu}$ has no elements of order p, necessarily $\psi = 0$.

(b) Assume first that Γ is a finite group of p-power order q. Then for every $\psi \in \chi_{\bar{\mu}}(\Gamma)$ and every $\alpha \in \Gamma$ we have

$$(q\psi)(\alpha) = \psi(q\alpha) = \varphi(0) = 0,$$

i.e., $q\psi = 0$. In light of (a), this implies that $\psi = 0$. Thus $\chi_{\bar{\mu}}(\Gamma) = 0$ in this case.

In the general case we write Γ as a direct limit $\varinjlim \Gamma_i$ of finite p-groups. By what we have just seen, $\chi_{\bar{\mu}}(\Gamma) = \varprojlim \chi_{\bar{\mu}}(\Gamma_i) = 0$.

(c) By the left-exactness of $\chi_{\bar{\mu}}$, the exact sequence

$$\Gamma_p \to \Gamma \to \Gamma/\Gamma_p \to 0$$

gives rise to an exact sequence

$$0 \to \chi_{\bar{\mu}}(\Gamma/\Gamma_p) \to \chi_{\bar{\mu}}(\Gamma) \to \chi_{\bar{\mu}}(\Gamma_p).$$

By (b), $\chi_{\bar{\mu}}(\Gamma_p) = 0$. Hence

$$|\chi_{\bar{\mu}}(\Gamma/\Gamma_p)| = |\chi_{\bar{\mu}}(\Gamma)| = |\Gamma| < \infty.$$

But by Corollary 1.4.6, $|\chi_{\bar{\mu}}(\Gamma/\Gamma_p)| \le |\Gamma/\Gamma_p|$. We conclude that $\Gamma_p = 0$. □

LEMMA 1.4.8. *Suppose that p is a prime number such that $\bar{\mu}$ does not contain elements of order p.*
(a) *If $\varphi\colon \Gamma \to \Gamma'$ is an epimorphism of abelian groups with p-primary kernel, then the induced map $\chi_{\bar{\mu}}(\Gamma') \to \chi_{\bar{\mu}}(\Gamma)$ is an isomorphism.*
(b) *If $\varphi\colon \Gamma \to \Gamma'$ is a homomorphism of abelian groups with p-primary cokernel, then the induced map $\chi_{\bar{\mu}}(\Gamma') \to \chi_{\bar{\mu}}(\Gamma)$ is injective.*

PROOF. (a) Consider the exact sequence

$$\mathrm{Ker}(\varphi) \to \Gamma \xrightarrow{\varphi} \Gamma' \to 0.$$

The left-exactness of $\chi_{\bar{\mu}}$ gives an exact sequence

$$0 \to \chi_{\bar{\mu}}(\Gamma') \to \chi_{\bar{\mu}}(\Gamma) \to \chi_{\bar{\mu}}(\mathrm{Ker}(\varphi)).$$

By Lemma 1.4.7(b) and the assumption, $\chi_{\bar{\mu}}(\mathrm{Ker}(\varphi)) = 0$.

(b) Consider the exact sequence

$$\Gamma \xrightarrow{\varphi} \Gamma' \to \mathrm{Coker}(\varphi) \to 0.$$

The left-exactness of $\chi_{\bar{\mu}}$ gives an exact sequence

$$\chi_{\bar{\mu}}(\mathrm{Coker}(\varphi)) \to \chi_{\bar{\mu}}(\Gamma') \to \chi_{\bar{\mu}}(\Gamma).$$

By Lemma 1.4.7(b) and the assumption, $\chi_{\bar{\mu}}(\mathrm{Coker}(\varphi)) = 0$. □

CHAPTER 2

Ordered Abelian Groups

As already explained in the Introduction, Krull's definition of valuations involves abelian groups equipped with a total order which is compatible with the operation of the group. This chapter studies the category of such groups in general, and focuses on a quantitative estimate of their complexity, called the *rank*. Of special importance are the ordered abelian groups of rank 1, which are studied in §2.5. The last three sections are devoted to the construction of fields of generalized formal power series and some related fields. These will be major sources of examples later on.

2.1 Basic properties and examples

An **ordered abelian group** is a pair $(\Gamma, \leq)$ consisting of an abelian group Γ and a total order $\leq$ on Γ such that for every $\alpha, \beta, \gamma \in \Gamma$, if $\alpha \leq \beta$, then $\alpha+\gamma \leq \beta+\gamma$. We list a few simple consequences of this condition:

LEMMA 2.1.1. *Let $(\Gamma, \leq)$ be an ordered abelian group.*
(a) *For $\alpha \in \Gamma$ with $\alpha > 0$ and for $n \in \mathbb{Z}$ one has $n \geq 0$ if and only if $n\alpha \geq 0$.*
(b) *If $\alpha, \beta \in \Gamma$ and $n, m \in \mathbb{Z}$ satisfy $0 \leq \alpha \leq \beta$ and $0 \leq n \leq m$, then $n\alpha \leq m\beta$.*
(c) *Γ is torsion-free.*
(d) *For every $n > 0$ the map $\Gamma \to \Gamma$, $\alpha \mapsto n\alpha$, is order-preserving.*

PROOF. (a) The "only if" part is proved by induction. For the "if" part suppose that $n\alpha \geq 0$. Since $\alpha > 0$ we have $0 > -\alpha = n\alpha - (n+1)\alpha$, which can happen only if $-(n+1)\alpha < 0$. By the "only if" part, this implies $-(n+1) < 0$, i.e., $n \geq 0$.

(b) By (a) we have

$$n\alpha \leq n\alpha + (m-n)\alpha + m(\beta - \alpha) = m\beta.$$

(c) This follows from (a).

(d) This follows from the previous assertions. □

A **morphism** of ordered abelian groups is an order-preserving group homomorphism.

Clearly, a subgroup of an ordered abelian group is an ordered abelian group with respect to the induced order. Likewise, the direct product of ordered abelian groups is an ordered abelian group with respect to the lexicographic order. A third general method to construct new ordered abelian groups from existing ones is given by the next result.

PROPOSITION 2.1.2. *Let $(\Gamma, \leq)$ be an ordered abelian group. There is a unique total order $\leq_{\text{div}}$ on Γ_{div} such that $(\Gamma_{\text{div}}, \leq_{\text{div}})$ is an ordered abelian group and $\iota_\Gamma : \Gamma \to \Gamma_{\text{div}}$ is a morphism of ordered abelian groups (the notation being as in §1.2).*

PROOF. Recall that every element of Γ_{div} can be written as $\alpha \otimes (1/m)$ with $\alpha \in \Gamma$ and $m > 0$. For $\alpha, \beta \in \Gamma$ and for $n, m > 0$ we set $\alpha \otimes (1/n) \leq_{\text{div}} \beta \otimes (1/m)$ if and only if $m\alpha \leq n\beta$.

To show that this relation is well defined, take $\alpha', \beta' \in \Gamma$ and $n', m' > 0$, and suppose that $\alpha \otimes (1/n) = \alpha' \otimes (1/n')$ and $\beta \otimes (1/m) = \beta' \otimes (1/m')$. Since Γ is torsion-free (Lemma 2.1.1(c)), ι_Γ is injective (Proposition 1.2.4(a)). We obtain that $n'\alpha = n\alpha'$ and $m'\beta = m\beta'$. Then $m\alpha \leq n\beta$ implies that

$$nmm'\alpha' = n'mm'\alpha \leq nn'm'\beta = nn'm\beta'.$$

Since the map $\Gamma \to \Gamma$, $\gamma \mapsto nm\gamma$, is order-preserving (Lemma 2.1.1(d)), $m'\alpha' \leq n'\beta'$, as required.

It is straightforward to verify that $\leq_{\text{div}}$ is a total order on Γ_{div} and that it is compatible with the group operation of Γ_{div}, thus making $(\Gamma_{\text{div}}, \leq_{\text{div}})$ into an ordered abelian group. By definition, if $\alpha \leq \beta$ in Γ, then

$$\iota_\Gamma(\alpha) = \alpha \otimes 1 \leq_{\text{div}} \beta \otimes 1 = \iota_\Gamma(\beta)$$

in Γ_{div}. Consequently, $\iota_\Gamma : (\Gamma, \leq) \to (\Gamma_{\text{div}}, \leq_{\text{div}})$ is a morphism of ordered abelian groups.

For the uniqueness, suppose that $\leq'_{\text{div}}$ is another total order on Γ_{div} with the same properties. When $0 \leq \alpha, \beta \in \Gamma$ and $n, m > 0$, Lemma 2.1.1(d) implies that $\alpha \otimes (1/n) \leq'_{\text{div}} \beta \otimes (1/m)$ if and only if $m\alpha \otimes 1 \leq'_{\text{div}} n\beta \otimes 1$. Since ι_Γ should be a morphism and is injective, this happens if and only if $m\alpha \leq n\beta$. Consequently, $\leq'_{\text{div}}$ coincides with $\leq_{\text{div}}$. □

Quotients by special subgroups are another way to construct ordered abelian groups:

PROPOSITION 2.1.3. *Let $(\Gamma, \leq)$ be an ordered abelian group and let Δ be a subgroup of Γ. The following conditions are equivalent:*

(a) *$\leq$ induces on Γ/Δ a structure of an ordered abelian group;*
(b) *Δ is the kernel of a morphism from $(\Gamma, \leq)$ to an ordered abelian group;*
(c) *if $\gamma \in \Gamma$, $\delta \in \Delta$ and $0 \leq \gamma \leq \delta$, then $\gamma \in \Delta$.*

PROOF. Straightforward. □

A subgroup Δ satisfying the equivalent conditions of Proposition 2.1.3 is said to be **convex** in $(\Gamma, \leq)$ (some texts call such subgroups "isolated" or "distinguished").

PROPOSITION 2.1.4. *Let Δ be a convex subgroup of an ordered abelian group $(\Gamma, \leq)$ and let Δ_0 be a subset of Γ. The following conditions are equivalent:*

(a) *Δ is contained in any convex subgroup of $(\Gamma, \leq)$ which contains Δ_0;*
(b) *every morphism from $(\Gamma, \leq)$ to another ordered abelian group which is zero on Δ_0 is zero on Δ;*
(c) *Δ is the set of all $\gamma \in \Gamma$ for which there exists $\delta \in \Delta_0$ with $\delta \geq 0$ and a positive integer n such that $-n\delta \leq \gamma \leq n\delta$.*

PROOF. (a)⇒(b): Let $\varphi: (\Gamma, \leq) \to (\Gamma', \leq')$ be a morphism of ordered abelian groups with $\Delta_0 \leq \mathrm{Ker}(\varphi)$. By Proposition 2.1.3, $\mathrm{Ker}(\varphi)$ is convex. By (a), it must contain Δ.

(b)⇒(a): Suppose that Δ' is a convex subgroup of $(\Gamma, \leq)$ such that $\Delta_0 \leq \Delta'$. By Proposition 2.1.3, the projection $\varphi: \Gamma \to \Gamma/\Delta'$ can be completed into a morphism of ordered abelian groups. It is zero on Δ_0, so by (b), also on Δ. Thus $\Delta \leq \mathrm{Ker}(\varphi) = \Delta'$.

(a)⇔(c): Straightforward. □

Notice that condition (c) uniquely determines Δ. We call Δ the **convex hull** of Δ_0 in $(\Gamma, \leq)$. The following observation is immediate.

LEMMA 2.1.5. *Let $(\Gamma, \leq)$ be an ordered abelian group and let $\alpha, \beta \in \Gamma$. The convex hull of $\{\alpha\}$ is contained in the convex hull of $\{\beta\}$ if and only if there exists a positive integer n such that $-n\beta \leq \alpha \leq n\beta$.*

LEMMA 2.1.6. *Let $(\Gamma, \leq)$ be an ordered abelian group, let Σ be a subgroup of Γ, and suppose that $0 \leq \alpha \in \Gamma$. Also, assume that every $\gamma \in \Gamma$ such that $0 \leq \gamma \leq \alpha$ is contained Σ. Then the convex hull of $\{\alpha\}$ is also contained in Σ.*

PROOF. For γ in the convex hull of $\{\alpha\}$ there exists $n \in \mathbb{Z}$ with $n\alpha \leq \gamma < (n+1)\alpha$. Then $0 \leq \gamma - n\alpha < \alpha$, so $\gamma - n\alpha \in \Sigma$. The assumption also implies that $\alpha \in \Sigma$. Hence $\gamma \in \Sigma$. □

REMARK 2.1.7. Suppose that $\varphi: (\Gamma, \leq) \to (\bar{\Gamma}, \leq)$ is an epimorphism of ordered abelian groups which is split by a morphism $\sigma: (\bar{\Gamma}, \leq) \to (\Gamma, \leq)$, i.e., $\varphi \circ \sigma = \mathrm{Id}_{\bar{\Gamma}}$. Let $\Delta = \mathrm{Ker}(\varphi)$. As abelian groups, $\Gamma \cong \bar{\Gamma} \oplus \Delta$ via the map

$$\gamma \mapsto (\varphi(\gamma), \gamma - \sigma(\varphi(\gamma))).$$

Moreover, this is an isomorphism of ordered abelian groups, where the direct sum is endowed with the (left-to-right) lexicographic order.

2.2 Ranks

We first note the following simple, yet fundamental, fact.

PROPOSITION 2.2.1. *The convex subgroups of an ordered abelian group $(\Gamma, \leq)$ form a chain with respect to inclusion.*

PROOF. Let Δ_1, Δ_2 be convex subgroups of $(\Gamma, \leq)$ and suppose that $\Delta_1 \not\leq \Delta_2$. Then there exists $\delta_1 \in \Delta_1 \setminus \Delta_2$, $\delta_1 > 0$. For every $\delta_2 \in \Delta_2$ with $\delta_2 \geq 0$, the convexity of Δ_2 implies that $\delta_2 < \delta_1$. It now follows from the convexity of Δ_1 that $\delta_2 \in \Delta_1$. We conclude that $\Delta_2 \leq \Delta_1$. □

The **rank** (also called **height**) of an ordered abelian group $(\Gamma, \leq)$ is the supremum of all non-negative integers n for which there exists a chain

$$\Delta_0 < \Delta_1 < \cdots < \Delta_n$$

of convex subgroups of $(\Gamma, \leq)$. We denote it by $\mathrm{rank}(\Gamma)$ (possibly $\mathrm{rank}(\Gamma) = \infty$).

LEMMA 2.2.2. *Let Δ be a convex subgroup of an ordered abelian group $(\Gamma, \leq)$ and consider Δ and Γ/Δ as ordered abelian groups with the induced orders. Then*

$$\operatorname{rank}(\Gamma) = \operatorname{rank}(\Delta) + \operatorname{rank}(\Gamma/\Delta) \quad .$$

PROOF. The map $\Delta' \mapsto \Delta'/\Delta$ is an order-preserving bijection between the convex subgroups Δ' of $(\Gamma, \leq)$ containing Δ and the convex subgroups of Γ/Δ. Furthermore, by Proposition 2.2.1, every convex subgroup of $(\Gamma, \leq)$ either contains, or is contained in, Δ. The assertion follows. □

2.2.3 EXAMPLES.

(1) The rank of $\mathbb{Z}$ with its usual order is 1.

(2) Let $(\Delta_i, \leq_i)$, $i = 1, 2$, be ordered abelian groups, and equip $\Gamma = \Delta_1 \times \Delta_2$ with the lexicographic order. Then $\{0\} \times \Delta_2$ is a convex subgroup of Γ, and $\Gamma/(\{0\} \times \Delta_2) \cong \Delta_1$ as ordered abelian groups. By Lemma 2.2.2,

$$\operatorname{rank}(\Gamma) = \operatorname{rank}(\Delta_1) + \operatorname{rank}(\Delta_2) \quad .$$

(3) Consider the lexicographically ordered abelian groups $\mathbb{Z}^n$, where $n \geq 0$, and $\mathbb{Z}^{\mathbb{N}}$. From examples (1) and (2) we obtain inductively that $\operatorname{rank}(\mathbb{Z}^n) = n$ for every n. Since there is a surjective morphism $\mathbb{Z}^{\mathbb{N}} \to \mathbb{Z}^n$ of ordered abelian groups, necessarily $\operatorname{rank}(\mathbb{Z}^{\mathbb{N}}) = \infty$.

PROPOSITION 2.2.4. *Let $(\Gamma', \leq)$ be an ordered abelian group and let Γ be a subgroup of Γ'.*

(a) *The set of all convex subgroups of Γ' is mapped surjectively onto the set of all convex subgroups of Γ by $\Delta' \mapsto \Gamma \cap \Delta'$.*

(b) *If Γ'/Γ is a torsion group, then the map in (a) is bijective. In particular, $\operatorname{rank}(\Gamma) = \operatorname{rank}(\Gamma')$ in this case.*

PROOF. (a) It is immediate to verify that if Δ' is a convex subgroup of Γ', then $\Gamma \cap \Delta'$ is a convex subgroup of Γ.

For the surjectivity, suppose that Δ is a convex subgroup of Γ. Let Δ' be its convex hull in $(\Gamma', \leq)$. The convexity of Δ in Γ implies that $\Gamma \cap \Delta' = \Delta$.

(b) The surjectivity has already been established. For the injectivity, suppose that Δ_1', Δ_2' are convex subgroups of Γ' satisfying $\Gamma \cap \Delta_1' = \Gamma \cap \Delta_2'$. For every $\gamma' \in \Delta_1'$ there exists a positive integer n such that $n\gamma' \in \Gamma$. Then

$$n\gamma' \in \Gamma \cap \Delta_1' = \Gamma \cap \Delta_2' \leq \Delta_2' \quad .$$

The convexity of Δ_2' implies that $\gamma' \in \Delta_2$. Thus $\Delta_1' \leq \Delta_2'$. By symmetry, $\Delta_1' = \Delta_2'$, as desired. □

From this and from Lemma 1.2.5(b) we deduce:

COROLLARY 2.2.5. *Let $(\Gamma, \leq)$ be an ordered abelian group and extend it to an ordered abelian group $(\Gamma_{\text{div}}, \leq_{\text{div}})$ as in Proposition 2.1.2. Then*

$$\operatorname{rank}(\Gamma) = \operatorname{rank}(\Gamma_{\text{div}}) \quad .$$

For non-convex subgroups of an ordered abelian group we have the following variant of Lemma 2.2.2.

PROPOSITION 2.2.6. *Let $(\Gamma, \leq)$ be an ordered abelian group and let Γ_0 be a subgroup of Γ with the induced order. Then*

$$\mathrm{rank}(\Gamma) \leq \mathrm{rank}(\Gamma_0) + \mathrm{rr}(\Gamma/\Gamma_0).$$

PROOF. The proof is based on [**Bou1**, Ch. VI, §10.2], with a simplification suggested by a referee.

Let $\Delta_0 < \Delta_1 < \cdots < \Delta_n$ be a strictly increasing sequence of convex subgroups of $(\Gamma, \leq)$. We need to show that $n \leq \mathrm{rank}(\Gamma_0) + \mathrm{rr}(\Gamma/\Gamma_0)$. To this end consider the set I of all indices $1 \leq i \leq n$ for which $\Delta_{i-1} \cap \Gamma_0 = \Delta_i \cap \Gamma_0$.

One first observes that the intersections $\Delta_i \cap \Gamma_0$ for $i \in \{1, \ldots, n\} \setminus I$ form a strictly increasing sequence of convex subgroups of Γ_0. Hence $n - |I| \leq \mathrm{rank}(\Gamma_0)$.

It therefore remains to show that $|I| \leq \mathrm{rr}(\Gamma/\Gamma_0)$. For each $i \in I$ we choose $\delta_i \in \Delta_i \setminus \Delta_{i-1}$ and show that the cosets of the δ_i in $(\Gamma/\Gamma_0)_{\mathrm{div}}$ are $\mathbb{Q}$-linearly independent. Indeed, suppose that $\sum_{i \in I} r_i \delta_i \in \Gamma_0$, where $r_i \in \mathbb{Q}$, $i \in I$, are not all zero. Choose $i^* \in I$ maximal with $r_{i^*} \neq 0$. Also, let m be the largest common denominator of the r_i. Then

$$\begin{aligned} m r_{i^*} \delta_{i^*} &= \sum_{i \in I} m r_i \delta_i - \sum_{i^* \neq i \in I} m r_i \delta_i \\ &\in (\Delta_{i^*} \cap \Gamma_0) + \Delta_{i^*-1} \\ &= (\Delta_{i^*-1} \cap \Gamma_0) + \Delta_{i^*-1} = \Delta_{i^*-1}. \end{aligned}$$

Since Δ_{i^*-1} is convex in $(\Gamma, \leq)$, this implies that $\delta_{i^*} \in \Delta_{i^*-1}$, a contradiction. □

Taking in Proposition 2.2.6 $\Gamma_0 = \{0\}$, we obtain in particular:

COROLLARY 2.2.7. *For an ordered abelian group $(\Gamma, \leq)$ one has*

$$\mathrm{rank}(\Gamma) \leq \mathrm{rr}(\Gamma).$$

EXAMPLE 2.2.8. The inequality in Corollary 2.2.7 may be strict. For instance, let $\Gamma = \mathbb{Z} + \mathbb{Z}\sqrt{2}$, with the order induced from $\mathbb{R}$. One has $\mathrm{rank}(\Gamma) = 1$ and $\mathrm{rr}(\Gamma) = 2$.

2.3 Cores

Let Σ be a subgroup of an ordered abelian group $(\Gamma, \leq)$. By Proposition 2.2.1, the convex subgroups of $(\Gamma, \leq)$ which are contained in Σ form a chain under inclusion. There is a maximal subgroup Γ_Σ in this chain, namely, its union. We call Γ_Σ the Σ-**core** of $(\Gamma, \leq)$.

2.3.1 EXAMPLES.

(1) Let $\Gamma = \mathbb{Z}$ with its usual order, and let $\Sigma = m\mathbb{Z}$, where $m \geq 2$. Then $\Gamma_\Sigma = \{0\}$.
(2) Consider $\Gamma = \mathbb{Z} \oplus \mathbb{Z}$ with the lexicographic order, and take $\Sigma = (m\mathbb{Z}) \oplus \mathbb{Z}$, where again $m \geq 2$. Then $\Gamma_\Sigma = \{0\} \oplus \mathbb{Z}$.

More generally, we have:

LEMMA 2.3.2. *Let Δ be a convex subgroup of an ordered abelian group $(\Gamma, \leq)$. Let Σ be a subgroup of Γ containing Δ. The following conditions are equivalent:*
(a) *Δ is the Σ-core of Γ;*
(b) *The Σ/Δ-core of Γ/Δ (with respect to the induced order) is trivial.*

PROOF. There is an obvious inclusion-preserving bijection between the following two sets:
(1) the convex subgroups Δ' of Γ such that $\Delta \leq \Delta' \leq \Sigma$; and
(2) the convex subgroup $\bar{\Delta}'$ of Γ/Δ such that $\bar{\Delta}' \leq \Sigma/\Delta$.
The equivalence follows. □

For later use we prove the following technical fact.

LEMMA 2.3.3. *Let $(\Gamma, \leq)$ be an ordered abelian group and let Σ be a subgroup of Γ such that $\Gamma_\Sigma = 0$. Let $\sigma \in \Sigma$ and let*

$$\Gamma_0 = \{\gamma \in \Gamma \mid 0 \leq \gamma \notin \Sigma\}.$$

Then $\sigma + \Gamma_0 \subseteq \Gamma_0$ if and only if $0 \leq \sigma$.

PROOF. The "if" part is obvious.
For the "only if" part suppose that $\sigma + \Gamma_0 \subseteq \Gamma_0$ but $\sigma < 0$. For every $\gamma \in \Gamma$ with $0 \leq \gamma \leq -\sigma$ we have $\sigma + \gamma \notin \Gamma_0$, so necessarily $\gamma \notin \Gamma_0$. Hence $\gamma \in \Sigma$. By Lemma 2.1.6, the convex hull of $\{-\sigma\}$ is contained in Σ. This contradicts our assumption that no non-zero convex subgroup of $(\Gamma, \leq)$ is contained in Σ. □

2.4 Cofinality and infinitesimals

We say that an ordered abelian group is **discrete** if it is cyclic and non-trivial. Being torsion-free, it must therefore be infinite, i.e., isomorphic to $\mathbb{Z}$.

PROPOSITION 2.4.1. *Let Γ be a subgroup of an ordered abelian group $(\Gamma', \leq)$. Suppose that $(\Gamma' : \Gamma) < \infty$. Then Γ is discrete if and only if Γ' is discrete.*

PROOF. For $n = (\Gamma' : \Gamma)$ we have a well defined monomorphism $\Gamma' \to \Gamma$, $\gamma' \mapsto n\gamma'$. Thus each of the groups Γ, Γ' embeds in the other. Since $(\Gamma' : \Gamma) < \infty$ and Γ' is torsion-free, if one of Γ, Γ' is trivial, then so is the other. Now use the fact that every non-trivial subgroup of $\mathbb{Z}$ is isomorphic to $\mathbb{Z}$. □

A subgroup Δ of an ordered abelian group $(\Gamma, \leq)$ is **cofinal in** $(\Gamma, \leq)$ if for every $\gamma \in \Gamma$ there exists $\delta \in \Delta$ such that $-\delta \leq \gamma \leq \delta$. We call $\alpha \in \Gamma$ **infinitesimal** if either $\alpha = 0$ or $\mathbb{Z}\alpha$ is not cofinal; equivalently, there exists $\gamma \in \Gamma$ such that $-\gamma \leq n\alpha \leq \gamma$ for all $n \in \mathbb{Z}$. We denote the set of all infinitesimal elements in $(\Gamma, \leq)$ by Γ_{inf}.

EXAMPLE 2.4.2. Denote the usual order on $\mathbb{R}$ by $\leq_{\mathbb{R}}$. Thus $\alpha \leq_{\mathbb{R}} \beta$ if and only if $\beta - \alpha$ is a square in $\mathbb{R}$. One has $\mathbb{R}_{\mathrm{inf}} = \{0\}$ (*Archimedes' Axiom*). More generally, for every subgroup Γ of $\mathbb{R}$ one has $\Gamma_{\mathrm{inf}} = \{0\}$ with respect to the induced order.

PROPOSITION 2.4.3. *Let $(\Gamma, \leq)$ be an ordered abelian group. Then:*
(a) *Γ_{inf} is a convex subgroup of $(\Gamma, \leq)$;*
(b) *for every surjective non-zero morphism $\varphi\colon (\Gamma, \leq) \to (\Gamma', \leq')$ of ordered abelian groups one has $\Gamma_{\mathrm{inf}} = \varphi^{-1}(\Gamma'_{\mathrm{inf}})$;*

(c) $\leq$ *induces on* $\bar{\Gamma} = \Gamma/\Gamma_{\text{inf}}$ *a structure of an ordered abelian group with* $\bar{\Gamma}_{\text{inf}} = \{0\}$.

PROOF. (a) Given non-zero $\alpha, \beta \in \Gamma_{\text{inf}}$, there exist $\gamma, \delta \in \Gamma$ such that $-\gamma \leq n\alpha \leq \gamma$ and $-\delta \leq n\beta \leq \delta$ for every integer n. Then also $-\gamma - \delta \leq n(\alpha \pm \beta) \leq \gamma + \delta$ for all n, so $\alpha \pm \beta \in \Gamma_{\text{inf}}$. It follows that Γ_{inf} is a group. Its convexity is immediate.

(b) Let $\alpha \in \Gamma$. If $\alpha \in \Gamma_{\text{inf}}$, then there exists $\gamma \in \Gamma$ such that for every $n \in \mathbb{Z}$ one has $-\gamma \leq n\alpha \leq \gamma$, whence also $-\varphi(\gamma) \leq n\varphi(\alpha) \leq \varphi(\gamma)$. We conclude that $\varphi(\alpha) \in \Gamma'_{\text{inf}}$.

Conversely, suppose that $\alpha \in \Gamma$ and $\varphi(\alpha) \in \Gamma'_{\text{inf}}$. Since φ is surjective and non-zero, there exists $\gamma \in \Gamma$ such that for every $n \in \mathbb{Z}$ one has $-\varphi(\gamma) < n\varphi(\alpha) < \varphi(\gamma)$, whence also $-\gamma < n\alpha < \gamma$. Therefore $\alpha \in \Gamma_{\text{inf}}$.

(c) Use Proposition 2.1.3, (a) and (b). □

Co-torsion subgroups are an important source of cofinal subgroups:

LEMMA 2.4.4. *Let* Γ *be a subgroup of an ordered abelian group* $(\Gamma', \leq)$ *such that* Γ'/Γ *is a torsion group. Then* Γ *is cofinal in* $(\Gamma', \leq)$.

PROOF. Given $\gamma' \in \Gamma'$ we take $n > 0$ with $n\gamma' \in \Gamma$. When $\gamma' \geq 0$ we have $-n\gamma' \leq \gamma' \leq n\gamma'$. When $\gamma' \leq 0$ we similarly have $n\gamma' \leq \gamma' \leq -n\gamma'$. □

For the proofs in the next section we will need the following technical lemma.

LEMMA 2.4.5. *Let* $(\Gamma, \leq)$ *be a non-discrete ordered abelian group. Suppose that* $0 < \alpha \in \Gamma$ *and that* $\mathbb{Z}\alpha$ *is cofinal. Then for every positive integer* n *there exists a positive* $\beta \in \Gamma$ *such that* $n\beta \leq \alpha$.

PROOF. Since $n < 2^n$, it suffices to prove the lemma when n is a power of 2. For this, it is enough to consider the case $n = 2$.

By assumption, $\Gamma \neq \mathbb{Z}\alpha$, so we choose $\gamma \in \Gamma \setminus \mathbb{Z}\alpha$. Therefore there exists $n \in \mathbb{Z}$ with $n\alpha < \gamma < (n+1)\alpha$. We may replace γ by $\gamma - n\alpha$ to assume that $0 < \gamma < \alpha$. Now $2\gamma + 2(\alpha - \gamma) = 2\alpha$, so either $2\gamma \leq \alpha$ or $2(\alpha - \gamma) \leq \alpha$. We take $\beta = \gamma$ or $\beta = \alpha - \gamma$, respectively. □

2.5 Ordered abelian groups of rank 1

Somewhat surprisingly, the notion of the rank of an ordered abelian group is closely related to the ordered structure of $\mathbb{R}$. This will be shown in the following two theorems. The main argument of the proof essentially goes back to O. Hölder [**Hö**].

THEOREM 2.5.1. *The following conditions on an ordered abelian group* $(\Gamma, \leq)$ *are equivalent:*

(a) *there is a non-zero morphism from* $(\Gamma, \leq)$ *to* $(\mathbb{R}, \leq_{\mathbb{R}})$;

(b) $\Gamma \neq \Gamma_{\text{inf}}$.

PROOF. (a)⇒(b): Let $\varphi\colon (\Gamma, \leq) \to (\mathbb{R}, \leq_{\mathbb{R}})$ be a non-zero morphism. Let $\Gamma' = \varphi(\Gamma)$. Then $\Gamma'_{\text{inf}} = \{0\}$ (see Example 2.4.2). By Proposition 2.4.3(b),

$$\Gamma_{\text{inf}} = \varphi^{-1}(\Gamma'_{\text{inf}}) = \varphi^{-1}(\{0\}) \neq \Gamma.$$

(b)⇒(a): Let $\bar{\Gamma} = \Gamma/\Gamma_{\text{inf}}$. By Proposition 2.4.3(c), $\bar{\Gamma}_{\text{inf}} = \{0\} \neq \bar{\Gamma}$. Since there is always a surjective morphism $\Gamma \to \bar{\Gamma}$ (by Proposition 2.4.3(a)), we may therefore replace Γ by $\bar{\Gamma}$ to assume, without loss of generality, that $\Gamma_{\text{inf}} = \{0\}$.

Now if Γ is discrete then (a) is clear.

So suppose that Γ is not discrete. We define a map $\varphi\colon \Gamma \to \mathbb{R}$ as follows: fix $\alpha \in \Gamma$ with $\alpha > 0$. By assumption, $\mathbb{Z}\alpha$ is cofinal in $(\Gamma, \leq)$. Thus for every $\gamma \in \Gamma$ there exists $k \in \mathbb{Z}$ such that $-k\alpha \leq \gamma \leq k\alpha$. Hence the subset

$$D_\gamma = \left\{ \frac{m}{l} \;\middle|\; m, l \in \mathbb{Z},\ l > 0,\ m\alpha \leq l\gamma \right\}$$

of $\mathbb{Q}$ is bounded from above by k (see Lemma 2.1.1). Moreover, taking $m = -k$ and $l = 1$, we see that D_γ contains $-k$, and in particular is non-empty. We may therefore set $\varphi(\gamma) = \sup(D_\gamma) \in \mathbb{R}$. Note that $\varphi(\alpha) = 1$, so φ is non-zero.

We claim that φ is order-preserving. Indeed, suppose that $\gamma_1, \gamma_2 \in \Gamma$ and $\gamma_1 \leq \gamma_2$. If $m, l \in \mathbb{Z}$, $l > 0$, and $m/l \in D_{\gamma_1}$, then $m\alpha \leq l\gamma_1 \leq l\gamma_2$, so $m/l \in D_{\gamma_2}$. Thus $D_{\gamma_1} \subseteq D_{\gamma_2}$, and the claim follows.

Next we observe that for every positive integer n and every $\gamma \in \Gamma$ one has $m/l \in D_{n\gamma}$ if and only if $m/ln \in D_\gamma$. Therefore $\varphi(n\gamma) = n\varphi(\gamma)$.

Finally we show that φ is a group homomorphism. To this end we take $\gamma_1, \gamma_2 \in \Gamma$ and a positive integer n. Lemma 2.4.5 yields a positive $\beta \in \Gamma$ such that $n\beta \leq \alpha$. The cofinality of $\mathbb{Z}\beta$ in $(\Gamma, \leq)$ yields $k_1, k_2 \in \mathbb{Z}$ such that

$$k_1\beta \leq \gamma_1 < (k_1 + 1)\beta, \quad k_2\beta \leq \gamma_2 < (k_2 + 1)\beta.$$

Consequently,

$$(k_1 + k_2)\beta \leq \gamma_1 + \gamma_2 < (k_1 + k_2 + 2)\beta.$$

Since φ is order-preserving, these inequalities give

$$|\varphi(\gamma_1) - k_1\varphi(\beta)|, |\varphi(\gamma_2) - k_2\varphi(\beta)| \leq \varphi(\beta)$$

and

$$|\varphi(\gamma_1 + \gamma_2) - (k_1 + k_2)\varphi(\beta)| \leq 2\varphi(\beta).$$

It follows that

$$|\varphi(\gamma_1 + \gamma_2) - \varphi(\gamma_1) - \varphi(\gamma_2)| \leq 4\varphi(\beta) = 4\varphi(n\beta)/n \leq 4\varphi(\alpha)/n.$$

But n was arbitrary, so necessarily $\varphi(\gamma_1 + \gamma_2) = \varphi(\gamma_1) + \varphi(\gamma_2)$. □

Combining this with the discussion of the previous section, we deduce:

THEOREM 2.5.2. *The following conditions on a non-zero ordered abelian group* $(\Gamma, \leq)$ *are equivalent:*

(a) $(\Gamma, \leq)$ *has rank* 1;
(b) $(\Gamma, \leq)$ *has no convex subgroups except* Γ *and* $\{0\}$;
(c) *every non-zero morphism from* $(\Gamma, \leq)$ *to an ordered abelian group is injective;*
(d) $\Gamma_{\text{inf}} = \{0\}$;
(e) *for every* $0 \neq \alpha \in \Gamma$ *the subgroup* $\mathbb{Z}\alpha$ *is cofinal;*
(f) *there is an injective morphism from* $(\Gamma, \leq)$ *to* $(\mathbb{R}, \leq_{\mathbb{R}})$.

PROOF. (a)⇔(b): This is an immediate consequence of the definition of the rank.

(b)⇔(c): Use Proposition 2.1.3.

(d)⇔(e): Immediate.

(b)⇒(e): The convex hull of $\mathbb{Z}\alpha$ is non-zero. By (b) it is all of Γ, i.e., $\mathbb{Z}\alpha$ is cofinal.

(e)⇒(b): Let Δ be a non-zero convex subgroup of Γ. Choose $0 \neq \alpha \in \Delta$. The convex hull of $\mathbb{Z}\alpha$ is contained in Δ. By (e) it is all of Γ. We conclude that $\Delta = \Gamma$.

(f)⇒(e): This follows from the fact that every non-zero subgroup of $\mathbb{R}$ is cofinal.

(c)+(d)⇒(f): (d) and Theorem 2.5.1 yield a non-zero morphism $(\Gamma, \leq) \to (\mathbb{R}, \leq_{\mathbb{R}})$. By (c), it is injective. □

Ordered abelian groups satisfying the equivalent conditions of Theorem 2.5.2 are also called **Archimedean**.

COROLLARY 2.5.3. *Let $(\Gamma, \leq)$ be an ordered abelian group of rank* 1. *The following conditions are equivalent:*
(a) Γ *is discrete;*
(b) *there exists $\alpha \in \Gamma$, $\alpha > 0$, such that there is no $\beta \in \Gamma$ with $0 < \beta < \alpha$.*

PROOF. (a)⇒(b): Choose a positive $\alpha \in \Gamma$ with $\Gamma = \mathbb{Z}\alpha$.

(b)⇒(a): By condition (e) of Theorem 2.5.2, $\mathbb{Z}\alpha$ is cofinal. Given $\gamma \in \Gamma$, we can therefore find $n \in \mathbb{Z}$ such that $n\alpha \leq \gamma < (n+1)\alpha$. Then $0 \leq \gamma - n\alpha < \alpha$, so $\gamma = n\alpha$. We conclude that $\Gamma = \mathbb{Z}\alpha$. □

COROLLARY 2.5.4. *Let $(\Gamma, \leq)$ be an ordered abelian group of rank* 1, *let Σ be a subgroup of Γ, and let $\alpha \in \Gamma$, $\alpha > 0$. Suppose that every $\beta \in \Gamma$ satisfying $0 \leq \beta < \alpha$ is contained in Σ. Then either Γ is discrete or $\Gamma = \Sigma$.*

PROOF. If there are no $\beta \in \Gamma$ such that $0 < \beta < \alpha$, then by Corollary 2.5.3, Γ is discrete.

If there exists such β, then by Lemma 2.1.6, the convex hull Δ of $\{\beta\}$ is contained in Σ. Since $(\Gamma, \leq)$ has rank 1, necessarily $\Gamma = \Delta$. Hence also $\Gamma = \Sigma$. □

2.6 Push-downs

The following fact about existence and properties of push-downs of ordered abelian groups will be needed in the study of extensions of valuations in §20.1.

PROPOSITION 2.6.1. *Let $\varphi\colon (\Gamma, \leq) \to (\Gamma_1 \leq)$ and $\eta\colon (\Gamma, \leq) \to (\bar{\Gamma}, \leq)$ be morphisms of ordered abelian groups, with η surjective. Then there is an ordered abelian group $(\bar{\Gamma}_1, \leq)$, a morphism $\varphi_1\colon (\bar{\Gamma}, \leq) \to (\bar{\Gamma}_1, \leq)$ and a surjective morphism $\eta_1\colon (\Gamma_1, \leq) \to (\bar{\Gamma}_1, \leq)$ satisfying:*
(a) $\varphi_1 \circ \eta = \eta_1 \circ \varphi$;
(b) *if φ is injective, then so is φ_1;*
(c) *if φ is surjective, then so is φ_1;*

(d) *if* $\varphi_2\colon (\bar\Gamma, \le) \to (\bar\Gamma_2, \le)$ *and* $\eta_2\colon (\Gamma_1, \le) \to (\bar\Gamma_2, \le)$ *are morphisms with* $\varphi_2 \circ \eta = \eta_2 \circ \varphi$, *then there is a unique morphism* $\theta\colon (\bar\Gamma_1, \le) \to (\bar\Gamma_2, \le)$ *such that* $\varphi_2 = \theta \circ \varphi_1$ *and* $\eta_2 = \theta \circ \eta_1$.

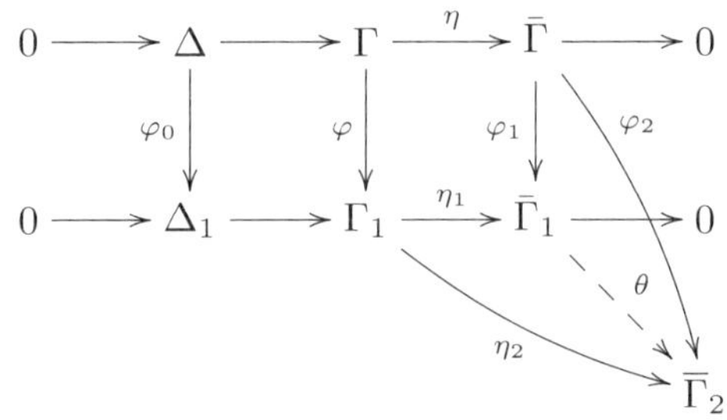

PROOF. Let $\Delta = \mathrm{Ker}(\eta)$. It is a convex subgroup of $(\Gamma, \le)$. Let Δ_1 be the convex hull of $\varphi(\Delta)$ in $(\Gamma_1, \le)$, let φ_0 be the restriction of φ to Δ and let $\bar\Gamma_1 = \Gamma_1/\Delta_1$. By Proposition 2.1.3, $(\Gamma_1, \le)$ induces on $\bar\Gamma_1$ a structure of an ordered abelian group, and the projection $\eta_1\colon \Gamma_1 \to \bar\Gamma_1$ is a surjective morphism of ordered abelian groups. Since the morphism $\eta_1 \circ \varphi$ is trivial on Δ, there is a unique morphism φ_1 making the upper part of the diagram above commutative (with exact rows). In particular, this gives (a).

(b) By the snake lemma (Lemma 1.1.3), we need to show that the map $\mathrm{Coker}(\varphi_0) \to \mathrm{Coker}(\varphi)$ is injective, i.e., that $\varphi(\Gamma) \cap \Delta_1 \le \varphi(\Delta)$. So suppose that $0 \le \gamma \in \Gamma$ and $\varphi(\gamma) \in \Delta_1$. Then there exists $\delta \in \Delta$ such that $0 \le \varphi(\gamma) \le \varphi(\delta)$. The injectivity of the morphism φ implies that $0 \le \gamma \le \delta$. Since Δ is convex, $\gamma \in \Delta$, whence $\varphi(\gamma) \in \varphi(\Delta)$.

(c) is immediate from (a).

(d) The assumption implies that η_2 is trivial on $\varphi(\Delta)$, whence on its convex hull Δ_1. Therefore it induces a morphism $\theta\colon (\bar\Gamma_1, \le) \to (\bar\Gamma_2, \le)$ such that $\eta_2 = \theta \circ \eta_1$. Then

$$\varphi_2 \circ \eta = \eta_2 \circ \varphi = \theta \circ \eta_1 \circ \varphi = \theta \circ \varphi_1 \circ \eta.$$

The surjectivity of η now implies that $\varphi_2 = \theta \circ \varphi_1$.

The uniqueness of θ follows from the equality $\eta_2 = \theta \circ \eta_1$ and the surjectivity of η_1. □

2.7 Well-ordered sets

Recall that a totally ordered set $(A, \le)$ is **well-ordered** if every non-empty subset of A has a minimal element. Equivalently, A has no infinite (strictly) decreasing sequence.

LEMMA 2.7.1. *An infinite totally ordered set must contain an infinite sequence that is either increasing or decreasing.*

PROOF. Suppose that $(A, \le)$ is an infinite totally ordered set that contains no infinite decreasing sequence. Then it has a smallest element a_1. The subset $A \setminus \{a_1\}$ has the same properties, and therefore has a smallest element a_2. We continue inductively, and define a_i to be the smallest element of $A \setminus \{a_1, \dots, a_{i-1}\}$. Then the infinite sequence $\{a_i\}_{i=1}^{\infty}$ is increasing. □

LEMMA 2.7.2. *Let $(\Gamma, \leq)$ be an ordered abelian group.*

(a) *A subset of a well-ordered subset of Γ is well-ordered.*

(b) *The union of finitely many well-ordered subsets of Γ is well-ordered.*

(c) *If A_1, A_2 are well-ordered subsets of Γ, then so is $A_1 + A_2$.*

(d) *If A_1, A_2 are well-ordered subsets of Γ and $\gamma \in \Gamma$, then $(\gamma - A_1) \cap A_2$ is finite.*

PROOF. (a) and (b) are obvious.

For (c) suppose that $a_{1,i} \in A_1$, $a_{2,i} \in A_2$, $i = 1, 2, \dots$, and that the sequence $\{a_{1,i} + a_{2,i}\}_{i=1}^{\infty}$ is decreasing. Then at least one of the sequences $\{a_{1,i}\}_{i=1}^{\infty}$, $\{a_{2,i}\}_{i=1}^{\infty}$ contains an infinite decreasing subsequence, contrary to our assumption.

(d) follows from Lemma 2.7.1. □

For an element γ of an ordered abelian group $(\Gamma, \leq)$ let Δ_γ be the convex hull of $\mathbb{Z}\gamma$. If $\gamma, \gamma' \in \Gamma$ and $0 \leq \gamma \leq \gamma'$, then $\Delta_\gamma \leq \Delta_{\gamma'}$ (see Lemma 2.1.5).

LEMMA 2.7.3. *Let $(\Gamma, \leq)$ be an ordered abelian group. Let A be a non-empty well-ordered set of positive elements of Γ. Let S be the set of all sums of finitely many elements of A.*

(a) *For every $\sigma \in S$ there exists $\alpha \in A$ such that $\sigma - \alpha \in S \cup \{0\}$ and $\Delta_\sigma = \Delta_\alpha$.*

(b) *Let $\sigma_i \in S$, $i = 1, 2, \dots$, and suppose that $\Delta_{\sigma_1} \geq \Delta_{\sigma_2} \geq \dots$. Then $\Delta_{\sigma_i} = \Delta_{\sigma_{i+1}}$ for all sufficiently large i.*

PROOF. (a) Write $\sigma = \alpha_1 + \dots + \alpha_n$, with $\alpha_1 \geq \dots \geq \alpha_n > 0$ in A. For $\alpha = \alpha_1$ we have

$$\sigma - \alpha = \alpha_2 + \dots + \alpha_n \in S \cup \{0\}.$$

Further, $0 < \alpha \leq \sigma \leq n\alpha$, so by Lemma 2.1.5, $\Delta_\sigma = \Delta_\alpha$.

(b) For each i part (a) gives rise to $\alpha_i \in A$ with $\Delta_{\sigma_i} = \Delta_{\alpha_i}$. If the Δ_{σ_i} do not stabilize as claimed, then there is an infinite decreasing subsequence $\Delta_{\sigma_{i_1}} > \Delta_{\sigma_{i_2}} > \dots$. Since $(\Gamma, \leq)$ is totally ordered and the map $\alpha \mapsto \Delta_\alpha$ is order-preserving, $\alpha_{i_1} > \alpha_{i_2} > \dots$. But this contradicts our assumption that A is well-ordered. □

THEOREM 2.7.4. *Let $(\Gamma, \leq)$ be an ordered abelian group. Let A be a well-ordered set of positive elements of Γ. Then the set S of all sums of finitely many elements of A is well-ordered.*

PROOF. Consider the collection $\mathcal{C}$ of all pairs $(\alpha, k) \in A \times \mathbb{N}$ for which there exists a decreasing sequence $\sigma_1 > \sigma_2 > \dots$ in S such that $\Delta_\alpha = \Delta_{\sigma_i}$ and $\sigma_i \leq k\alpha$ for all sufficiently large i. We order $\mathcal{C}$ lexicographically with respect to the given order on A and the standard order on $\mathbb{N}$. Note that this is a well-ordering.

Every infinite decreasing sequence $\sigma_1 > \sigma_2 > \dots$ in S gives rise in the following manner to a pair in $(\alpha, k) \in \mathcal{C}$ such that $\Delta_\alpha = \Delta_{\sigma_i}$ for sufficiently large i: clearly $\Delta_{\sigma_1} \geq \Delta_{\sigma_2} \geq \dots$, so Lemma 2.7.3(b) yields a convex subgroup Δ of Γ and $i_0 \geq 0$ such that $\Delta_{\sigma_i} = \Delta$ for all $i \geq i_0$. By Lemma 2.7.3(a) there exists $\alpha \in A$ with $\Delta = \Delta_\alpha$. For $i \geq i_0$ we therefore have

$$\sigma_i \in \Delta_{\sigma_i} = \Delta = \Delta_\alpha.$$

In particular, there exists a positive integer k such that $\sigma_{i_0} \leq k\alpha$. Therefore $\sigma_i \leq k\alpha$ for all $i \geq i_0$. Thus $(\alpha, k) \in \mathcal{C}$, as required.

Now assume that there exists an infinite decreasing sequence in S. In particular we obtain from the preceding construction that $\mathcal{C} \neq \emptyset$. Being well-ordered, $\mathcal{C}$ then has a minimal element (α^*, k^*). Let $\sigma_1 > \sigma_2 > \ldots$ be an infinite decreasing sequence in S that corresponds to (α^*, k^*), as in the definition of $\mathcal{C}$. Thus $\Delta_{\alpha^*} = \Delta_{\sigma_i}$ and $\sigma_i \leq k^*\alpha^*$ for all sufficiently large i. For each i, Lemma 2.7.3(a) yields $\alpha_i \in A$ such that $\sigma_i - \alpha_i \in S \cup \{0\}$ and $\Delta_{\sigma_i} = \Delta_{\alpha_i}$. Since A is well-ordered, at most finitely many σ_i can be in A. Therefore for all sufficiently large i we have $\sigma_i - \alpha_i \in S$.

The sequence $\{\sigma_i\}$ is decreasing, while the sequence $\{\alpha_i\}$ does not contain an infinite decreasing subsequence. Hence some subsequence of $\{\sigma_i - \alpha_i\}_{i=1}^{\infty}$ is decreasing. We may pass to subsequences to assume, without loss of generality, that $\{\sigma_i - \alpha_i\}_{i=1}^{\infty}$ is decreasing. Let (α, k) be a pair in $\mathcal{C}$ corresponding to this subsequence as above. Then $(\alpha^*, k^*) \leq (\alpha, k)$.

Since $\sigma_i - \alpha_i \leq \sigma_i$ we must have $\Delta_\alpha = \Delta_{\sigma_i - \alpha_i} \leq \Delta_{\sigma_i} = \Delta_{\alpha^*}$ for some i. As $\alpha^* \leq \alpha$ we deduce that $\Delta_\alpha = \Delta_{\alpha^*}$.

For all sufficiently large i we now have

$$\Delta_{\alpha^*} = \Delta_{\sigma_i} = \Delta_{\sigma_i - \alpha_i} = \Delta_{\alpha_i}.$$

Hence there exists a positive integer m_i such that $\alpha^* \leq m_i\alpha_i$. Therefore $\sigma_i \leq k^* m_i \alpha_i$ for large i. It follows that the pair $(\alpha_i, k^* m_i)$ is in $\mathcal{C}$, with $\{\sigma_j\}_{j=1}^{\infty}$ as the corresponding decreasing sequence. By the minimality, $(\alpha^*, k^*) \leq (\alpha_i, k^* m_i)$, and therefore $\alpha^* \leq \alpha_i$.

For large i we now get

$$\sigma_i - \alpha_i \leq \sigma_i - \alpha^* \leq (k^* - 1)\alpha^*.$$

Consequently $(\alpha^*, k^* - 1) \in \mathcal{C}$, with $\{\sigma_i - \alpha_i\}$ as the corresponding decreasing sequence. But this contradicts the minimality of (α^*, k^*). □

PROPOSITION 2.7.5. *Let $(\Gamma, \leq)$ be an ordered abelian group. Let A be a well-ordered set of positive elements of Γ. Then an element of Γ can be written as a sum of finitely many elements of A in only finitely many ways.*

PROOF. Let S be the set of all sums of finitely many elements of A, and fix $\sigma \in S$. We show that σ has only finitely many presentations as a sum of elements of A. By Theorem 2.7.4, S is well-ordered, so we may assume that this is known for all smaller elements of S. Let A_0 be the set of all $\alpha \in A$ which appear in such presentations of σ. In this case $\sigma - \alpha \in S \cup \{0\}$.

CLAIM I: *A_0 is finite.*

Since A_0 is a subset of the well-ordered set A, it does not contain an infinite decreasing sequence. On the other hand, if $\{\alpha_i\}_{i=1}^{\infty}$ is an infinite increasing sequence in A_0, then $\{\sigma - \alpha_i\}_{i=1}^{\infty}$ is an infinite decreasing sequence in $S \cup \{0\}$, which is impossible, by Theorem 2.7.4 again. The finiteness of A_0 now follows from Lemma 2.7.1.

CLAIM II: *$\alpha \in A_0$ appears in at most finitely many presentations of σ.*

Indeed, there is an obvious bijective correspondence between:

(i) the presentations of σ as a sum of at least two elements of A and in which α appears; and
(ii) the presentations of $\sigma - \alpha$ as a sum of elements of A.

By assumption, there are only finitely many presentations as in (ii). Therefore there are only finitely many presentation as in (i). Also, there is clearly at most one presentation of σ using a single element of A. The claim follows.

By combining these two claims we obtain the assertion. □

In general, the union of an infinite collection of well-ordered subsets of an ordered group $(\Gamma, \leq)$ obviously need not be well-ordered. Lemma 2.7.6 below shows that this is nevertheless so in a special situation which will occur when we study generalized power series fields (see §2.8 and §18.4). This situation is the following one:

Let K be a set with a distinguished element 0 (in the application, K will be taken to be a field, and 0 its zero element). For an ordered abelian group $(\Gamma, \leq)$ let K^Γ be the set of all maps $f: \Gamma \to K$. Given $f \in K^\Gamma$ we define its **support** to be

$$\mathrm{Supp}(f) = \{\gamma \in \Gamma \mid f(\gamma) \neq 0\}.$$

We also define the **cut** Γ_f of f to be the set of all $\delta \in \Gamma$ for which there exists $\gamma \in \mathrm{Supp}(f)$ with $\delta \leq \gamma$. In particular, $\mathrm{Supp}(f) \subseteq \Gamma_f$.

For $f, f' \in K^\Gamma$ we write $f \preceq f'$ if f and f' coincide on Γ_f. Then f and f' coincide on $\mathrm{Supp}(f)$, so $\mathrm{Supp}(f) \subseteq \mathrm{Supp}(f')$ and therefore $\Gamma_f \subseteq \Gamma_{f'}$. It follows that $\preceq$ is a partial order on K^Γ.

LEMMA 2.7.6. *Let C be a non-empty chain in $(K^\Gamma, \preceq)$. Then:*
(a) *C has a unique least upper bound f^* in $(K^\Gamma, \preceq)$;*
(b) $\mathrm{Supp}(f^*) = \bigcup_{f \in C} \mathrm{Supp}(f)$;
(c) *if the supports* $\mathrm{Supp}(f)$, *$f \in C$, are well-ordered with respect to $\leq$, then so is* $\mathrm{Supp}(f^*)$.

PROOF. For $\gamma \in \Gamma$ we set:

$$f^*(\gamma) = \begin{cases} f(\gamma) & \text{if } \gamma \in \mathrm{Supp}(f) \text{ for some } f \in C, \\ 0 & \text{if } \gamma \notin \bigcup_{f \in C} \mathrm{Supp}(f). \end{cases}$$

Observe that since C is a chain, f^* is well defined. The verification of (a) and (b) is straightforward.

To obtain (c), assume that $\gamma_1 > \gamma_2 > \dots$ is an infinite strictly decreasing sequence in $\mathrm{Supp}(f^*)$. By (b), $\gamma_1 \in \mathrm{Supp}(f_1)$ for some $f_1 \in C$. Since $\mathrm{Supp}(f_1)$ is well-ordered, it cannot contain all the γ_i. Hence $\gamma_i \notin \mathrm{Supp}(f_1)$ for some $i \geq 2$. By (b) again, we can find $f_2 \in C$ with $\gamma_i \in \mathrm{Supp}(f_2)$. Then $\mathrm{Supp}(f_2) \not\subseteq \mathrm{Supp}(f_1)$, so $f_2 \not\preceq f_1$. But C is a chain, so necessarily $f_1 \preceq f_2$, i.e., f_1 and f_2 coincide on Γ_{f_1}. As $\gamma_i < \gamma_1 \in \mathrm{Supp}(f_1)$ we have $\gamma_i \in \Gamma_{f_1}$. This gives the contradiction $0 = f_1(\gamma_i) = f_2(\gamma_i) \neq 0$. □

2.8 Formal power series

In this section we give two general constructions of fields from ordered abelian groups. These constructions will serve as key examples of valued fields. The first one, of generalized power series, goes back to H. Hahn [**Ha**].

Let $(\Gamma, \leq)$ be an ordered abelian group and let K be a field.

DEFINITION 2.8.1. We denote the set of all maps $f\colon \Gamma \to K$ with a well-ordered support $\mathrm{Supp}(f)$ by $K((\Gamma))$.

Given $f, g \in K((\Gamma))$, define a map $f + g\colon \Gamma \to K$ by

$$(f+g)(\gamma) = f(\gamma) + g(\gamma)$$

for $\gamma \in \Gamma$. We have $\mathrm{Supp}(f+g) \subseteq \mathrm{Supp}(f) \cup \mathrm{Supp}(g)$. In light of Lemma 2.7.2(a)(b), $f + g \in K((\Gamma))$.

Further, for each $\gamma \in \Gamma$ the set $(\gamma - \mathrm{Supp}(f)) \cap \mathrm{Supp}(g)$ is finite (Lemma 2.7.2(d)). Hence we may define a map $f \cdot g\colon \Gamma \to K$ by the convolution

$$(f \cdot g)(\gamma) = \sum_{\delta \in \Gamma} f(\gamma - \delta) g(\delta)$$

for $\gamma \in \Gamma$. Then $\mathrm{Supp}(f \cdot g) \subseteq \mathrm{Supp}(f) + \mathrm{Supp}(g)$, so by Lemma 2.7.2(a)(c), $f \cdot g \in K((\Gamma))$.

It is convenient to write f formally as $\sum_{\gamma \in \Gamma} f(\gamma) t^{\gamma}$, where t is a symbol. Then the operations $+, \cdot$ on $K((\Gamma))$ defined above correspond to the formal addition and multiplication of power series.

THEOREM 2.8.2. *$K((\Gamma))$ is a field with respect to these operations.*

PROOF. The verification of the commutative, associative and distributive laws is routine. Also, the constant map 0 is neutral with respect to addition, and the map e given by $e(0) = 1$ and $e(\gamma) = 0$ for $\gamma \neq 0$ is neutral with respect to multiplication. The additive inverse of $f \in K((\Gamma))$ is the map $-f \in K((\Gamma))$ defined by $(-f)(\gamma) = -f(\gamma)$ for $\gamma \in \Gamma$.

It remains to show that every non-zero $f \in K((\Gamma))$ has a multiplicative inverse. Indeed, the set $\mathrm{Supp}(f)$ is non-empty and well-ordered. We may therefore set $\gamma = \min(\mathrm{Supp}(f))$. Define $g \in K((\Gamma))$ by

$$g(\gamma') = \begin{cases} f(\gamma)^{-1} & \gamma' = -\gamma, \\ 0 & \gamma' \neq -\gamma \end{cases}$$

(i.e., g is the formal power series $f(\gamma)^{-1} t^{-\gamma}$). Let $m = e - f \cdot g$. For $n \geq 0$ we also set $h_n = \sum_{k=0}^{n} m^k$. Then $m, h_n \in K((\Gamma))$, and $\mathrm{Supp}(m)$ is strictly positive.

Let S be the set of all sums of finitely many elements of $\mathrm{Supp}(m)$. By Theorem 2.7.4, S is well-ordered. For every non-negative integer k, the support $\mathrm{Supp}(m^k)$ consists of sums of k elements of $\mathrm{Supp}(m)$, and therefore is a subset of S. Hence $\mathrm{Supp}(h_n) \subseteq S$ for each n. Moreover, in light of Proposition 2.7.5, an element of Γ can be in the support of at most finitely many m^k. Consequently, for each $\gamma \in \Gamma$ the sequence $\{h_n(\gamma)\}_{n=0}^{\infty}$ stabilizes to some $h(\gamma) \in K$ (by abuse of notation, $h = \sum_{k=0}^{\infty} m^k$). For the map $h\colon \Gamma \to K$ we thus obtain $\mathrm{Supp}(h) \subseteq S$, so by Lemma 2.7.2(a), $h \in K((\Gamma))$.

We claim that $(f \cdot g) \cdot h = e$. Indeed, let $\gamma \in \Gamma$ and let

$$S_0 = (\gamma - \mathrm{Supp}(f \cdot g)) \cap S.$$

One has

$$((f \cdot g) \cdot h)(\gamma) = \sum_{\delta \in \Gamma} \Big((f \cdot g)(\gamma - \delta) \cdot h(\delta)\Big) = \sum_{\delta \in S_0} \Big((f \cdot g)(\gamma - \delta) \cdot h(\delta)\Big)$$

and for every n also

$$\big((f\cdot g)\cdot h_n\big)(\gamma)=\sum_{\delta\in\Gamma}\Big((f\cdot g)(\gamma-\delta)\cdot h_n(\delta)\Big)=\sum_{\delta\in S_0}\Big((f\cdot g)(\gamma-\delta)\cdot h_n(\delta)\Big).$$

By Lemma 2.7.2(d), S_0 is finite. Therefore for n sufficiently large, h and h_n coincide on it. It follows that

$$\big((f\cdot g)\cdot h\big)(\gamma)=\big((f\cdot g)\cdot h_n\big)(\gamma)=\Big((e-m)\cdot\sum_{k=0}^{n}m^k\Big)(\gamma)=(e-m^{n+1})(\gamma).$$

Taking n even larger, if necessary, we get $(e-m^{n+1})(\gamma)=e(\gamma)$, proving our claim.

It follows that

$$f\cdot\big(g\cdot h\big)=\big(f\cdot g\big)\cdot h=e,$$

so f is invertible in $K((\Gamma))$. □

We call $K((\Gamma))$ the field of **generalized formal power series** over K with exponents in Γ. When $\Gamma=\mathbb{Z}$ with the usual order, $K((\mathbb{Z}))$ is the field $K((t))$ of **Laurent series** over K. We embed K in $K((\Gamma))$ by identifying $a\in K$ with the map which is a at 0 and zero elsewhere. In this manner K becomes a subfield of $K((\Gamma))$.

For the next proof we observe that the subset $K[[\Gamma]]$ of $K((\Gamma))$ consisting of all $f\in K((\Gamma))$ with non-negative support (and including the zero map) forms a subring of $K((\Gamma))$. The map

$$\psi\colon K[[\Gamma]]\to K,\quad f\mapsto f(0),$$

is a ring homomorphism.

PROPOSITION 2.8.3. *K is algebraically closed in $K((\Gamma))$.*

PROOF. Suppose that $G(X)$ is an irreducible polynomial in $K[X]$, and that $0\neq f\in K((\Gamma))$ satisfies $G(f)=0$. Let $n=\deg(G)\,(\geq 1)$ and let $\gamma=\min(\mathrm{Supp}(f))$. If $\gamma<0$, then $(G(f))(n\gamma)\neq 0$, a contradiction. We conclude that $f\in K[[\Gamma]]$. For the K-homomorphism ψ as above, we therefore get

$$G(\psi(f))=\psi(G(f))=\psi(0)=0.$$

Since $G(X)$ is irreducible, this can happen only when $n=1$. Thus $f\in K$. □

The construction above is functorial in the following sense: suppose that $K\subseteq K'$ are fields and that Γ,Γ' are ordered abelian groups with $\Gamma\leq\Gamma'$. Identify each map $f\colon\Gamma\to K$ with the map $\Gamma'\to K'$ which is f on Γ and is 0 elsewhere, to obtain an embedding of $K((\Gamma))$ as a subfield of $K'((\Gamma'))$.

PROPOSITION 2.8.4. $[K'((\Gamma')):K((\Gamma))]=(\Gamma':\Gamma)[K':K]$.

PROOF. Take a system of representatives γ_i', $i\in I$, for the cosets of Γ in Γ'. Also take a K-linear basis α_j', $j\in J$, of K'. Then every $f'\in K'((\Gamma'))$ can be uniquely written as a formal sum

$$f'=\sum_{i\in I}\sum_{j\in J}f_{ij}\cdot\alpha_j'\cdot t^{\gamma_i'}$$

for some $f_{ij} \in K((\Gamma))$. Thus, when $(\Gamma' : \Gamma), [K' : K] < \infty$ the monomials $\alpha'_j t^{\gamma'_i}$, $i \in I$, $j \in J$, form a linear basis of $K'((\Gamma'))$ over $K((\Gamma))$. When $(\Gamma' : \Gamma) = \infty$ or $[K' : K] = \infty$ we obtain that $[K'((\Gamma')) : K((\Gamma))] = \infty$. □

Let K and $(\Gamma, \leq)$ be as before, and consider the collection of all intermediate groups $\Gamma \leq \Gamma' \leq \Gamma_{\rm div}$ with $(\Gamma' : \Gamma) < \infty$. It forms a direct system with respect to inclusion. We define the field of **Γ-Puiseux series over** K to be the direct limit

$$K_{\rm Puis}((\Gamma)) = \varinjlim K((\Gamma')).$$

In light of Proposition 2.8.4, it is an algebraic extension of $K((\Gamma))$. Obviously, it is a subfield of $K((\Gamma_{\rm div}))$. By Proposition 2.8.3, K is algebraically closed in $K_{\rm Puis}((\Gamma))$. In the special case where $\Gamma = \mathbb{Z}$ we obtain the familiar field

$$K_{\rm Puis}((\mathbb{Z})) = \bigcup_{n \in \mathbb{N}} K((\tfrac{1}{n}\mathbb{Z}))$$

of classical Puiseux series over K.

In general, the field $K_{\rm Puis}((\Gamma))$ is *strictly* contained in $K((\Gamma_{\rm div}))$. For instance, the formal power series $\sum_{k=2}^{\infty} t^{k+k^{-1}}$ is in $K((\mathbb{Q}))$, but not in any $K((\frac{1}{n}\mathbb{Z}))$, hence not in $K_{\rm Puis}((\mathbb{Z}))$.

2.9 Generalized rational functions

As a third general construction of this type, we imitate the construction of the field of rational functions in one variable over a given field. To this end let as before K be a field and $(\Gamma, \leq)$ an ordered abelian group. Let R be the subset of $K((\Gamma))$ consisting of all maps $f\colon \Gamma \to K$ with *finite* support. It forms a ring with respect to the addition and multiplication of $K((\Gamma))$. The quotient field $K(\Gamma)$ of R is a subfield of $K((\Gamma))$. We call it the field of **generalized rational functions** over K with exponents in Γ. We identify again the map $f \in R$ with the (finite) formal linear combination $\sum_{\gamma \in \Gamma} f(\gamma) t^\gamma$, and view the elements of K as the constant formal polynomials in $K(\Gamma)$. For example, when $\Gamma = \mathbb{Z}$ with its standard order, $K(\mathbb{Z})$ is the usual field $K(t)$ of rational functions in one variable t over K. More generally, we have:

PROPOSITION 2.9.1. *The field K is algebraically closed in $K(\Gamma)$. The transcendence degree of $K(\Gamma)$ over K is the rational rank*

$$\operatorname{rr}(\Gamma) = \dim_{\mathbb{Q}}(\Gamma \otimes_{\mathbb{Z}} \mathbb{Q}).$$

PROOF. The first assertion follows from Proposition 2.8.3.

To compute the transcendence degree, choose $\gamma_i \in \Gamma$, $i \in I$, such that $\gamma_i \otimes 1$, $i \in I$, form a $\mathbb{Q}$-linear basis of $\Gamma_{\rm div} = \Gamma \otimes_{\mathbb{Z}} \mathbb{Q}$. Consider the monomials t^{γ_i} (i.e., the maps $\Gamma \to K$ which are 1 at γ_i and 0 elsewhere). Then each monomial t^γ, $\gamma \in \Gamma$, is algebraic over $K(t^{\gamma_i} \mid i \in I)$. Since the monomials t^γ, $\gamma \in \Gamma$, generate $K(\Gamma)$ over K, the extension $K(\Gamma)/K(t^{\gamma_i} \mid i \in I)$ is therefore algebraic.

It remains to show that t^{γ_i}, $i \in I$, are algebraically independent over K. To this end let $i_1, \ldots, i_m$ be distinct elements of I. Let

$$H(X_1, \ldots, X_m) = \sum_{r_1, \ldots, r_m \in \mathbb{N}} a_{r_1, \ldots, r_m} X_1^{r_1} \cdots X_n^{r_m}$$

be a polynomial in $K[X_1, \dots, X_m]$ such that $H(t^{\gamma_{i_1}}, \dots, t^{\gamma_{i_m}}) = 0$, i.e.,

$$\sum_{r_1,\dots,r_m \in \mathbb{N}} a_{r_1,\dots,r_m} t^{r_1\gamma_{i_1}+\cdots+r_n\gamma_{i_m}} = 0.$$

The combinations $r_1\gamma_{i_1}+\cdots+r_n\gamma_{i_m}$ are distinct for distinct m-tuples $(r_1, \dots, r_m) \in \mathbb{N}^m$. Consequently, $H = 0$, as desired. □

REMARK 2.9.2. When Γ is a free $\mathbb{Z}$-module, the extension $K(\Gamma)/K$ is even purely transcendental. Indeed, one can take in the previous argument the γ_i, $i \in I$, to be a $\mathbb{Z}$-basis of Γ. Then $K(\Gamma) = K(t^{\gamma_i} \mid i \in I)$. In general, this need not be the case, e.g., $K(\mathbb{Q})/K$ is not purely transcendental.

As an application of these constructions we obtain:

COROLLARY 2.9.3 (Hilbert [**Hi**]). *Every ordered abelian group embeds in the multiplicative group of some field.*

PROOF. Choose an arbitrary field K and let $F = K(\Gamma)$. To any $\gamma \in \Gamma$ we associate a map $f_\gamma : \Gamma \to K$ in $K(\Gamma)$ by setting

$$f_\gamma(\gamma') = \begin{cases} 1 & \text{if } \gamma = \gamma', \\ 0 & \text{if } \gamma \neq \gamma' \end{cases}$$

(in formal power series notation, f_γ corresponds to t^γ). We observe that the map $\gamma \to f_\gamma$ is a group monomorphism $\Gamma \to K(\Gamma)^\times$. □

Part II

VALUATIONS AND ORDERINGS

In this part of the book we define and study the basic notions associated with valuations and orderings on a field F (§3, §6). In §4 we present the main examples of valued fields – mostly those arising from unique factorization domains and generalized power series fields. An additional important class of examples – namely, quotient fields of rings of Witt vectors – will be postponed to §12, due to its technical complexity. In §8 we discuss the interplay between the arithmetic and topological aspects of valuations and orderings, which lead to the various approximation theorems (§10). For rank 1 valuations and Archimedean orderings these topologies are metric, leading to natural notions of completeness (§9).

Generally speaking, the field F will be kept fixed, and we will study the collection of all valuations and orderings on it. This collection has a natural tree structure, which is compatible with the operation of residue field in a natural sense (§5, §7).

Our main focus will be on the multiplicative group of the field and its various subgroups. Thus our point of view here will be *K-theoretic*, as discussed in the Introduction. A main notion which then arises is that of valuations and orderings that are *compatible* with such a multiplicative subgroup (see §11.1); this is an analog of Henselity, to be discussed in Part III. Restricting the above tree structure to compatible valuations leads to explicit constructions of canonical valuations on the field (§§11.2–4).

CHAPTER 3

Valuations

Krull valuations can be described in several equivalent ways: as certain subrings of the field, as homomorphisms from its multiplicative group onto an ordered abelian group (satisfying the ultrametric inequality), and as places on the field. This chapter presents these three approaches and the connections between them.

3.1 Valuation rings

Let F be a field. A **valuation ring** on F is a subring O of F such that for every $x \in F^\times$ at least one of x, x^{-1} is in O. As usual, we denote the group of all $x \in F^\times$ such that both x and x^{-1} are in O by $O^\times$, and call it the **group of units** of O. Note that every subring O' of F containing the valuation ring O is also a valuation ring on F.

LEMMA 3.1.1. *Let O be a valuation ring. Then $\mathfrak{m} = O \setminus O^\times$ is the unique maximal ideal in O.*

PROOF. Clearly, $0 \in \mathfrak{m}$. To show that $O\mathfrak{m} = \mathfrak{m}$, take $x \in O$ and $0 \neq y \in \mathfrak{m}$. Then $xy \in O$ and $y^{-1} \notin O$. As $xO \subseteq O$ also $y^{-1} \notin xO$, i.e., $(xy)^{-1} \notin O$, showing that $xy \in \mathfrak{m}$.

Next we show that for $0 \neq x, y \in \mathfrak{m}$ also $x + y \in \mathfrak{m}$. By symmetry, we may assume that $xy^{-1} \in O$. Then $(x+y)y^{-1} = xy^{-1} + 1 \in O$. Since $y^{-1} \notin O$ this implies that $x + y \notin O^\times$, as desired.

Consequently, $\mathfrak{m}$ is an ideal in O.

Finally, if I is an ideal in O, then $(O \setminus \mathfrak{m}) \cap I = O^\times \cap I = \emptyset$, so $I \subseteq \mathfrak{m}$. □

In the notation of Lemma 3.1.1, we call the field $\bar{F} = O/\mathfrak{m}$ the **residue field** of O. We observe that $\operatorname{char} \bar{F} = p$ for some prime number p if and only if $p \in \mathfrak{m}$. Consequently, $\operatorname{char} \bar{F} = 0$ if and only if $\mathbb{Q} \subseteq O$.

The set $1 + \mathfrak{m}$ is a subgroup of $O^\times$, called the group of **principal units** (or **1-units**) of O. We observe that

$$O = O^\times \cup (1 - O^\times), \quad \mathfrak{m} = (1 - O^\times) \setminus O^\times.$$

The valuation ring O is called **trivial** if $O = F$.

EXAMPLE 3.1.2. Consider $F = \mathbb{Q}$, let p be a prime number, and let $O = O_p$ be the set consisting of 0 as well as of all rational numbers $p^r n/m$, where $n, m \in \mathbb{Z}$ are not divisible by p, and $r \geq 0$. It is a valuation ring, called the **p-adic** valuation ring on $\mathbb{Q}$. Its maximal ideal $\mathfrak{m}$ consists of 0 as well as all such fractions with $r \geq 1$. The residue field $O/\mathfrak{m}$ may be identified with $\mathbb{F}_p$.

Many more examples of valuation rings will be given in §4.

PROPOSITION 3.1.3. *A valuation ring O on a field F is integrally closed in F.*

PROOF. Let $a \in F$, let $c_1, \dots, c_n \in O$, and suppose that

$$a^n + c_1 a^{n-1} + c_2 a^{n-2} + \cdots + c_{n-1} a + c_n = 0.$$

We need to show that $a \in O$. If not, then $a \neq 0$ and $a^{-1} \in O$. Dividing by a^{n-1}, we get

$$a = -c_1 - c_2 a^{-1} - \cdots - c_{n-1} a^{-(n-2)} - c_n a^{-(n-1)} \in O,$$

a contradiction. □

3.2 Valuations

We fix again a field F. A **valuation** on F is a group homomorphism v from $F^\times$ into an ordered abelian group $(\Gamma, \leq)$ such that for every $x, y \in F^\times$ with $x \neq -y$ the following **ultrametric inequality** holds:

$$v(x+y) \geq \min\{v(x), v(y)\}.$$

As already mentioned in the Introduction, sometimes one uses the term **"Krull valuation"** for a valuation, to distinguish it from discrete valuations, or valuations of rank 1 (see below).

Of course, one can replace Γ in the definition above by its subgroup $v(F^\times)$, to assume that $v\colon F^\times \to \Gamma$ is surjective (here and in the sequel we suppress the total order $\leq$ on Γ). In this case we call Γ the **value group** of the valuation v. It is often convenient to extend v to all of F by setting $v(0)$ to be a formal element ∞, which by definition is strictly larger than all the elements of Γ. One then sets $\gamma + \infty = \infty + \infty = \infty$ for every $\gamma \in \Gamma$.

LEMMA 3.2.1. *For a valuation v on F one has:*
(a) $v(-1) = 0$;
(b) $v(-x) = v(x)$ *for every* $x \in F^\times$;
(c) *if* $x, y \in F^\times$ *satisfy* $v(x) < v(y)$, *then* $v(x+y) = v(x)$;
(d) *if* $x_1, \dots, x_n \in F^\times$ *and* $v(x_1), \dots, v(x_n)$ *are distinct, then*

$$v\Big(\sum_{i=1}^n x_i\Big) = \min_{1\leq i\leq n} v(x_i);$$

(e) *If* $x, y \in F^\times$ *satisfy* $x^m - x = y$, *where* $m \geq 2$ *is an integer and* $v(y) < 0$, *then* $mv(x) = v(y)$.

PROOF. (a) follows from $(-1)^2 = 1$ and from the fact that an ordered abelian group is torsion-free (Lemma 2.1.1(c)). (b) follows from (a).

For (c) suppose that $v(x) < v(y)$ but $v(x+y) \neq v(x)$. By the ultrametric inequality, this implies that $v(x+y) > v(x)$. The ultrametric inequality once again gives

$$v(x) = v((x+y) - y) \geq \min\{v(x+y), v(y)\} > v(x),$$

a contradiction.

(d) follows from (c) by induction on n.

For (e) we first observe that $v(x) < 0$, by the ultrametric inequality. Hence $v(x^m) = mv(x) < v(x)$, so (c) gives

$$mv(x) = v(x^m) = v(x^m - x) = v(y),$$

as required. □

When v is a valuation on the field F we call the pair (F, v) a **valued field**. We say that a valuation v is **trivial** if the value group $v(F^\times)$ is trivial, i.e., v is the zero homomorphism. Likewise we define the **rank** of v to be

$$\operatorname{rank}(v) = \operatorname{rank}(v(F^\times))$$

in the sense of §2.2. Valuations of rank 1 are also called **Archimedean** valuations; however, to avoid confusion with Archimedean absolute values (which correspond to orderings rather than to valuations), we will not use this terminology here.

3.2.2 EXAMPLES.

(1) Let F be an algebraic extension of a finite field $\mathbb{F}_p$. Then any valuation v on F must be trivial. Indeed, for every $a \in F^\times$ there is a positive integer n with $a^n = 1$, whence $nv(a) = 0$. Since $v(F^\times)$ is torsion free, $v(a) = 0$.

We shall see later (in Example 14.1.3) that all other fields do have non-trivial valuations.

(2) As a first example of a non-trivial valuation, consider $F = \mathbb{Q}$ and let p be a prime number. The ***p*-adic valuation** $v = v_p$ on $\mathbb{Q}$ is defined by

$$v_p(p^r n/m) = r$$

for $n, m \in \mathbb{Z}$ are not divisible by p. It is straightforward to verify that it is indeed a valuation (we shall prove a more general result in §4.1).

Valuations v, u on F are called **equivalent** if there is an isomorphism of ordered abelian groups $\varphi: v(F^\times) \to u(F^\times)$ such that $u = \varphi \circ v$. This is obviously an equivalence relation. One usually does not distinguish between equivalent valuations. As the following proposition shows, valuation rings and equivalence classes of valuations are essentially the same objects.

PROPOSITION 3.2.3.

(a) *If $v: F^\times \to \Gamma$ is a valuation on F, then*

$$O_v = \{x \in F \mid v(x) \geq 0\}$$

is a valuation ring on F.

(b) *Let O be a valuation ring on F, let S be a subgroup of $F^\times$, and define a binary relation $\leq$ on $S/(S \cap O^\times)$ by setting $x(S \cap O^\times) \leq y(S \cap O^\times)$ if and only if $y/x \in O$. Then $(S/(S \cap O^\times), \leq)$ is an ordered abelian group.*

(c) *Let v be a valuation on F and let S be a subgroup of $F^\times$. Then v induces an isomorphism $S/(S \cap O_v^\times) \cong v(S)$ of ordered abelian groups (for the order defined in (b)).*

(d) *For a valuation ring O on F, the projection $v_O: F^\times \to F^\times/O^\times$ is a valuation on F (for the order on $F^\times/O^\times$ defined in (b) with $S = F^\times$).*

(e) *The maps*

$$v \mapsto O_v, \quad O \mapsto v_O$$

give an inverse bijective correspondence between the equivalence classes of valuations on F and the valuation rings on F.

PROOF. (a) We have $v(1) = 0$, $v(0) = \infty > 0$, and by Lemma 3.2.1(a), $v(-1) = 0$. Therefore $1, 0, -1 \in O_v$. If $x, y \in O_v$, then

$$v(xy) = v(x) + v(y) \geq 0, \quad v(x+y) \geq \min\{v(x), v(y)\} \geq 0.$$

Thus $xy, x + y \in O_v$, so O_v is a ring. Clearly, it contains one of x, x^{-1} for every $x \in F^\times$.

(b) Since $1 \in O$, the relation $\leq$ is reflexive. By the definition of $O^\times$ it is asymmetric. Since O is closed under multiplication, it is transitive. For $x, y \in S$ either x/y or its inverse y/x are in O, showing that $\leq$ is a total order.

To show that the order $\leq$ is compatible with the group operation, take $x, y, z \in S$ with $x(S \cap O^\times) \leq y(S \cap O^\times)$. Then $yz/xz = y/x \in O$, so

$$x(S \cap O^\times) \cdot z(S \cap O^\times) \leq y(S \cap O^\times) \cdot z(S \cap O^\times) \quad ,$$

as desired.

(c) As $O_v^\times = \mathrm{Ker}(v)$, the valuation v induces a group isomorphism $S/(S \cap O_v^\times) \cong v(S)$. If $x, y \in S$ and $x(S \cap O_v^\times) \leq y(S \cap O_v^\times)$, then $y/x \in O_v$, so $v(x) \leq v(y)$. Therefore this is an isomorphism of ordered abelian groups.

(d) We only need to verify the ultrametric inequality. To this end we take $x, y \in F^\times$ with $x \neq -y$. By symmetry we may assume that $v_O(x) \leq v_O(y)$. Then $y/x \in O$, so $(x+y)/x = 1 + y/x \in O$, i.e., $v_O(x+y) \geq v_O(x)$.

(e) Suppose that O is a valuation ring on F. As $v_O(1) = 0$ we have $0 \leq v_O(x)$ if and only if $x \in O$. Therefore $O_{v_O} = O$.

Now let v and u be equivalent valuations on F. Thus there is an isomorphism $\varphi\colon v(F^\times) \to u(F^\times)$ of ordered abelian groups such that $u = \varphi \circ v$. For $x \in F^\times$ we then have $v(x) \geq 0$ if and only if $u(x) \geq 0$. Thus $O_v = O_u$.

Conversely, suppose that v and u are valuations with $O_v = O_u$. Then $F^\times/O_v^\times = F^\times/O_u^\times$ as ordered abelian groups (with respect to the orders defined in (b)). By (c), v is equivalent to the valuation $v_{O_v}\colon F^\times \to F^\times/O_v^\times$ induced by O_v, and similarly for u. Hence v and u are equivalent. □

For example, for a prime number p, the p-adic valuation on $\mathbb{Q}$ corresponds in this way to the p-adic valuation ring on $\mathbb{Q}$ (see Examples 3.1.2 and 3.2.2(2)).

NOTATION. Let v be a valuation on F and let O_v be the corresponding valuation ring, as in Proposition 3.2.3(a). We denote the maximal ideal and the residue field of O_v by $\mathfrak{m}_v$ and $\bar{F}_v$, respectively. We call them the **valuation ideal** and the **residue field of** v. Likewise we denote the group $1 + \mathfrak{m}_v$ of **principal units** of O_v by G_v. Thus

$$\mathfrak{m}_v = \{x \in F \mid v(x) > 0\}$$

and

$$G_v = \{x \in F \mid v(1-x) > 0\}.$$

We further have

$$O_v^\times = \{x \in F \mid v(x) = 0\}.$$

The projection

$$\pi_v\colon O_v \to \bar{F}_v = O_v/\mathfrak{m}_v$$

is called the **place associated with** v. It is a ring epimorphism, and it induces a group epimorphism $O_v^\times \to \bar{F}_v^\times$ with kernel G_v. Notice that v is a trivial valuation if and only if O_v is the trivial valuation ring (i.e., $O_v = F$).

Given a subgroup S of $F^\times$, we call the subgroup

$$\bar{S}_v = \pi_v(S \cap O_v^\times)$$

of $\bar{F}_v^\times$ the **push-down** of S under v. The maps v and π_v induce short exact sequences of abelian groups

$$(3.2.1) \qquad 1 \to S \cap O_v^\times \to S \xrightarrow{v} v(S) \to 0$$

$$(3.2.2) \qquad 1 \to S \cap G_v \to S \cap O_v^\times \xrightarrow{\pi_v} \bar{S}_v \to 1.$$

In particular, for $S = F^\times$ we obtain the exact sequences

$$(3.2.3) \qquad 1 \to O_v^\times \to F^\times \xrightarrow{v} v(F^\times) \to 0$$

$$(3.2.4) \qquad 1 \to G_v \to O_v^\times \xrightarrow{\pi_v} \bar{F}_v^\times \to 1.$$

From (3.2.1)–(3.2.4) and the snake lemma (Lemma 1.1.3) we deduce the following canonical exact sequences:

$$(3.2.5) \qquad 1 \to O_v^\times/(S \cap O_v^\times) \to F^\times/S \to v(F^\times)/v(S) \to 0,$$

$$(3.2.6) \qquad 1 \to G_v/(S \cap G_v) \to O_v^\times/(S \cap O_v^\times) \to \bar{F}_v^\times/\bar{S}_v \to 1.$$

An important situation (which will be studied in more detail in §11.1) is when $G_v \leq S$. Then the sequences (3.2.5)–(3.2.6) combine to the following canonical exact sequence:

$$(3.2.7) \qquad 1 \to \bar{F}_v^\times/\bar{S}_v \to F^\times/S \to v(F^\times)/v(S) \to 0.$$

The sequences (3.2.5) and (3.2.6) are especially useful when $S = (F^\times)^m$ for a positive integer m. Then $O_v^\times/(S \cap O_v^\times) = O_v^\times/m$ and $\bar{F}_v^\times/\bar{S}_v = \bar{F}_v^\times/m$. In some cases we also have $G_v/(S \cap G_v) = G_v/m$, as the next lemma shows.

LEMMA 3.2.4. *Let (F, v) be a valued field. Let m be a positive integer such that the map $\pi_v\colon {}_m(O_v^\times) \to {}_m(\bar{F}_v^\times)$ is surjective. Then $G_v^m = (F^\times)^m \cap G_v$.*

PROOF. Consider the commutative diagrams of multiplications by m:

$$\begin{array}{ccccccccc}
1 & \longrightarrow & O_v^\times & \longrightarrow & F^\times & \longrightarrow & v(F^\times) & \longrightarrow & 0 \\
 & & \downarrow m & & \downarrow m & & \downarrow m & & \\
1 & \longrightarrow & O_v^\times & \longrightarrow & F^\times & \longrightarrow & v(F^\times) & \longrightarrow & 0
\end{array}$$

$$\begin{array}{ccccccccc}
1 & \longrightarrow & G_v & \longrightarrow & O_v^\times & \longrightarrow & \bar{F}_v^\times & \longrightarrow & 1 \\
 & & \downarrow m & & \downarrow m & & \downarrow m & & \\
1 & \longrightarrow & G_v & \longrightarrow & O_v^\times & \longrightarrow & \bar{F}_v^\times & \longrightarrow & 1.
\end{array}$$

By Lemma 2.1.1(c), $v(F^\times)$ is torsion-free. Therefore the snake lemma (Lemma 1.1.3) shows that the induced homomorphisms

$$O_v^\times/m \to F^\times/m, \quad G_v/m \to O_v^\times/m$$

are injective. Therefore so is their composition, i.e., $G_v^m = (F^\times)^m \cap G_v$. □

We say that a valued field (E, u) **extends** a valued field (F, v) if F is a subfield of E and u extends v. Just as the standard notation E/F for a field extension, we write $(E, u)/(F, v)$ for a valued field extension. Then

$$\text{(3.2.8)} \qquad O_v = O_u \cap F, \quad O_v^\times = O_u^\times \cap F, \quad \mathfrak{m}_v = \mathfrak{m}_u \cap F, \quad G_v = G_u \cap F.$$

By Proposition 3.2.3 (for the subgroup $F^\times$ of $E^\times$), $v(F^\times) \cong F^\times/O_v^\times$ and $u(E^\times) \cong E^\times/O_u^\times$ as ordered abelian groups. Hence, when (3.2.8) holds, we may canonically identify $v(F^\times)$ as a subgroup of $u(E^\times)$ and view (E, u) as a valued field extension of (F, v). Also, (3.2.8) gives a canonical embedding of $\bar{F}_v$ as a subfield of $\bar{E}_u$.

We further say that a valued field extension $(E, u)/(F, v)$ is **immediate** if $v(F^\times) = u(E^\times)$ and $\bar{F}_v = \bar{E}_u$. From the exact sequence (3.2.7) (with F, S replaced by $E, F^\times G_u$, respectively) we obtain:

LEMMA 3.2.5. *An extension $(E, u)/(F, v)$ of valued fields is immediate if and only if $E^\times = F^\times G_u$.*

3.3 Places

While we have associated to any valuation v its place $\pi_v: O_v \to \bar{F}_v$, it is also possible to go the other way around. To see this let $F, \bar{F}$ be fields and let ∞ be a formal symbol. We extend the addition and multiplication on $\bar{F}$ into partial binary operations on $\bar{F} \cup \{\infty\}$ by setting $\infty + \infty = \infty$, $\infty \cdot \infty = \infty$, $a + \infty = \infty$ for all $a \in \bar{F}$, and $a \cdot \infty = \infty$ for all $a \in \bar{F} \setminus \{0\}$. A **place** from F to $\bar{F}$ is a map $\pi: F \to \bar{F} \cup \{\infty\}$ satisfying $\pi(1) = 1$ and such that

$$\pi(a + b) = \pi(a) + \pi(b), \quad \pi(ab) = \pi(a)\pi(b)$$

whenever $a, b \in F$ and the right-hand sides are well defined.

EXAMPLE 3.3.1. Let $F = \mathbb{Q}$ and $\bar{F} = \mathbb{F}_p$, where p is a prime number. Let $a = p^r n/m$ be a non-zero rational number, where $n, m \in \mathbb{Z}$ are not divisible by p. We define $\pi(a)$ to be the residue of n/m in $\mathbb{F}_p$, if $r \geq 0$, and $\pi(a) = \infty$ if $r < 0$. Also let $\pi(0) = 0$. This gives a place on $\mathbb{Q}$, called the **p-adic place**. It is related to the p-adic valuation ring and the p-adic valuation on $\mathbb{Q}$ as in the following proposition.

PROPOSITION 3.3.2. *Let π be a place from the field F to the field $\bar{F}$.*

(a) *The set*

$$O = \{a \in F \mid \pi(a) \neq \infty\}$$

is a valuation ring on F with maximal ideal

$$\mathfrak{m} = \{a \in F \mid \pi(a) = 0\}.$$

(b) *If v is the valuation on F corresponding to O, then there is an embedding $\bar{F}_v \to \bar{F}$ which commutes with $\pi_v: O = O_v \to \bar{F}_v$ and $\pi: O \to \bar{F}$.*

PROOF. (a) By definition, $1 \in O$, and O is closed under addition and multiplication. As

$$1 = \pi(1) = \pi(1) + \pi(0) = 1 + \pi(0),$$

we have $\pi(0) = 0$, whence $0 \in O$. Similarly,

$$0 = \pi(0) = \pi(1 + (-1)) = \pi(1) + \pi(-1) = 1 + \pi(-1)$$

implies that $-1 \in O$, so O is a ring.

It is immediate to see that $\mathfrak{m}$ is subset of O that contains 0 and is additively closed. If $a \in O$ and $b \in \mathfrak{m}$, then $\pi(ab) = \pi(a) \cdot 0 = 0$, so $ab \in \mathfrak{m}$. Thus $\mathfrak{m}$ is an ideal in O.

If $a \in O \setminus \mathfrak{m}$ and $\pi(a^{-1}) = \infty$, then $\pi(a) \in F \setminus \{0\}$ and

$$1 = \pi(a \cdot a^{-1}) = \pi(a) \cdot \pi(a^{-1}) = \infty,$$

a contradiction. Therefore $a \in O \setminus \mathfrak{m}$ implies that $a^{-1} \in O$. This shows that $\mathfrak{m}$ is the unique maximal ideal of O.

It remains to show that if $a \in F \setminus O$, then $a^{-1} \in O$. Otherwise $\pi(a) = \pi(a^{-1}) = \infty$, whence

$$1 = \pi(1) = \pi(a) \cdot \pi(a^{-1}) = \infty \cdot \infty = \infty,$$

a contradiction.

(b) The restriction of π to O is a ring homomorphism into the field $\bar{F}$ and has kernel $\mathfrak{m}$. Hence it induces a field homomorphism $\bar{F}_v = O/\mathfrak{m} \to \bar{F}$ as required. □

Conversely, suppose that O is a valuation ring on the field F and let $\mathfrak{m}$ and $\bar{F}$ be its maximal ideal and residue field, respectively. Define a map $\pi\colon F \to \bar{F} \cup \{\infty\}$ by

$$\pi(a) = \begin{cases} a + \mathfrak{m} & a \in O, \\ \infty & a \notin O. \end{cases}$$

It is straightforward to verify that π is a place from F to $\bar{F}$. Here

$$O = \{a \in F \mid \pi(a) \neq \infty\},$$

so this map $O \mapsto \pi$ is inverse to the map $\pi \mapsto O$ of Proposition 3.3.2. Thus the notions of a place from F and of a valuation ring on F are essentially equivalent.

3.4 Discrete valuations

A valuation v on the field F is called **discrete** (more precisely, **discrete of rank** 1) if $v(F^\times) \cong \mathbb{Z}$. In this case we say that (F, v) is a **discrete valued field**. Likewise, a **discrete valuation ring** is a domain which is a valuation ring on its quotient field F and such that the corresponding valuation on F is discrete. In this section we characterize these valuation rings (Proposition 3.4.3), and prove some special properties of discrete valuations, which will be used later on. First we make the following useful observation.

LEMMA 3.4.1. *Let R be a domain with quotient field F and let Γ be an ordered abelian group. Let $v\colon R \setminus \{0\} \to \Gamma$ be a map satisfying:*

(i) *$v(ab) = v(a) + v(b)$ for all non-zero $a, b \in R$;*
(ii) *$v(a + b) \geq \min\{v(a), v(b)\}$ for all non-zero $a, b \in R$ with $a \neq -b$.*

Then v extends uniquely to a valuation $\hat{v}\colon F^\times \to \Gamma$. It is given by

$$\hat{v}(a/b) = v(a) - v(b)$$

for $a, b \in R \setminus \{0\}$.

PROOF. By (i), the map $\hat{v}\colon F^\times \to \Gamma$ defined as in the last sentence of the lemma is a well defined group homomorphism. By (i) again, $v(1) = 0$, so $\hat{v}$ extends v. To prove the ultrametric inequality, let $a, a', b, b' \in R \setminus \{0\}$ and suppose that $a/b \neq -a'/b'$. Then

$$\begin{aligned}\hat{v}\left(\frac{a}{b} + \frac{a'}{b'}\right) &= v(ab' + a'b) - v(bb') \\ &\geq \min\Big\{v(ab'), v(a'b)\Big\} - v(b) - v(b') \\ &= \min\Big\{v(a) - v(b), v(a') - v(b')\Big\} = \min\Big\{\hat{v}\Big(\frac{a}{b}\Big), \hat{v}\Big(\frac{a'}{b'}\Big)\Big\}.\end{aligned}$$

The uniqueness is immediate. □

3.4.2 REMARKS.

(a) A discrete valuation ring O is a principal ideal domain. Indeed, let I be an ideal in O. There exists $a \in I$ with $v_O(a)$ minimal, where v_O is the valuation corresponding to O. Since I is an ideal, $aO \subseteq I$. Conversely, for every $b \in I$ one has $v_O(a) \leq v_O(b)$, i.e., $b \in aO$. Thus $I = aO$ is a principal ideal.

(b) In particular, for every discrete valuation ring O on F there exists $\lambda \in F$ such that the maximal ideal of O is λO. Equivalently, $v(F^\times) = v(\lambda)\mathbb{Z}$, where $v = v_O$ is the corresponding valuation. We call such an element λ a **uniformizer** of O. It is often convenient to identify $v(F^\times)$ with $\mathbb{Z}$ and $v(\lambda)$ with 1.

PROPOSITION 3.4.3. *A ring O is a discrete valuation ring if and only if there exists $\lambda \in O$ such that:*

(i) *λ is not a zero divisor in O;*
(ii) *$\mathfrak{m} = \lambda O$ is a maximal ideal of O;*
(iii) $\bigcap_{k=0}^{\infty} \lambda^k O = \{0\}$.

Moreover, in this case the valuation $\hat{v}$ corresponding to O is given for a non-zero $a \in O$ by $\hat{v}(a) = k$, where k is the maximal non-negative integer such that $a \in \lambda^k O$.

PROOF. If O is a discrete valuation ring on a field, then (i)–(iii) obviously hold with $\mathfrak{m} = \lambda O$ for a uniformizer λ of O.

Conversely, suppose that λ satisfies (i)–(iii). Given $a \in O \setminus \{0\}$ we define $v(a) \in \mathbb{N}$ as in the assertion (by (iii), v is well defined). We show that O is a domain and the map $v\colon O \setminus \{0\} \to \mathbb{Z}$ satisfies conditions (i) and (ii) of Lemma 3.4.1 for $R = O$.

To this end take non-zero $a, b \in O$ with $v(a) = k$ and $v(b) = l$. Write $a = \lambda^k a_1$ and $b = \lambda^l b_1$, with $a_1, b_1 \in O \setminus \mathfrak{m}$. Then $ab = \lambda^{k+l} a_1 b_1 \in \lambda^{k+l} O$. Further, if $ab \in \lambda^{k+l+1} O$, then by (i), $a_1 b_1 \in \lambda O = \mathfrak{m}$. Since $\mathfrak{m}$ is a prime ideal, $a_1 \in \mathfrak{m}$ or $b_1 \in \mathfrak{m}$, a contradiction. We conclude that $v(ab) = k + l$. In particular, this shows that O is a domain. Furthermore, setting $r = \min\{k, l\}$ we have $a + b \in \lambda^r O$. Hence, if $a \neq -b$, then

$$v(a+b) \geq r = \min\{v(a), v(b)\}.$$

Next let F be the quotient field of O. It follows from Lemma 3.4.1 that v extends to a valuation $\hat{v}: F^\times \to \mathbb{Z}$ such that $\hat{v}(a/b) = v(a) - v(b)$ for $a, b \in O \setminus \{0\}$. By the construction, $O \subseteq O_{\hat{v}}$.

For the converse inclusion take a non-zero $x \in O_{\hat{v}}$, and write $x = a/b$, with $a, b \in O \setminus \{0\}$. As before, there exist $a_1, b_1 \in O \setminus \mathfrak{m}$ and non-negative integers k, l such that $a = \lambda^k a_1$ and $b = \lambda^l b_1$, so $x = \lambda^{k-l} a_1/b_1$. By (ii), $O \setminus \mathfrak{m} \subseteq O^\times$. Hence $a_1, b_1 \in O^\times \subseteq O_{\hat{v}}^\times$. Now

$$0 \leq \hat{v}(x) = v(\lambda^k a_1) - v(\lambda^l b_1) = k - l,$$

so $x \in O$. We conclude that $O_{\hat{v}} = O$. □

Proposition 3.4.4. *Let (F, v) be a discrete valued field. For every $i \geq 1$ there is a group isomorphism*

$$(1 + \mathfrak{m}_v^i)/(1 + \mathfrak{m}_v^{i+1}) \cong \bar{F}_v.$$

Proof. Choose a uniformizer λ of O_v. Then $\mathfrak{m}_v^i = \lambda^i O_v$. Define a map $\psi: 1 + \mathfrak{m}_v^i \to \bar{F}_v$ by $\psi(1 + \lambda^i b) = \pi_v(b)$ for $b \in O_v$. Given $b, c \in O_v$ we have

$$\begin{aligned} \psi((1+\lambda^i b)(1+\lambda^i c)) &= \psi(1 + \lambda^i(b + c + \lambda^i bc)) \\ &= \pi_v(b + c + \lambda^i bc) = \pi_v(b) + \pi_v(c) \\ &= \psi(1 + \lambda^i b) + \psi(1 + \lambda^i c). \end{aligned}$$

Hence ψ is a homomorphism. It is clearly surjective and its kernel is $1 + \lambda^i \mathfrak{m}_v = 1 + \mathfrak{m}_v^{i+1}$. □

For discrete valuations we have the following refinement of Lemma 3.2.5.

Proposition 3.4.5. *Let (F, v), (E, u) be discrete valued fields such that (E, u) is an immediate extension of (F, v). Then*

$$(1 + \mathfrak{m}_u^i)/(1 + \mathfrak{m}_v^i) \cong E^\times/F^\times$$

canonically for all $i \geq 1$. In particular, $E^\times = F^\times(1 + \mathfrak{m}_u^i)$.

Proof. For $i = 1$ this follows from the exact sequences (3.2.5)–(3.2.6) for the subgroup $F^\times$ of $E^\times$. For $i \geq 1$ Proposition 3.4.4 gives a commutative diagram with exact rows:

$$\begin{array}{ccccccccc} 0 & \longrightarrow & 1 + \mathfrak{m}_v^{i+1} & \longrightarrow & 1 + \mathfrak{m}_v^i & \longrightarrow & \bar{F}_v & \longrightarrow & 1 \\ & & \downarrow & & \downarrow & & \downarrow \cong & & \\ 0 & \longrightarrow & 1 + \mathfrak{m}_u^{i+1} & \longrightarrow & 1 + \mathfrak{m}_u^i & \longrightarrow & \bar{E}_u & \longrightarrow & 1. \end{array}$$

The snake lemma (Lemma 1.1.3) gives rise to a natural isomorphism

$$(1 + \mathfrak{m}_u^{i+1})/(1 + \mathfrak{m}_v^{i+1}) \xrightarrow{\sim} (1 + \mathfrak{m}_u^i)/(1 + \mathfrak{m}_v^i).$$

The desired isomorphism now follows from the case $i = 1$ by induction. □

CHAPTER 4

Examples of Valuations

We now illustrate the notions introduced in the previous chapter by several examples of valuations. Main sources of examples will be the field of generalized formal power series, as constructed in §2.8, and the fields of generalized Puiseux series and generalized rational functions with exponents in a given ordered abelian group. Another important example – namely, quotient fields of rings of Witt vectors – will be studied in §12. We start however with some fundamental examples of valuations on quotient fields of unique factorization domains.

4.1 Valuations from unique factorization domains

Let R be a unique factorization domain with quotient field F. Let $\mathcal{P}$ be the collection of all prime elements in R (up to multiplication by a unit). Thus every element x of $F^\times$ can be uniquely written as

$$x = u \prod_{\pi \in \mathcal{P}} \pi^{v_\pi(x)},$$

with $u \in R^\times$ and integers $v_{\pi(x)}$ which are 0 for all but finitely many $\pi \in \mathcal{P}$. For each $\pi \in \mathcal{P}$ the localization $R_{\pi R}$ at πR and its maximal ideal $\pi R_{\pi R}$ satisfy conditions (i)–(iii) of Proposition 3.4.3. Hence $R_{\pi R}$ is a discrete valuation ring on F. The corresponding valuation is $v_\pi \colon F^\times \to \mathbb{Z}$. We call v_π the **π-adic** valuation on F. Its residue field is $R_{\pi R}/\pi R_{\pi R} \cong R/\pi R$. We have the following converse result:

LEMMA 4.1.1. *Let R be a unique factorization domain with quotient field F. Let O be a non-trivial valuation ring on F containing R. Then O is the localization of R at πR for some prime element π in R.*

PROOF. Let $\mathfrak{m}$ be the maximal ideal of O. Since O is non-trivial, we can find a non-zero $x \in \mathfrak{m}$. Write $x = a/b$ with $0 \neq a, b \in R$. Then $a = bx \in R\mathfrak{m} \subseteq O\mathfrak{m} = \mathfrak{m}$. The ideal $R \cap \mathfrak{m}$ of R is prime and contains a, hence is non-zero. Since R is a principal ideal domain, there exists a prime element $\pi \in R$ with $R \cap \mathfrak{m} = \pi R$. Thus

$$R \setminus \pi R \subseteq O \setminus \mathfrak{m} = O^\times,$$

so $R_{\pi R} \subseteq O$. Furthermore, $\pi R_{\pi R} \subseteq \pi O \subseteq \mathfrak{m}$.

Conversely, take $x \in O$. Then $x^{-1} \notin \mathfrak{m}$, and in particular $x^{-1} \notin \pi R_{\pi R}$. As we have observed, $R_{\pi R}$ is a valuation ring with maximal ideal $\pi R_{\pi R}$. Hence $x \in R_{\pi R}$. We conclude that $O = R_{\pi R}$. □

We now consider two important special cases.

EXAMPLE 4.1.2 (Ostrowski [**O2**]): *Valuations on $\mathbb{Q}$.*

Apply the preceding discussion to the unique factorization domain $R = \mathbb{Z}$. For every prime number p we obtain as in Example 3.2.2(2) the p-adic valuation v_p on

$\mathbb{Q}$, given by $v_p(p^n a/b) = n$ for $n, a, b \in \mathbb{Z}$ with $p \not| a, b$. It is a discrete valuation with residue field $\mathbb{F}_p$. Furthermore, since every valuation ring on $\mathbb{Q}$ contains $\mathbb{Z}$ and in light of Lemma 4.1.1, the p-adic valuations are the only non-trivial valuations on $\mathbb{Q}$ (up to equivalence).

EXAMPLE 4.1.3: *K-trivial valuations on $K(t)$ for a field K.*

Take in the preceding discussion $R = K[t]$ and $F = K(t)$. For every non-constant irreducible polynomial $f(t) \in K[t]$ we obtain an **$f(t)$-adic valuation** on $K(t)$. Its value group is $\mathbb{Z}$ and its residue field is isomorphic to $K[t]/f(t)K[t]$, i.e., to the extension of K obtained by adjoining a zero of $f(t)$. By Lemma 4.1.1, up to equivalence these are the only non-trivial valuations on $K(t)$ which are non-negative on $K[t] \setminus \{0\}$.

Next suppose that v is valuation on $K(t)$ which is trivial on K but which is negative on some non-zero polynomial in $K[t]$. Then necessarily $v(t) < 0$. For every polynomial $\sum_{i=0}^{n} a_i t^i$ in $K[t]$ with $a_n \neq 0$ we obtain that

$$v\Big(\sum_{i=0}^{n} a_i t^i\Big) = \min_{1 \leq i \leq n} iv(t) = nv(t)$$

(see Lemma 3.2.1(d)). Hence v is a discrete valuation with value group $\mathbb{Z}v(t)$. It is equivalent to the **degree valuation** $v_\infty\colon K(t)^\times \to \mathbb{Z}$, defined by

$$v_\infty(f) = -\deg(f)$$

for a non-zero polynomial $f \in K[t]$. We note that v_∞ is also the s-adic valuation on $K(t)$ constructed as above from the ring $K[s]$, where $s = 1/t$.

The valuation ring O_{v_∞} (resp., the maximal ideal $\mathfrak{m}_{v_\infty}$) of v_∞ consists of all rational functions f/g, with $f, g \in K[t]$, $g \neq 0$, and $\deg(f) \leq \deg(g)$ (resp., $\deg(f) < \deg(g)$). Note that when $\deg(f) \leq \deg(g)$ there exist unique $a \in K$ and $r \in K[t]$ such that $f = ag + r$ and $\deg(r) < \deg(g)$. Then $f/g \in a + \mathfrak{m}_{v_\infty}$. Therefore the residue field $\bar{F}_{v_\infty}$ of v_∞ is isomorphic to K via the canonical map $a + \mathfrak{m}_{v_\infty} \mapsto a$.

Another basic example follows.

EXAMPLE 4.1.4: *The p-adic numbers.*

Let $\mathbb{Z}_p = \varprojlim \mathbb{Z}/p^k$ be the ring of p-adic integers. The ring $\mathbb{Z}/p^k$ has a unique maximal ideal $p\mathbb{Z}/p^k\mathbb{Z}$. Therefore $\mathbb{Z}_p$ has a unique maximal ideal $p\mathbb{Z}_p$. Furthermore, $\bigcap_{k=0}^{\infty} p^k\mathbb{Z}_p = \{0\}$. We conclude from Proposition 3.4.3 that $\mathbb{Z}_p$ is a discrete valuation ring with residue field $\mathbb{Z}_p/p \cong \mathbb{Z}/p = \mathbb{F}_p$. The quotient field $\mathbb{Q}_p$ of $\mathbb{Z}_p$ is the field of **p-adic numbers**, discovered by K. Hensel [**He**].

Every non-zero element a of $\mathbb{Q}_p$ can be uniquely written as $a = p^k b$, where k is an integer and $b \in \mathbb{Z}_p^\times$. Denoting the valuation on $\mathbb{Q}_p$ with valuation ring $\mathbb{Z}_p$ by v_p we then have $v_p(a) = k$. Thus for $a \in \mathbb{Q}_p$ one has $v_p(a) = k$ if and only if $\alpha \in p^k\mathbb{Z}_p \setminus p^{k+1}\mathbb{Z}_p$.

4.2 Valuations on power series fields

EXAMPLE 4.2.1: *Generalized formal power series.*

Let $(\Gamma, \leq)$ be an ordered abelian group and let (K, w) be a valued field. Let $K((\Gamma))$ be the field of generalized power series with coefficients in K and exponents in Γ, as constructed in §2.8. Recall that it consists of all maps $f\colon \Gamma \to K$ with a well-ordered support $\mathrm{Supp}(f)$. Equip $\Gamma \oplus w(K^\times)$ with the (left-to-right) lexicographic order. Given a non-zero $f \in K((\Gamma))$ let $\gamma = \min(\mathrm{Supp}(f))$ and define

$$v(f) = \big(\gamma, w(f(\gamma))\big).$$

This gives a well-defined map

$$v\colon K((\Gamma))^\times \to \Gamma \oplus w(K^\times).$$

We show that v is a valuation on $K((\Gamma))$. Thus, for non-zero $f, g \in K((\Gamma))$ with

$$\gamma_1 = \min(\mathrm{Supp}(f)), \quad \gamma_2 = \min(\mathrm{Supp}(g)),$$

we need to verify:

(i) $v(f \cdot g) = v(f) + v(g)$; and

(ii) if $f \neq -g$, then $v(f+g) \geq \min(v(f), v(g))$.

PROOF OF (i). We have $\min(\mathrm{Supp}(f \cdot g)) = \gamma_1 + \gamma_2$ and $(f \cdot g)(\gamma_1 + \gamma_2) = f(\gamma_1)g(\gamma_2)$. Hence

$$\begin{aligned} v(f \cdot g) &= (\gamma_1 + \gamma_2, w(f(\gamma_1)) + w(g(\gamma_2))) \\ &= (\gamma_1, w(f(\gamma_1))) + (\gamma_2, w(g(\gamma_2))) = v(f) + v(g), \end{aligned}$$

as desired.

PROOF OF (ii). Let $\gamma = \min(\mathrm{Supp}(f+g))$. Note that $\min\{\gamma_1, \gamma_2\} \leq \gamma$, and this is an equality when $\gamma_1 \neq \gamma_2$. We break the discussion into four cases.

CASE (1): $\gamma_1 < \gamma_2$. Then $\gamma = \gamma_1$ and $g(\gamma) = 0$, so

$$v(f+g) = (\gamma, w((f+g)(\gamma))) = (\gamma_1, w(f(\gamma_1))) = v(f).$$

CASE (2): $\gamma_1 > \gamma_2$. We similarly obtain that $v(f+g) = v(g)$.

CASE (3): $\gamma_1 = \gamma_2 < \gamma$. By the definition of the lexicographic order,

$$\begin{aligned} v(f+g) &= (\gamma, w(f(\gamma) + g(\gamma))) \\ &> \min\{(\gamma_1, w(f(\gamma_1))), (\gamma_2, w(g(\gamma_2)))\} \\ &= \min\{v(f), v(g)\}. \end{aligned}$$

CASE (4): $\gamma_1 = \gamma_2 = \gamma$. Then

$$\begin{aligned} v(f+g) &= (\gamma, w(f(\gamma) + g(\gamma))) \\ &\geq (\gamma, \min\{w(f(\gamma)), w(g(\gamma))\}) \\ &= \min\{(\gamma, w(f(\gamma))), (\gamma, w(g(\gamma)))\} \\ &= \min\{v(f), v(g)\} \end{aligned}$$

in this case as well.

Altogether, this proves that v is a valuation on $K((\Gamma))$ with value group $\Gamma \oplus w(K^\times)$. Its valuation ring O_v consists of all maps $f\colon \Gamma \to K$ such that either $f = 0$, or $\gamma = \min(\mathrm{Supp}(f)) > 0$, or $\gamma = 0$ and $w(f(0)) \geq 0$. The maximal ideal $\mathfrak{m}_v$ of O_v consists of all maps $f\colon \Gamma \to K$ such that either $f = 0$, or $\gamma = \min(\mathrm{Supp}(f)) > 0$, or $\gamma = 0$ and $w(f(0)) > 0$.

To compute the residue field $\bar{F}_v = O_v/\mathfrak{m}_v$, we identify the elements of O_w as power series in O_v with support in $\{0\}$. This gives a ring homomorphism $O_w \to O_v$, which in turn induces an isomorphism $O_w/\mathfrak{m}_w \xrightarrow{\sim} O_v/\mathfrak{m}_v$. Thus $\bar{K}_w \cong \bar{F}_v$ canonically.

In particular, when w is the trivial valuation we get $O_v = K[[\Gamma]]$ (in the sense of §2.8) and $\bar{F}_v = K$.

EXAMPLE 4.2.2: *Laurent series.*

Take in the previous example $\Gamma = \mathbb{Z}$ with its usual order, and take w to be the trivial valuation on K. We obtain a valuation v on the field of Laurent series $K((\mathbb{Z})) = K((t))$ over K in the formal variable t. Recall that $K((t))$ consists of all formal series $\sum_{i=N}^{\infty} a_i t^i$ where $N \in \mathbb{Z}$ and $a_i \in K$ for all $i \geq N$. When $a_N \neq 0$ one has $v(\sum_{i=N}^{\infty} a_i t^i) = N$. Thus v is a discrete valuation with residue field K.

EXAMPLE 4.2.3: *Γ-Puiseux series.*

Let (K, w) be a valued field and Γ an ordered abelian group. Consider the field

$$K_{\mathrm{Puis}}((\Gamma)) = \varinjlim K((\Gamma'))$$

of Γ-Puiseux series over K in the sense of §2.8. Recall that here Γ' ranges over all subgroups of Γ_{div} that contain Γ as a subgroup of finite index, and for such subgroups $\Gamma' \leq \Gamma''$ one identifies a map $f \in K((\Gamma'))$ with the map in $K((\Gamma''))$ which coincides with f on Γ' and is 0 elsewhere. Denote the canonical valuation on $K((\Gamma'))$ (in the sense of Example 4.2.1) by $v_{\Gamma'}$. When $\Gamma' \leq \Gamma''$ we observe that $v_{\Gamma''}$ extends $v_{\Gamma'}$. Therefore the $v_{\Gamma'}$ combine to a valuation v on $K_{\mathrm{Puis}}((\Gamma))$ with residue field $\bar{K}_w$ and value group

$$\Gamma_{\mathrm{div}} \oplus w(K^\times) = (\varinjlim \Gamma') \oplus w(K^\times) \quad .$$

In particular, take $\Gamma = \mathbb{Z}$ and $w = 0$. Then

$$K_{\mathrm{Puis}}((\mathbb{Z})) = \bigcup_{n=1}^{\infty} K((\tfrac{1}{n}\mathbb{Z}))$$

is the field of classical Puiseux series over K. The valuation constructed above then has residue field K and value group $\mathbb{Q}$.

EXAMPLE 4.2.4: *Generalized rational functions.*

Let $(\Gamma, \leq)$ be an ordered abelian group, let (K, w) be a valued field, and let $K(\Gamma)$ be the field of generalized rational functions in one variable over K with exponents in Γ, as in §2.9. Construct a valuation v on $K((\Gamma))$ as in Example 4.2.1. The restriction of v to $K(\Gamma)$ is also a valuation with the same value group Γ and the same residue field $\bar{K}_w$.

For instance, take $\Gamma = \mathbb{Z}$ and let w be the trivial valuation on K. Then $K(\Gamma)$ is the field $K(t)$ of rational functions over K, and the valuation v is its t-adic valuation.

4.3 Gauss valuations

Next we present a construction of valuations which is in some sense opposite to the constructions of §4.2.

EXAMPLE 4.3.1: *Extended Gauss valuations.*

Let (K, w) be a valued field and let $(\Gamma, \leq)$ be an ordered abelian group. For convenience we set as before $w(0) = \infty$. We construct a valuation on the field $K(\Gamma)$ of generalized rational functions over K with exponents in Γ as follows.

Order $w(K^\times) \oplus \Gamma$ lexicographically (note that this is the opposite lexicographic order to the one used in Examples 4.2.1–4.2.4). Let R be the ring of all $f \in K((\Gamma))$ with finite support. Define a map

$$v\colon R \setminus \{0\} \to w(K^\times) \oplus \Gamma$$

by setting

$$v(f) = \min_{\gamma \in \Gamma}\Big(w(f(\gamma)), \gamma\Big)$$

for $f \in R$ (note that it is enough to consider $\gamma \in \mathrm{Supp}(f)$, so v is well defined). We verify conditions (i) and (ii) of Lemma 3.4.1.

PROOF OF (I). For non-zero $f, g \in R$, let $h = f \cdot g$. Recall that

$$h(\gamma) = \sum_{\delta' + \delta'' = \gamma} f(\delta')g(\delta'')$$

for $\gamma \in \Gamma$. We need to show that $v(h) = v(f) + v(g)$. To this end choose $\delta_0', \delta_0'' \in \Gamma$ with

$$v(f) = (w(f(\delta_0')), \delta_0'), \quad v(g) = (w(g(\delta_0'')), \delta_0'').$$

Now consider arbitrary $\delta', \delta'' \in \Gamma$. If $\delta' < \delta_0'$, then $w(f(\delta')) > w(f(\delta_0'))$. As $w(g(\delta'')) \geq w(g(\delta_0''))$ this implies that $w(f(\delta')g(\delta'')) > w(f(\delta_0')g(\delta_0''))$. Similarly $\delta'' < \delta_0''$ implies that $w(f(\delta')g(\delta'')) > w(f(\delta_0')g(\delta_0''))$. Setting $\gamma_0 = \delta_0' + \delta_0''$, we deduce from Lemma 3.2.1(c) that

$$w(h(\gamma_0)) = w\Big(f(\delta_0')g(\delta_0'') + \sum_{\substack{\delta' + \delta'' = \gamma_0 \\ (\delta', \delta'') \neq (\delta_0', \delta_0'')}} f(\delta')g(\delta'')\Big) = w(f(\delta_0')g(\delta_0'')).$$

For every $\gamma \in \Gamma$ the ultrametric inequality for w now gives

$$\begin{aligned} w(h(\gamma)) &= w\Big(\sum_{\delta' + \delta'' = \gamma} f(\delta')g(\delta'')\Big) \\ &\geq \min_{\delta' + \delta'' = \gamma} w(f(\delta')g(\delta'')) \\ &\geq w(f(\delta_0')g(\delta_0'')) = w(h(\gamma_0)). \end{aligned}$$

If this is an equality, then for some δ', δ'' with $\delta' + \delta'' = \gamma$ we have

$$w(f(\delta')) = w(f(\delta_0')), \quad w(g(\delta'')) = w(g(\delta_0'')).$$

The minimality conditions on δ_0', δ_0'' then imply that $\delta' \geq \delta_0'$ and $\delta'' \geq \delta_0''$, so $\gamma \geq \gamma_0$. Therefore $(w(h(\gamma)), \gamma) \geq (w(h(\gamma_0)), \gamma_0)$.

Altogether we obtain:

$$\begin{aligned} v(h) = \min_{\gamma \in \Gamma}\big(w(h(\gamma)), \gamma\big) &= \big(w(h(\gamma_0)), \gamma_0\big) \\ &= (w(f(\delta_0')g(\delta_0'')), \delta_0' + \delta_0'') = v(f) + v(g), \end{aligned}$$

as desired.

PROOF OF (ii). Consider non-zero $f, g \in R$ with $f \neq -g$. We need to show the ultrametric inequality $v(f+g) \geq \min\{v(f), v(g)\}$. Take $\gamma_0, \gamma_1, \gamma_2 \in \Gamma$ with

$$v(f+g) = (w(f(\gamma_0)+g(\gamma_0)), \gamma_0),\ v(f) = (w(f(\gamma_1)), \gamma_1),\ v(g) = (w(g(\gamma_2)), \gamma_2).$$

Without loss of generality, $w(f(\gamma_0)) \leq w(g(\gamma_0))$. The ultrametric inequality for w and the choice of γ_1 then give

$$w(f(\gamma_0)+g(\gamma_0)) \geq w(f(\gamma_0)) \geq w(f(\gamma_1)).$$

If $w(f(\gamma_0)+g(\gamma_0)) > w(f(\gamma_1))$, then we are done.

If $w(f(\gamma_0)+g(\gamma_0)) = w(f(\gamma_1))$, then also $w(f(\gamma_0)) = w(f(\gamma_1))$. By the choice of γ_1 we must have $\gamma_0 \geq \gamma_1$. Hence $v(f+g) \geq v(f)$ in this case as well.

Now by Lemma 3.4.1, v extends in a natural way to a valuation $\hat{v}$ on the quotient field $K(\Gamma)$ of R. Specifically, $\hat{v}(f/g) = v(f) - v(g)$ for $0 \neq f, g \in R$.

The value group of $\hat{v}$ is $w(K^\times) \oplus \Gamma$. To compute its residue field note that $O_{\hat{v}} \cap R$ consists of all $f \in R$ such that $f(\gamma) \in O_w$ for all $\gamma \in \Gamma$, and $f(\gamma) \in \mathfrak{m}_w$ for all $0 > \gamma \in \Gamma$. Likewise, $\mathfrak{m}_{\hat{v}} \cap R$ consists of all $f \in R$ such that $f(\gamma) \in O_w$ for all $\gamma \in \Gamma$ and $f(\gamma) \in \mathfrak{m}_w$ for all $0 \geq \gamma \in \Gamma$. Hence $(O_{\hat{v}} \cap R)/(\mathfrak{m}_{\hat{v}} \cap R) \cong \bar{K}_w$ canonically. It follows that

$$\overline{K(\Gamma)}_{\hat{v}} = O_{\hat{v}}/\mathfrak{m}_{\hat{v}} \cong \bar{K}_w.$$

EXAMPLE 4.3.2: *Restricted Gauss valuations.*

Consider a valued field (K, w) and an ordered abelian group $(\Gamma, \leq)$. We construct the extended Gauss valuation $\hat{v}\colon K(\Gamma)^\times \to w(K^\times) \oplus \Gamma$ as in Example 4.3.1. Since the projection $w(K^\times) \oplus \Gamma \to w(K^\times)$ is a morphism of ordered abelian groups, its composition with $\hat{v}$ gives rise to a valuation $\bar{v}\colon K(\Gamma)^\times \to w(K^\times)$. Let R be as before the ring of all maps in $K((\Gamma))$ with finite support. Then

$$\bar{v}(f) = \min_{\gamma \in \Gamma} w(f(\gamma))$$

for $f \in R \setminus \{0\}$.

To compute the residue field of $(K(\Gamma), \bar{v})$, we observe that $O_{\bar{v}} \cap R$ consists of all maps $f \in R$ such that $f(\gamma) \in O_w$ for all $\gamma \in \Gamma$. Similarly, $\mathfrak{m}_{\bar{v}} \cap R$ consists of all maps $f \in R$ such that $f(\gamma) \in \mathfrak{m}_w$ for all $\gamma \in \Gamma$. Hence $(O_{\bar{v}} \cap R)/(\mathfrak{m}_{\bar{v}} \cap R) \cong \bar{K}_w(\Gamma)$. It follows that

$$\overline{K(\Gamma)}_{\bar{v}} \cong \bar{K}_w(\Gamma).$$

EXAMPLE 4.3.3: *Classical Gauss valuations.*

Let (K, w) again be a valued field. We take in Examples 4.3.1 and 4.3.2 $\Gamma = \mathbb{Z}$ to obtain the extended (resp., restricted) Gauss valuation $\hat{v}$ (resp., $\bar{v}$) on the field $K(t)$ of all rational functions over K in the variable t. More specifically, $\hat{v}\colon K(t)^\times \to w(K^\times) \oplus \mathbb{Z}$ is given on non-zero polynomials by

$$\hat{v}\Big(\sum_{i=0}^{n} a_i t^i\Big) = \min_{0 \leq i \leq n} (w(a_i), i)$$

and $\bar{v}\colon K(t)^\times \to w(K^\times)$ is given by

$$\bar{v}\Big(\sum_{i=0}^{n} a_i t^i\Big) = \min_{0 \leq i \leq n} w(a_i)$$

(where as before we use the convention $w(0) = \infty$). The residue fields of $\hat{v}$ and $\bar{v}$ are $\bar{K}_w$ and $\bar{K}_w(t)$, respectively. We call $\hat{v}$ (resp., $\bar{v}$) the **classical extended** (resp., **restricted**) **Gauss valuation** on $K(t)$.

EXAMPLE 4.3.4: *Contents.*
Let R be a unique factorization domain with quotient field K and let $F = K(t)$. Given a prime element π of R, let w_π be the corresponding π-adic valuation on K (Example 4.1.1). Let $v_{\pi,\mathrm{cont}}$ be the classical restricted Gauss valuation on F which corresponds to w_π, as in Example 4.3.3. Thus

$$v_{\pi,\mathrm{cont}}\Big(\sum_{i=0}^{n} a_i t^i\Big) = \min_{0\le i\le n} w_\pi(a_i)$$

for $a_0, \dots, a_n \in K$. We call $v_{\pi,\mathrm{cont}}$ the **π-content valuation** on F. Notice that for given $a_0, \dots, a_n$, this can be non-zero for only finitely many π. We may therefore define the **content** of $f \in F^\times$ to be the product

$$\mathrm{cont}(f) = \prod_{\pi} \pi^{v_{\pi,\mathrm{cont}}(f)},$$

where π ranges over all primes in R (up to a multiplication by a unit). For non-zero $f, g \in F$ we have $v_{\pi,\mathrm{cont}}(fg) = v_{\pi,\mathrm{cont}}(f) + v_{\pi,\mathrm{cont}}(g)$ for all primes π, whence

$$\mathrm{cont}(fg) = \mathrm{cont}(f) \cdot \mathrm{cont}(g).$$

When restricted to polynomials in $K[t]$, this is the classical *Gauss lemma* [**Lg**, Ch. V, §6]. This observation motivated the name "Gauss valuation" in Examples 4.3.1–4.3.3.

EXAMPLE 4.3.5: *Mirror valuations on rational function fields.*
Let K be a field and let $(\Gamma, \le)$ be an ordered abelian group. Let R again be the subring of $K((\Gamma))$ consisting of all maps with finite support. Given a map $f\colon \Gamma \to K$, define a map $f^\#\colon \Gamma \to K$ by

$$f^\#(\gamma) = f(-\gamma).$$

The involution $f \mapsto f^\#$ is an automorphism of R. It extends uniquely to an automorphism $^\#$ of the quotient field $F = K(\Gamma)$ of R. From every valuation v on F we therefore obtain a new valuation $v^\#$ on F, by setting

$$v^\#(f) = v(f^\#).$$

We call $v^\#$ the **mirror valuation** of v.
For instance, suppose that w is a valuation on K. Ordering $\Gamma \oplus w(K^\times)$ lexicographically, we have the valuation v_1 on F given by $v_1(f) = (\gamma, w(f(\gamma)))$, where $\gamma = \min(\mathrm{Supp}(f))$ for non-zero $f \in R$ (see Example 4.2.4). It gives rise to the valuation

$$v_1^\#(f) = v_1(f^\#) = (-\delta, w(f(\delta))),$$

where $\delta = \max(\mathrm{Supp}(f))$. When w is the trivial valuation on K and $\Gamma = \mathbb{Z}$, this valuation is equivalent to the degree valuation v_∞ on $K(\mathbb{Z}) = K(t)$ (in the sense of Example 4.1.3).

Similarly, order $w(K^\times) \oplus \Gamma$ lexicographically, and consider the extended Gauss valuation v_2 on F (Example 4.3.1). Recall that $v_2(f) = \min_{\gamma\in\Gamma}(w(f(\gamma)), \gamma)$ for a non-zero $f \in R$. It gives rise to the mirror valuation $v_2^\#$ on F defined by

$$v_2^\#(f) = v_2(f^\#) = \min_{\gamma\in\Gamma}(w(f(-\gamma)), \gamma) = \min_{\delta\in\Gamma}(w(f(\delta)), -\delta).$$

It is equivalent to the extended Gauss valuation corresponding to w and to the ordered abelian group $(\Gamma, \leq^\#)$, where $\gamma \leq^\# \gamma'$ if and only if $\gamma \geq \gamma'$.

CHAPTER 5

Coarsenings of Valuations

We now fix the field F, and consider the collection of all valuations on it. This collection has a natural structure of a partially ordered set. In fact, we shall see later that it is even a tree. Comparable valuations give rise in a natural way to a quotient valuation on the residue field of the smaller (coarser) valuation. This results in a rich structure theory relating these partially ordered sets on various fields. Moreover, in §10 we will see that this partial order can be also interpreted topologically.

5.1 Coarser and finer

We denote as before the localization of a subring O of the field F at a prime ideal $\mathfrak{p}$ by $O_{\mathfrak{p}}$.

PROPOSITION 5.1.1. *The following conditions on valuations v and u on F are equivalent:*

(a) $G_u \leq G_v$;
(b) $\mathfrak{m}_u \subseteq \mathfrak{m}_v$;
(c) $O_u \supseteq O_v$;
(d) $O_u^\times \geq O_v^\times$;
(e) *there exists a valuation ring $\bar{O}$ on $\bar{F}_u$ such that $O_v = \pi_u^{-1}(\bar{O})$;*
(f) *there exists a prime ideal $\mathfrak{p}$ in O_v such that $O_u = (O_v)_{\mathfrak{p}}$;*
(g) *there exists a morphism $\varphi\colon v(F^\times) \to u(F^\times)$ of ordered abelian groups with $u = \varphi \circ v$;*
(h) *there exists a convex subgroup Δ of $v(F^\times)$ and an isomorphism*

$$u(F^\times) \cong v(F^\times)/\Delta$$

of ordered abelian groups which commutes with u and v.

PROOF. (a)$\Leftrightarrow$(b)$\Leftrightarrow$(c), (e)$\Rightarrow$(c), (f)$\Rightarrow$(c), (g)$\Rightarrow$(c), (b)+(c)$\Rightarrow$(d): Immediate.

(d)$\Rightarrow$(c): We have

$$O_u = O_u^\times \cup (1 - O_u^\times) \supseteq O_v^\times \cup (1 - O_v^\times) = O_v.$$

(g)$\Leftrightarrow$(h): Use Proposition 2.1.3.

(b)+(c)$\Rightarrow$(e): Take $\bar{O} = \pi_u(O_v)$. It is a subring of $\bar{F}_u$. Moreover, if $a \in O_u$ and $\pi_u(a) \notin \bar{O}$, then $a \notin O_v$. Hence $a \neq 0$ and $a^{-1} \in O_v \subseteq O_u$, so $\pi_u(a)^{-1} = \pi_u(a^{-1}) \in \bar{O}$. Thus $\bar{O}$ is a valuation ring on $\bar{F}_u$.

Trivially $O_v \subseteq \pi_u^{-1}(\bar{O})$. Conversely, take $a \in \pi_u^{-1}(\bar{O})$. There exists $b \in O_v$ with $\pi_u(a) = \pi_u(b)$, i.e.,

$$a - b \in \mathfrak{m}_u \subseteq \mathfrak{m}_v \subseteq O_v,$$

so $a \in O_v$, as desired.

(b)+(c)$\Rightarrow$(f): We take $\mathfrak{p} = \mathfrak{m}_u$. By (b) and (c), $\mathfrak{p} \subseteq \mathfrak{m}_v \subseteq O_v \subseteq O_u$, so $\mathfrak{p}$ is a prime ideal of O_v. Further, $(O_v)_\mathfrak{p} \subseteq O_u$. Conversely, suppose that $a \in O_u$. If $a \in O_v$, then trivially $a \in (O_v)_\mathfrak{p}$. If $a \in O_u \setminus O_v$, then $a \neq 0$ and $a^{-1} \in O_v \setminus \mathfrak{p}$, so $a = 1/(a^{-1}) \in (O_v)_\mathfrak{p}$.

(c)$\Rightarrow$(g): Define a binary relation $\leq_u$ on $F^\times/O_u^\times$ by setting $xO_u^\times \leq_u yO_u^\times$ if and only if $y/x \in O_u$. Define $\leq_v$ similarly. By Proposition 3.2.3(b), $(F^\times/O_v^\times, \leq_v)$ and $(F^\times/O_u^\times, \leq_u)$ are ordered abelian groups. The natural projection $F^\times/O_v^\times \to F^\times/O_u^\times$ is then a surjective morphism of ordered abelian groups. By Proposition 3.2.3(c)(d), the projection $F^\times \to F^\times/O_v^\times$ is a valuation which is equivalent to v, and similarly for u. Condition (g) now follows from the commutative diagram:

$$\begin{array}{ccccccc} & & F^\times & = & F^\times & & \\ & \swarrow^{v} & \downarrow & & \downarrow & \searrow^{u} & \\ v(F^\times) & \xleftarrow{\sim} & F^\times/O_v^\times & \longrightarrow & F^\times/O_u^\times & \xrightarrow{\sim} & u(F^\times). \end{array} \qquad \square$$

When the equivalent conditions of Proposition 5.1.1 are satisfied we say that the valuation u is **coarser** than v, and that v is **finer** than u. We then write $u \leq v$. Notice that this is a partial order on the collection of all valuations on F.

REMARK 5.1.2. Recall that the rank of a valuation v on F was defined in §3.2 to be the rank of its value group, i.e., the supremum of all non-negative integers n for which there exists a chain

$$0 = \Delta_0 < \Delta_1 < \cdots < \Delta_n$$

of convex subgroups of $v(F^\times)$. In view of condition (h) above, this is the supremum of all non-negative integers n for which there exist a sequence of valuations

$$v = v_0, v_1, \ldots, v_n$$

on F, where v_i is a proper coarsening of v_{i-1} for $i = 1, \ldots, n$. In particular, v has rank 1 if and only if it has no coarsenings except v itself and the trivial valuation.

5.2 Quotients and compositions of valuations

Consider valuations v and u on the field F with u coarser than v. As we have seen in Proposition 5.1.1, $O_v = \pi_u^{-1}(\bar{O})$ for some valuation ring $\bar{O}$ on $\bar{F}_u$. This valuation ring is uniquely determined by v and u, namely $\bar{O} = \pi_u(O_v) = O_v/\mathfrak{m}_u$. Its maximal ideal is $\pi_u(\mathfrak{m}_v) = \mathfrak{m}_v/\mathfrak{m}_u$, and its group of units is $\bar{O}^\times = \pi_u(O_v^\times)$. Consequently,

$$\pi_u(G_v) = 1 - \pi_u(\mathfrak{m}_v) = 1 - (\mathfrak{m}_v/\mathfrak{m}_u) \quad .$$

Furthermore, the epimorphisms $O_v^\times \to \bar{O}^\times$ and $G_v \to G_v/G_u$ induced by π_u give rise to isomorphisms $\bar{O}^\times \cong O_v^\times/G_u$ and $\pi_u(G_v) \cong G_v/G_u$. By Proposition 3.2.3(e) there is a unique valuation w on $\bar{F}_u$ (up to equivalence) such that $\bar{O} = O_w$. We call w the **quotient valuation** of v by u, and denote it by v/u. One has a commutative diagram of *additive* abelian groups with exact rows and columns:

$$
\begin{array}{ccccccccc}
 & & & & 0 & & 0 & & \\
 & & & & \downarrow & & \downarrow & & \\
0 & \longrightarrow & \mathfrak{m}_u & \longrightarrow & \mathfrak{m}_v & \xrightarrow{\pi_u} & \mathfrak{m}_w & \longrightarrow & 0 \\
 & & \| & & \downarrow & & \downarrow & & \\
0 & \longrightarrow & \mathfrak{m}_u & \longrightarrow & O_v & \xrightarrow{\pi_u} & O_w & \longrightarrow & 0 \\
 & & & & \downarrow{\scriptstyle \pi_v} & & \downarrow{\scriptstyle \pi_w} & & \\
 & & & & \bar{F}_v & \xrightarrow{\sim} & (\bar{F}_u)_w & & \\
 & & & & \downarrow & & \downarrow & & \\
 & & & & 0 & & 0 & &
\end{array}
\tag{5.2.1}
$$

where the isomorphism follows from the snake lemma (Lemma 1.1.3).

Next let S be a subgroup of $F^\times$ and write as before $\bar{S}_u = \pi_u(S \cap O_u^\times)$. As $G_u \le G_v \le O_v^\times \le O_u^\times$, one has

$$\bar{S}_u \cap G_w = \pi_u(S \cap O_u^\times) \cap \pi_u(G_v) = \pi_u(S \cap G_v),$$

$$\bar{S}_u \cap O_w^\times = \pi_u(S \cap O_u^\times) \cap \pi_u(O_v^\times) = \pi_u(S \cap O_v^\times). \tag{5.2.2}$$

We therefore obtain a commutative diagram of *multiplicative* abelian groups with exact rows and columns, and with an isomorphism at the lower row (by the snake lemma again):

$$
\begin{array}{ccccccccc}
 & & & & 1 & & 1 & & \\
 & & & & \downarrow & & \downarrow & & \\
1 & \longrightarrow & S \cap G_u & \longrightarrow & S \cap G_v & \xrightarrow{\pi_u} & \pi_u(S \cap G_v) & \longrightarrow & 1 \\
 & & \| & & \downarrow & & \downarrow & & \\
1 & \longrightarrow & S \cap G_u & \longrightarrow & S \cap O_v^\times & \xrightarrow{\pi_u} & \pi_u(S \cap O_v^\times) & \longrightarrow & 1 \\
 & & & & \downarrow{\scriptstyle \pi_v} & & \downarrow{\scriptstyle \pi_w} & & \\
 & & & & \bar{S}_v & \xrightarrow{\sim} & (\bar{S}_u)_w & & \\
 & & & & \downarrow & & \downarrow & & \\
 & & & & 1 & & 1 & &
\end{array}
\tag{5.2.3}
$$

We summarize:

PROPOSITION 5.2.1. *Let v and u be valuations on F, with u coarser than v, and let $w = v/u$. Let $S \le F^\times$. Then:*

(a) $O_w = \pi_u(O_v) = O_v/\mathfrak{m}_u$*;*

(b) $\mathfrak{m}_w = \pi_u(\mathfrak{m}_v) = \mathfrak{m}_v/\mathfrak{m}_u$*;*

(c) $O_w^\times = \pi_u(O_v^\times) \cong O_v^\times/G_u$ *canonically;*

(d) $G_w = \pi_u(G_v) \cong G_v/G_u$ *canonically;*

(e) $\bar{F}_v \cong (\bar{F}_u)_w$ *canonically;*
(f) $\bar{S}_v \cong (\bar{S}_u)_w$ *via the isomorphism of (e);*
(g) *there is a canonical short exact sequence of ordered abelian groups*

$$0 \to w(\bar{S}_u) \to v(S) \to u(S) \to 0$$

which is functorial in S*;*
(h) *there is a canonical short exact sequence of groups*

$$0 \to w(\bar{F}_u^\times)/w(\bar{S}_u) \to v(F^\times)/v(S) \to u(F^\times)/u(S) \to 0.$$

PROOF. We have already seen (a)–(f).

(g) Equality (5.2.2) and the snake lemma (Lemma 1.1.3) give rise to a commutative diagram with exact rows and columns:

$$\begin{array}{ccccccccc}
 & & & & 1 & & 1 & & \\
 & & & & \downarrow & & \downarrow & & \\
1 & \longrightarrow & S\cap G_u & \longrightarrow & S\cap O_v^\times & \xrightarrow{\pi_u} & \bar{S}_u\cap O_w^\times & \longrightarrow & 1 \\
 & & \| & & \downarrow & & \downarrow & & \\
1 & \longrightarrow & S\cap G_u & \longrightarrow & S\cap O_u^\times & \xrightarrow{\pi_u} & \bar{S}_u & \longrightarrow & 1 \\
 & & & & \downarrow & & \downarrow & & \\
 & & & & (S\cap O_u^\times)/(S\cap O_v^\times) & \xrightarrow{\sim} & \bar{S}_u/(\bar{S}_u\cap O_w^\times) & & \\
 & & & & \downarrow & & \downarrow & & \\
 & & & & 1 & & 1 & &
\end{array}$$

Moreover, if $a, b \in S \cap O_u^\times$ and $a/b \in O_v$, then $\pi_u(a)/\pi_u(b) \in O_w$. Therefore the group isomorphism in the lower row of the diagram is also an isomorphism of ordered abelian groups, with respect to the orders defined in Proposition 3.2.3(b). We thus get an exact sequence of ordered abelian groups

$$1 \to \bar{S}_u/(\bar{S}_u \cap O_w^\times) \to S/(S\cap O_v^\times) \to S/(S\cap O_u^\times) \to 1.$$

By Proposition 3.2.3(c), w, v, u induce isomorphisms of these ordered abelian groups with the groups in (g). All these isomorphisms are canonical, whence the asserted functoriality.

(h) Use (g) for the subgroups S and $F^\times$, and apply the snake lemma once again. □

LEMMA 5.2.2. *Let* u, v', v *be valuations on the field* F *with* u *coarser than* v' *and* v' *coarser than* v*. Then:*
(a) v'/u *is coarser than* v/u*;*
(b) *the quotient valuations* v/v' *and* $(v/u)/(v'/u)$ *are equivalent.*

PROOF. (a) By Proposition 5.2.1(a), $O_{v'/u} = \pi_u(O_{v'}) \supseteq \pi_u(O_v) = O_{v/u}$.

(b) For notational simplicity, we denote here the value group of a valuation w by Γ_w. Then Proposition 5.2.1(g) gives a commutative diagram of ordered abelian groups with exact rows and columns:

$$\begin{array}{ccccccccc}
 & & 0 & & 0 & & & & \\
 & & \downarrow & & \downarrow & & & & \\
 & & \Gamma_{(v/u)/(v'/u)} & & \Gamma_{v/v'} & & & & \\
 & & \downarrow & & \downarrow & & & & \\
0 & \longrightarrow & \Gamma_{v/u} & \longrightarrow & \Gamma_v & \longrightarrow & \Gamma_u & \longrightarrow & 0 \\
 & & \downarrow & & \downarrow & & \| & & \\
0 & \longrightarrow & \Gamma_{v'/u} & \longrightarrow & \Gamma_{v'} & \longrightarrow & \Gamma_u & \longrightarrow & 0 \\
 & & \downarrow & & \downarrow & & & & \\
 & & 0 & & 0 & & & &
\end{array}$$

As before, the snake lemma yields a canonical isomorphism $\Gamma_{(v/u)/(v'/u)} \xrightarrow{\sim} \Gamma_{v/v'}$ of ordered abelian groups. One can verify that it is compatible with the valuations $(v/u)/(v'/u)$ and v/v'. □

Opposite to the construction of quotient valuations, one has the following construction of composition of valuations. When u is a valuation on F and w is a valuation on $\bar{F}_u$, the inverse image $\pi_u^{-1}(O_w)$ is a valuation ring on F. The corresponding valuation v is then finer than u, and w is equivalent to v/u. For the places π_v, π_u and π_w associated with v, u and w, respectively, π_v is the composition $\pi_w \circ \pi_u$ on O_v. By abuse of notation, we call v the **composition** of w with u. In view of condition (e) of Proposition 5.1.1, every valuation on F which is finer than u is equivalent to the composition of some valuation on $\bar{F}_u$ with u.

We now demonstrate the situation discussed in this and the previous sections by some more concrete examples.

EXAMPLE 5.2.3. Let Γ be an ordered abelian group and let (K, w) be a valued field. Let $F = K((\Gamma))$ be the field of generalized formal power series over K with exponents in Γ (§2.8). We endow $\Gamma \oplus w(K^\times)$ with the lexicographic order. From w and from the trivial valuation on K we construct valuations

$$v\colon F^\times \to \Gamma \oplus w(K^\times), \quad u\colon F^\times \to \Gamma \oplus \{0\} \cong \Gamma,$$

respectively, as in Example 4.2.1. Thus for a non-zero $f \in F$ with $\gamma = \min(\mathrm{Supp}(f))$ we set $v(f) = (\gamma, w(f(\gamma)))$ and $u(f) = (\gamma, 0)$. Then v is finer than u.

Recall that O_v consists of 0 as well as of all non-zero elements $f \in F$ such that either $\min(\mathrm{Supp}(f)) > 0$ or both $\min(\mathrm{Supp}(f)) = 0$ and $w(f(0)) \geq 0$. Likewise, $\mathfrak{m}_u$ consists of 0 as well as of all non-zero $f \in F$ such that $\min(\mathrm{Supp}(f)) > 0$. Hence $O_{v/u} = O_v/\mathfrak{m}_u \cong O_w$ canonically. Thus v/u is equivalent to w, and v is equivalent to the composition of w with u.

EXAMPLE 5.2.4. Let (K,w) be a valued field, let Γ be an ordered abelian group, and order $w(K^\times)\oplus\Gamma$ lexicographically. Let $F = K(\Gamma)$ be the field of generalized rational functions over K with exponents in Γ (§2.9). Further, let

$$\hat{v}\colon F^\times \to w(K^\times)\oplus\Gamma, \quad \bar{v}\colon F^\times \to w(K^\times)$$

be the extended and restricted Gauss valuations, respectively (Examples 4.3.1 and 4.3.2). By definition, $\bar{v}$ is the composition of $\hat{v}$ with the projection $w(K^\times)\oplus\Gamma \to w(K^\times)$. Since this projection is a morphism of ordered abelian groups, $\hat{v}$ is finer than $\bar{v}$ (by condition (g) of Proposition 5.1.1).

Also recall that $\bar{F}_{\hat{v}} \cong \bar{K}_w$ and $\bar{F}_{\bar{v}} \cong \bar{K}_w(\Gamma)$ canonically. The valuation $\hat{v}/\bar{v}$ on $\bar{F}_{\bar{v}}$ corresponds under the latter isomorphism to the canonical valuation $u\colon \bar{K}_w(\Gamma)^\times \to \Gamma$, given by $u(f) = \min(\mathrm{Supp}(f))$ for $f\neq 0$ (see Example 4.2.4). Equivalently, $\hat{v}$ is the composition of w with $\bar{v}$.

EXAMPLE 5.2.5. Let (K,w) be a valued field and let $F = K(\mathbb{Z}) = K(t)$. We endow again $\mathbb{Z}\oplus w(K^\times)$ with the lexicographic order. Consider the mirror valuation

$$v^\#\colon F^\times \to \mathbb{Z}\oplus w(K^\times)$$

given for a non-zero $f\in F$ by

$$v^\#(f) = (-\delta, w(f(\delta))),$$

where $\delta = \max(\mathrm{Supp}(f))$ (Example 4.3.5). The composition of $v^\#$ with the projection $\mathbb{Z}\oplus w(K^\times)\to\mathbb{Z}$ is the degree valuation v_∞ on F (Example 4.1.3). Hence $v^\#$ is finer than v_∞. The canonical isomorphism $K\cong \bar{F}_{v_\infty}$ identifies w with $v^\#/v_\infty$. Equivalently, $v^\#$ is the composition of w with v_∞.

5.3 Coarsenings in the mixed characteristic case

Valued fields (F,v) with $\operatorname{char} F = 0$ and $\operatorname{char}\bar{F}_v = p > 0$ are relatively difficult to analyze. This is the so-called **mixed characteristic case**. The following useful *slicing lemma* reduces this situation to valuations of rank 1:

LEMMA 5.3.1. *Suppose that* $\operatorname{char} F = 0$ *and let* v *be a valuation on* F *with* $\operatorname{char}\bar{F}_v = p > 0$. *There exists a coarsening* u^+ *of* v *and a coarsening* u^- *of* u^+ *such that* $\bar{F}_{u^-}$ *has characteristic* 0, $\bar{F}_{u^+}\cong(\bar{F}_{u^-})_{u^+/u^-}$ *has characteristic* p, *and* u^+/u^- *has rank* 1.

PROOF. We denote again the value group of a valuation w by Γ_w. Proposition 5.1.1 gives a bijective correspondence between the following sets:

(1) the coarsenings u of v;
(2) the prime ideals $\mathfrak{p}$ in O_v;
(3) the convex subgroups Δ of Γ_v.

Moreover, these correspondences are order-preserving/inversing in the following sense: suppose that $u_i, \mathfrak{p}_i, \Delta_i$ correspond to each other via these bijections, $i=1,2$. Then

$$u_1 \le u_2 \iff \mathfrak{p}_1\subseteq\mathfrak{p}_2 \iff \Delta_1\ge\Delta_2.$$

Proposition 2.2.1 therefore implies that the sets (1)–(3) form chains in the respective senses.

We partition the set of all prime ideals $\mathfrak{p}$ in O_v into the subset A^+ consisting of all prime ideals $\mathfrak{p}$ with $p\in\mathfrak{p}$, and the subset A^- consisting of all prime ideals $\mathfrak{p}$ with

$p \notin \mathfrak{p}$. As $\operatorname{char} \bar{F}_v = p$ we have $\mathfrak{m}_v \in A^+$. As $\operatorname{char} F = 0$ we further have $\{0\} \in A^-$. Since A^+ is a chain, the intersection $\mathfrak{p}^+$ of all its ideals is a minimal element in A^+. Likewise, since A^- is a chain, the union $\mathfrak{p}^-$ of all its ideals is a maximal element in A^-. To $\mathfrak{p}^+$ (resp., $\mathfrak{p}^-$) we associate objects u^+, Δ^+ (resp., u^-, Δ^-) as above. We may identify u^+ (resp., u^-) with the morphism $F^\times \to \Gamma_v/\Delta^+$ (resp., $F^\times \to \Gamma_v/\Delta^-$) induced by v. Now $\mathfrak{p}^- \subset \mathfrak{p}^+$, so u^- is coarser than u^+ and $\Delta^- \supset \Delta^+$. As $p \in \mathfrak{p}^+ = \mathfrak{m}_{u^+}$ we have $\operatorname{char} \bar{F}_{u^+} = p$. As $p \notin \mathfrak{p}^- = \mathfrak{m}_{u^-}$ we further have $\operatorname{char} \bar{F}_{u^-} \neq p$. By Proposition 5.2.1(e), $\bar{F}_{u^+} = (\bar{F}_{u^-})_{u^+/u^-}$.

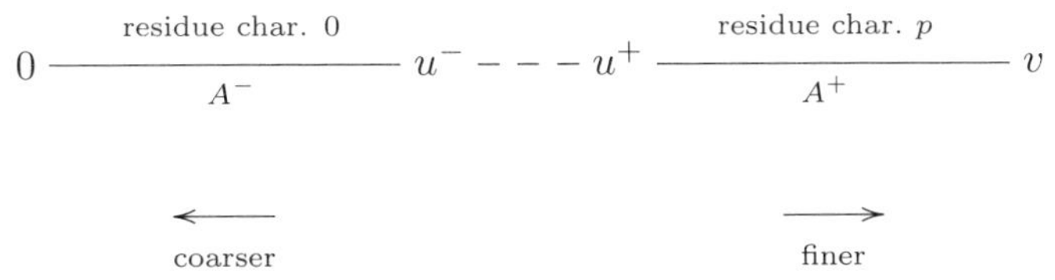

Finally, Proposition 5.2.1(g) (for $S = F^\times$) yields a short exact sequence of ordered abelian groups

$$0 \to \Gamma_{v^+/v^-} \to \Gamma_u/\Delta^+ \to \Gamma_u/\Delta^- \to 0 \quad .$$

Hence $\Gamma_{v^+/v^-} \cong \Delta^-/\Delta^+$ as ordered abelian groups. Since there are no prime ideals of O_v between $\mathfrak{p}^-$ and $\mathfrak{p}^+$, there are no convex subgroups of Γ_v between Δ^- and Δ^+. Therefore

$$\operatorname{rank}(\Gamma_{u^+/u^-}) = \operatorname{rank}(\Delta^-/\Delta^+) = 1$$

(see Lemma 2.2.2). $\square$

CHAPTER 6

Orderings

After studying valuations in the last three chapters, we now introduce the second main arithmetical object studied in this book, namely, orderings on fields. While their technical definition certainly looks quite different from that of valuations, both kinds of objects behave in a similar way in many respects, and combine to a unified and harmonic theory.

6.1 Ordered fields

An **ordering** on the field F is an additively closed subgroup P of $F^\times$ such that $(F^\times : P) = 2$. Note that as $1 \in P$ and $0 \notin P$, necessarily $-1 \notin P$, so in fact $F^\times = P \cup - P$. We then say that the pair (F, P) is an **ordered field**. This is related to the standard notion of order as follows:

Let $<$ be a strict total order on F (i.e., a binary relation which is anti-reflexive, anti-symmetric, and transitive, and in which any two elements are comparable). We say that $<$ is an **ordering relation** on F if:

(i) for every $a, b, c \in F$ such that $a < b$ one has $a + c < b + c$;
(ii) for every $a, b, c \in F$ such that $a < b$ and $0 < c$ one has $ac < bc$.

PROPOSITION 6.1.1.
(a) *Let P be an ordering on F and define a binary relation $<_P$ on F by setting $a <_P b$ if and only if $b - a \in P$. Then $<_P$ is an ordering relation on F.*
(b) *Let $<$ be an ordering relation on F. Then*

$$P_< = \{a \in F \mid 0 < a\}$$

is an ordering on F.
(c) *The maps $P \mapsto <_P$ and $< \mapsto P_<$ are converse one to each other.*

PROOF. The proofs of (a) and (c) are straightforward.

We verify (b). The fact that $P_<$ is additively (resp., multiplicatively) closed follows from (i) (resp., (ii)). To prove that $1 \in P_<$ assume that $1 < 0$. Then $0 < -1$, by (i), and therefore $0 = 0 \cdot (-1) < (-1)^2 = 1$ by (ii), a contradiction.

Next, if $0 < a$ and $a^{-1} < 0$, then (ii) would give $1 = a^{-1}a < 0 \cdot a = 0$, contrary to what we have just seen. This shows that $P_<$ is closed under inversion. Since $<$ is total, for every $a \in F^\times$ precisely one of $0 < a$, $a < 0$ holds, and the latter condition is equivalent to $0 < -a$, by (i). This shows that $F^\times = P_< \cup - P_<$. □

Given an ordering P on F, we also define a total ordering $\leq_P$ by setting $a \leq_P b$ if and only if either $a <_P b$ or $a = b$. Just as for the ordered set $\mathbb{R}$, one defines for $a, b \in F$ the open intervals

$$(a, b)_P = (a + P) \cap (b - P), \quad (a, \infty)_P = a + P, \quad (-\infty, b) = b - P.$$

We also write

$$[a, b]_P = (a, b)_P \cup \{a, b\},$$

etc. The absolute value $|\cdot|_P$ on F with respect to P is defined by

$$|a|_P = \begin{cases} a & \text{if } a \in P, \\ -a & \text{if } a \notin P. \end{cases}$$

We note the following standard facts:

PROPOSITION 6.1.2. *For every $a, b, c, d \in F$ one has:*
(a) $|a + b|_P \leq_P |a|_P + |b|_P$ **(the triangle inequality)**;
(b) $\big||a|_P - |b|_P\big|_P \leq_P |a - b|_P$;
(c) *if $|a|_P \leq_P |b|_P$ and $|c|_P \leq_P |d|_P$, then $|ac| \leq_P |bd|_P$;*
(d) *when $0 <_P a, b$ one has $a <_P b$ if and only $b^{-1} <_P a^{-1}$.*

Recall that $(F^\times)^2$ is the set of all squares in $F^\times$. A **preordering** on F is a subset T of $F^\times$ containing $(F^\times)^2$ which is additively and multiplicatively closed. Then $-1 \notin T$. For $t \in T$ we obtain $t^{-1} = (t^{-1})^2 t \in TT = T$. Therefore, T is a subgroup of $F^\times$. Note that an ordering on F is just a preordering of index 2 in $F^\times$.

LEMMA 6.1.3. *Let T be a preordering on F and let $a \in F^\times \setminus T$. Then*

$$S = T \cup -aT \cup (T - aT)$$

is also a preordering on F.

PROOF. The fact that S contains $(F^\times)^2$ and is additively and multiplicatively closed follows from the corresponding properties of T. As T is a subgroup of $F^\times$ not containing a, we have $0 \notin S$. □

PROPOSITION 6.1.4. *Every preordering on F extends to an ordering on F.*

PROOF. Let T be a preordering on F. Zorn's lemma yields a maximal preordering P on F containing T. If $a \in F^\times \setminus P$, then by Lemma 6.1.3 and the maximality,

$$P = P \cup -aP \cup (P - aP) \quad .$$

In particular, $-a \in P$. This show that $F^\times = P \cup -P$. Since P is an additively closed subgroup of $F^\times$ not containing -1, it is therefore an ordering on F. □

Given a subgroup S of $F^\times$, let $\sum S$ be the subset of $F^\times$ consisting of all sums of finitely many elements of S. The following observation essentially goes back to [**Se1**].

PROPOSITION 6.1.5. *Let S be a subgroup of $F^\times$ containing $(F^\times)^2$. The following conditions are equivalent:*
(a) *there is an ordering P on F containing S;*
(b) $-1 \notin \sum S$;
(c) *$\sum S$ is a preordering on F.*

PROOF. (a)⇒(b): Just note that an ordering is additively closed and does not contain -1.

(b)⇒(c): Obviously, $T = \sum S$ is multiplicatively closed. By (b), 0 is not a sum of elements of S. This implies that $0 \notin T + T$, so T is additively closed. Hence it is a preordering on F.

(c)⇒(a): The preordering $\sum S$ contains S. By Proposition 6.1.4, it extends to an ordering P on F. □

In the special case where $S = (F^\times)^2$, we abbreviate and write $\sum F^2$ for $\sum (F^\times)^2$. In this case Proposition 6.1.5 reads as follows:

COROLLARY 6.1.6 (Artin–Schreier [**AS1**]). *The following conditions on a field F are equivalent:*
(a) *there exists an ordering on F;*
(b) $-1 \notin \sum F^2$;
(c) $\sum F^2$ *is a preordering on F.*

A field F satisfying the equivalent conditions of Corollary 6.1.6 is said to be **formally real**.

COROLLARY 6.1.7. *If F has an ordering, then* $\operatorname{char} F = 0$.

PROOF. If $\operatorname{char} F = p > 0$, then $-1 = (p-1) \in \sum F^2$, so condition (b) of Corollary 6.1.6 does not hold. □

PROPOSITION 6.1.8. *A subset of the field F is a preordering if and only if it is the intersection of a non-empty collection of orderings on F.*

PROOF. As we have already observed, every ordering is a preordering. Furthermore, the intersection of a non-empty collection of preorderings is a preordering.

Conversely, let T be a preordering on F, and let T_1 be the intersection of all orderings P on F containing T. Clearly, $T \subseteq T_1$. Suppose that there exists $a \in T_1 \smallsetminus T$. Then Lemma 6.1.3 would give rise to a preordering S on F with $T \subseteq S$ and $-a \in S$. By Proposition 6.1.4, there is an ordering P on F with $S \subseteq P$. But then $-a \in S \subseteq P$ as well as $a \in T_1 \subseteq P$, a contradiction. □

COROLLARY 6.1.9 (Artin [**Ar**]). *Let F be a formally real field. Then $\sum F^2$ is the intersection of all the orderings on F.*

PROOF. By assumption, $\sum F^2$ is a preordering on F. Moreover, it is contained in every ordering on F. Now use Proposition 6.1.8. □

On the other hand, we have:

PROPOSITION 6.1.10. *If the field F is not formally real and if* $\operatorname{char} F \neq 2$, *then* $F^\times = \sum F^2$.

PROOF. By assumption, $-1 \in \sum F^2$. Hence, for every $a \in F$ we have

$$a = \left(\frac{a+1}{2}\right)^2 - \left(\frac{a-1}{2}\right)^2 \in (F^\times)^2 + \left(\sum F^2\right)(F^\times)^2 \subseteq \left(\sum F^2\right) \cup \{0\}. \quad \square$$

REMARK. The definitions in this section slightly deviate from the more standard terminology, in which an ordering is taken to be a set of the form $P \cup \{0\}$, where P is an ordering in our sense (and similarly for preorderings). Our terminology is however more natural from the current point of view, of studying the arithmetic of a field F by means of subgroups of $F^\times$.

6.2 Examples of orderings

EXAMPLE 6.2.1. The field $\mathbb{R}$ has only one ordering, namely, its set of non-zero squares.

EXAMPLE 6.2.2. The standard ordering on the field $\mathbb{Q}$ of rational numbers is its only ordering. Indeed, consider a non-zero rational number n/m, where $0 \neq n, m \in \mathbb{Z}$. If it is positive in the standard ordering, then nm is a sum of 1's, whence $n/m = nm/m^2 \in \sum \mathbb{Q}^2$. On the other hand, every ordering on $\mathbb{Q}$ contains $\sum \mathbb{Q}^2$, by Corollary 6.1.9.

EXAMPLE 6.2.3. The field $\mathbb{Q}(\sqrt{2})$ has two orderings, induced by its two embeddings into $\mathbb{R}$.

EXAMPLE 6.2.4. Let $\theta \in \mathbb{R}$ be transcendental over $\mathbb{Q}$ and let $<_{\mathbb{R}}$ be the usual ordering relation on $\mathbb{R}$. The set P of all $f \in \mathbb{Q}(t)$ such that $0 <_{\mathbb{R}} f(\theta)$ is an ordering on $\mathbb{Q}(t)$.

EXAMPLE 6.2.5. Let $(K, \bar{P})$ be an ordered field, let Γ be an ordered abelian group, and let $\chi\colon \Gamma \to \{\pm 1\}$ be a group homomorphism. Furthermore, let $F = K((\Gamma))$ and let $v\colon F^\times \to \Gamma$ be its canonical valuation corresponding to the trivial valuation on K, as in Example 4.2.1. Thus $v(f) = \min(\mathrm{Supp}(f))$ for $f \in F^\times$. Let P be the set of all $f \in F^\times$ such that

$$\chi(v(f)) \cdot f(v(f)) \in \bar{P}.$$

We claim that P is an ordering on F. To see this, let $f, g \in P$ and let $\gamma = v(f)$ and $\delta = v(g)$. Then

$$v(f \cdot g) = \gamma + \delta \quad , \quad (f \cdot g)(\gamma + \delta) = f(\gamma) g(\delta).$$

Hence

$$\begin{aligned} \chi(v(f \cdot g)) \cdot (f \cdot g)(v(f \cdot g)) &= \chi(\gamma + \delta) \cdot (f \cdot g)(\gamma + \delta) \\ &= \chi(\gamma) f(\gamma) \chi(\delta) g(\delta) \in \bar{P} \cdot \bar{P} = \bar{P}, \end{aligned}$$

so $f \cdot g \in P$, showing that P is multiplicatively closed.

To show that P is additively closed, let $f, g \in P$ and let γ, δ be as above. If $\gamma < \delta$, then $v(f+g) = \gamma$ and $(f+g)(\gamma) = f(\gamma)$. Hence

$$\chi(v(f+g)) \cdot (f+g)(v(f+g)) = \chi(\gamma) \cdot f(\gamma) \in \bar{P},$$

so $f + g \in P$ in this case. Likewise, if $\gamma > \delta$, then $f + g \in P$. Finally, suppose that $\gamma = \delta$. Since $\chi(\gamma) f(\gamma), \chi(\gamma) g(\gamma) \in \bar{P}$ and $\bar{P}$ is additively closed, $f(\gamma) + g(\gamma) \neq 0$. Hence

$$v(f+g) = \min(\mathrm{Supp}(f+g)) = \gamma = \delta.$$

It follows that

$$\chi(v(f+g)) \cdot (f+g)(\gamma) = \chi(\gamma) \cdot (f(\gamma) + g(\gamma)) \in \bar{P} + \bar{P} \subseteq \bar{P},$$

so again, $f+g\in P$.

Clearly, $F^\times = P \uplus -P$. It follows that P is an ordering on F. Viewing K as the subfield of F consisting of all constant power series, we obtain that $\bar{P} = P\cap K$.

6.3 Archimedean orderings

We conclude this chapter by studying orderings which satisfy *Archimedes' Axiom* (condition (a) of the following proposition). Recall that every ordered field contains $\mathbb{Q}$ (Corollary 6.1.7).

PROPOSITION 6.3.1. *The following conditions on an ordering P on a field F are equivalent:*

(a) *for every $x\in F^\times$ there exists a positive integer n with $|x|_P <_P n$;*
(b) *for every $x\in F^\times$ there exists $r\in\mathbb{Q}$ with $|x|_P <_P r$;*
(c) *$(0,y)_P\cap\mathbb{Q}\neq\emptyset$ for every $y\in P$;*
(d) *$\mathbb{Q}$ intersects every non-empty open interval $(z,t)_P$ in F;*
(e) *(F,P) embeds in $(\mathbb{R},(\mathbb{R}^\times)^2)$ as an ordered field.*

PROOF. (a)$\Leftrightarrow$(b): Just observe that the restriction of $<_P$ to $\mathbb{Q}$ is its unique order $<_\mathbb{Q}$.

(b)$\Rightarrow$(c): For $y\in P$ (b) yields $r\in\mathbb{Q}$ such that $y^{-1} <_P r$. Then $r^{-1}\in (0,y)_P\cap\mathbb{Q}$.

(c)$\Rightarrow$(b): For $x\in F^\times$ (c) yields $s\in(0,|x|_P^{-1})_P\cap\mathbb{Q}$. Then $|x|_P < s^{-1}\in\mathbb{Q}$.

(a)+(c)$\Rightarrow$(d): If one of the intervals $(z,t)_P$, $(-t,-z)_P$ intersects $\mathbb{Q}$, then so does the other. We may therefore assume that $0<_P t$. Further, in view of (c), we may assume that $0<_P z$. Now (c) yields $\epsilon\in(0,t-z)_P\cap\mathbb{Q}$. (a) gives rise to a minimal positive integer n such that $z\epsilon^{-1} <_P n$. Then $n-1\le_P z\epsilon^{-1}$, so

$$n\epsilon = (n-1)\epsilon+\epsilon \le_P z+\epsilon <_P t \quad ,$$

whence $n\epsilon\in(z,t)_P\cap\mathbb{Q}$.

(d)$\Rightarrow$(c): Trivial.

(d)$\Rightarrow$(e): We may define maps $f,g\colon F\to\mathbb{R}$ by setting for $a\in F$,

$$f(a)=\sup\{r\in\mathbb{Q}\mid r<_P a\},\quad g(a)=\inf\{r\in\mathbb{Q}\mid a<_P r\}.$$

For $a,b\in F$ one has

$$f(-a)=-g(a),\quad f(a)+f(b)\le f(a+b),\quad g(a)+g(b)\ge g(a+b).$$

When $0<_P a,b$ also

$$f(a)f(b)\le f(ab),\quad g(a)g(b)\ge g(ab).$$

Now by (d), for every $a\in F$ and $0<_P\epsilon\in\mathbb{Q}$ there exist rational numbers in $(a-\epsilon,a+\epsilon)_P$. It follows that $f(a)=g(a)$. We deduce that $f=g$ is a ring homomorphism, whose restriction to $\mathbb{Q}$ is the identity map. Moreover, it is order-preserving, and in particular injective.

(e)$\Rightarrow$(d): Immediate. □

When the equivalent conditions of Proposition 6.3.1 are satisfied one says that the ordering P is **Archimedean**.

CHAPTER 7

The Tree of Localities

We fix again a field F and study the valuations and orderings on F in a unified way, under the name "localities". We extend the partial order of coarsening of valuations to localities, define and study quotient localities, and prove the main properties of this partially ordered set. Of special importance will be the construction of the valuation ring associated with an ordering, which goes back to Artin, Schreier and Baer (§7.3).

7.1 Localities

A **locality** on the field F is either a valuation or an ordering on F. The **trivial locality** is the trivial valuation on F. Given a locality λ on F we define a subgroup G_λ of $F^\times$ as follows:

$$G_\lambda = \begin{cases} 1+\mathfrak{m}_v & \text{if } \lambda = v \text{ is a valuation,} \\ P & \text{if } \lambda = P \text{ is an ordering.} \end{cases}$$

In the valuation case this agrees with our earlier notation in §3.2. Likewise we extend the coarsening relation between valuations to arbitrary localities λ_1, λ_2 on F by saying that λ_1 is **coarser** than λ_2 (and that λ_2 is **finer** than λ_1) if $G_{\lambda_1} \leq G_{\lambda_2}$ – in other words, either:

(i) λ_1, λ_2 are valuations and λ_1 is coarser than λ_2 in the sense of §5.1;
(ii) λ_1 is a valuation, λ_2 is an ordering, and $1+\mathfrak{m}_{\lambda_1} \leq \lambda_2$; or
(iii) λ_1, λ_2 are equal orderings.

We denote this relation by $\lambda_1 \leq \lambda_2$. In case (ii) one has the following alternative definition:

Lemma 7.1.1. *Let P be an ordering on the field F and let u be a valuation on F. The following conditions are equivalent:*
(a) u is coarser than P;
(b) $(-1,1)_P \subseteq O_u$.

Proof. (a)⇒(b): Take $x \in (-1,1)_P$. To show that $x \in O_u$ we may assume that $x \in P$ (otherwise replace x by $-x$). Then

$$1 - x^{-1} = x^{-1}(x-1) \in P(-P) = -P.$$

As $1+\mathfrak{m}_u \leq P$, this implies that $x^{-1} \notin \mathfrak{m}_u$. Therefore $x \in O_u$.

(b)⇒(a): Consider a non-zero $m \in \mathfrak{m}_u$. We need to show that $1+m \in P$. Now $m^{-1} \notin O_u$, so $m^{-1} \notin (-1,1)_P$. As $m \neq \pm 1$, we deduce that $m \in (-1,1)_P$, whence $1+m \in P$, as desired. □

EXAMPLE 7.1.2. Let $\bar{P}$ be an ordering on a field K, let Γ be an ordered abelian group, and let $\chi:\Gamma \to \{\pm 1\}$ be a group homomorphism. Let v be the valuation on $F = K((\Gamma))$ given by $v(f) = \min(\mathrm{Supp}(f))$ for a non-zero $f \in F$ (Example 4.2.1). Construct an ordering P on F as in Example 6.2.5. Thus $f \in P$ if and only if $\chi(v(f)) \cdot f(v(f)) \in \bar{P}$. When $f \in G_v$ we have $\chi(v(f)) = \chi(0) = 1$ and $f(v(f)) = f(0) = 1$. Hence $G_v \leq P$, i.e., v is coarser than P.

As we have observed in §5, if a valuation v on F is finer than a valuation u, then $\pi_u(O_v)$ is a valuation ring on $\bar{F}_u$. The following lemma gives an analog of this fact for preorderings and orderings.

LEMMA 7.1.3. *Let u be a valuation on F, let T be a preordering on F, and suppose that $-1 \notin G_v T$. Then the push-down $\bar{T} = \pi_u(T \cap O_u^\times)$ is a preordering on $\bar{F}_u$. Moreover, if in addition T is an ordering on F, then $\bar{T}$ is an ordering on $\bar{F}_u$.*

PROOF. Clearly, $\bar{T}$ is a subgroup of $\bar{F}_u^\times$. As $(O_v^\times)^2 \leq T$, we have $(\bar{F}_u^\times)^2 \leq \bar{T}$. Next, if $-1 \in \bar{T}$, then

$$-1 \in G_v(T \cap O_v^\times) \leq G_v T,$$

a contradiction. Thus $-1 \notin \bar{T}$, and therefore $\bar{T} \cap -\bar{T} = \emptyset$. From this and from the fact that T is additively closed we deduce that so is $\bar{T}$.

Finally, suppose that T is an ordering on F, i.e., $F^\times = T \cup -T$. Then $\bar{F}_u^\times = \bar{T} \cup -\bar{T}$, so $\bar{T}$ is an ordering on $\bar{F}_u$. □

Notice that the assumption $-1 \in G_v T$ is in particular satisfied when $G_v \leq T$.

7.2 Localities on residue fields

For a locality λ on F which is finer than the valuation u, we define the **quotient locality** λ/u in $\bar{F}_u$ to be:

- the quotient valuation v/u, if $\lambda = v$ is a valuation;
- the push-down $\bar{P}_u = \pi_u(P \cap O_u^\times)$, if $\lambda = P$ is an ordering.

Note that λ/u is a locality on $\bar{F}_u$ of the same type (i.e., a valuation or an ordering) as λ.

PROPOSITION 7.2.1. *Let u be a valuation on the field F.*

(a) *For every locality λ on F which is finer than u one has*

$$G_{\lambda/u} = \pi_u(G_\lambda \cap O_u^\times).$$

(b) *If λ, λ' are localities on F finer than u and if λ' is finer than λ, then λ'/u is finer than λ'/u.*

(c) *If λ, λ' are localities on F which are finer than u and are not both orderings, and if λ'/u is finer than λ/u, then λ' is finer than λ.*

(d) *The map $v \mapsto v/u$ gives an order-preserving bijection between the set of all equivalence classes of valuations v on F finer than u and the set of all equivalence classes of valuations on $\bar{F}_u$.*

PROOF. (a) When λ is an ordering, is trivial from the definitions. When λ is a valuation, $G_\lambda \leq O_\lambda^\times \leq O_u^\times$, so this is Proposition 5.2.1(d).

(b) Apply (a).

(c) By (a) and the assumptions,

$$\pi_u(G_\lambda \cap O_u^\times) \le \pi_u(G_{\lambda'} \cap O_u^\times).$$

Since the kernel G_u of $\pi_u\colon O_u^\times \to \bar{F}_u^\times$ is contained in $G_{\lambda'} \cap O_u^\times$, we obtain $G_\lambda \cap O_u^\times \le G_{\lambda'} \cap O_u^\times$. Since λ, λ' are not both orderings, neither are $\lambda/u, \lambda'/u$. Therefore λ/u – whence also λ – is a valuation. Hence $G_\lambda \le O_\lambda^\times \le O_u^\times$, so $G_\lambda \le G_{\lambda'}$.

(d) The injectivity follows from (c).

To prove its surjectivity, let w be a valuation on $\bar{F}_u$. Let v be the composition of w and u, in the sense of §5.2. It is finer than u, and $O_{v/u} = O_w$.

The bijection is order-preserving by (b). □

7.3 The tree structure

Given an ordering P on the field F and a subring R of F, we define

$$O_R(P) = \bigcup_{a \in P \cap R} (-a, a)_P.$$

Notice that the following conditions on $x \in F$ are equivalent:

(i) $x \in O_R(P)$;

(ii) $|x|_P <_P a$ for some $a \in P \cap R$;

(iii) $|x|_P \le_P a$ for some $a \in P \cap R$.

PROPOSITION 7.3.1.

(a) *$O_R(P)$ is a valuation ring on F containing R, and is coarser than P.*

(b) *Every valuation v on F which is coarser than P is also coarser than the valuation corresponding to $O_\mathbb{Q}(P)$.*

(c) *P is Archimedean if and only if $F = O_\mathbb{Q}(P)$.*

PROOF. (a) By condition (iii) above, $R \subseteq O_R(P)$. In particular, $-1, 0, 1 \in O_R(P)$.

Next let $x, y \in O_R(P)$. Then there exist $a, b \in P \cap R$ such that $|x|_P \le_P a$ and $|y|_P \le_P b$. It follows from the triangle inequality (Proposition 6.1.2(a)) that $|x+y|_P \le_P a+b$, so $x+y \in O_R(P)$. By Proposition 6.1.2(c), $|xy|_P \le_P ab$, so $xy \in O_R(P)$. We conclude that $O_R(P)$ is a ring.

If $x \in F^\times \setminus O_R(P)$, then $1 <_P |x|_P$, so by Proposition 6.1.2(d), $|x^{-1}|_P <_P 1$. Hence $x^{-1} \in O_R(P)$. This shows that $O_R(P)$ is a valuation ring. As $(-1,1)_P \subseteq O_R(P)$, Lemma 7.1.1 shows that the corresponding valuation is coarser than P.

(b) By Lemma 7.1.3 and (a), $\bar{P}_v$ is an ordering on $\bar{F}_v$. It therefore follows from Corollary 6.1.7 that $\operatorname{Char} \bar{F}_v = 0$, whence $\mathbb{Q} \subseteq O_v$. By Lemma 7.1.1,

$$O_\mathbb{Q}(P) = \bigcup_{a \in P \cap \mathbb{Q}} a(-1,1)_P \subseteq \mathbb{Q} O_v = O_v.$$

(c) This is immediate. □

COROLLARY 7.3.2. *Any set of localities on the field F has a finest common coarsening (i.e., an infimum).*

PROOF. Let λ_i, $i \in I$, be distinct localities on F, with $|I| \geq 2$. If some λ_i is an ordering, then Proposition 7.3.1(b) allows us to replace it by the valuation corresponding to $O_{\mathbb{Q}}(\lambda_i)$. Thus we may assume that all the λ_i are valuations. The ring generated by all the valuation rings O_{λ_i}, $i \in I$, is then also a valuation ring on F. The corresponding valuation is the finest common coarsening of the λ_i. □

A more explicit description of the finest common coarsening v of a collection v_i, $i \in I$, of valuations can be given as follows. Let $\prod_{i \in I} O_{v_i}$ be the set of all products $\prod_{i \in I} z_i$, where $z_i \in O_{v_i}$ for all i, and $z_i = 1$ for all but finitely many i (this makes sense also when I is infinite).

PROPOSITION 7.3.3. $O_v = \prod_{i \in I} O_{v_i}$.

PROOF. We first show that $\prod_{i \in I} O_{v_i}$ is a ring. Indeed, consider $z = \prod_{i \in I} z_i$ and $t = \prod_{i \in I} t_i$, with $z_i, t_i \in O_{v_i}$ as above. Clearly,

$$zt = \prod_{i \in I} (z_i t_i) \in \prod_{i \in I} O_{v_i}.$$

Also take an arbitrary $i_0 \in I$. By the ultrametric inequality,

$$v_{i_0}(z + t) \geq \min\{v_{i_0}(z), v_{i_0}(t)\}$$

– say, $v_{i_0}(z + t) \geq v_{i_0}(z)$. Then

$$z + t \in zO_{i_0} \subseteq \Big(\prod_{i \in I} O_{v_i}\Big) O_{i_0} = \prod_{i \in I} O_{v_i},$$

proving our claim.

Next, $\prod_{i \in I} O_{v_i}$ contains the valuation rings O_{v_i}, $i \in I$. Hence it is itself a valuation ring. There are no proper subrings of $\prod_{i \in I} O_{v_i}$ containing every O_{v_i}, $i \in I$. Therefore it corresponds to the finest common coarsening v. □

PROPOSITION 7.3.4. *Let λ_i, $i \in I$, be localities on the field F which are not all orderings, and let u be their finest common coarsening. Then the finest common coarsening of λ_i/u, $i \in I$, is trivial.*

PROOF. Let $\bar{\mu}$ be a locality on $\bar{F}_u$ which is coarser than λ_i/u for every $i \in I$. Since λ_i are not all orderings, neither are the λ_i/u. Hence $\bar{\mu}$ is a valuation on $\bar{F}_u$. Proposition 7.2.1(d) yields a valuation v on F which is finer than u and such that $\bar{\mu} = v/u$. By Proposition 7.2.1(c), v is coarser than every λ_i. We conclude that $v = u$, so $\bar{\mu}$ is the trivial valuation. □

PROPOSITION 7.3.5. *Let λ, λ', μ be localities on a field F. Suppose that λ and λ' are coarser than μ. Then either λ is finer than λ' or λ' is finer than λ.*

PROOF. We may assume that $\lambda, \lambda' \neq \mu$. Thus $\lambda = v$ and $\lambda' = v'$ are valuations.

Assume first that $\mu = u$ is a valuation as well. In light of condition (h) of Proposition 5.1.1, there exist convex subgroups Δ, Δ' of $u(F^\times)$ such that v, v' are equivalent to the maps

$$F^\times \to u(F^\times)/\Delta, \quad F^\times \to u(F^\times)/\Delta',$$

respectively, induced by u. By Proposition 2.2.1, one of Δ, Δ' must contain the other – say, $\Delta \leq \Delta'$. Then the projection $u(F^\times)/\Delta \to u(F^\times)/\Delta'$ is a morphism of ordered abelian groups which commutes with the two maps above. We conclude again from condition (h) of Proposition 5.1.1 that v is finer than v'.

Next suppose that $\mu = P$ is an ordering. By Proposition 7.3.1(b), v and v' are then coarser than the valuation corresponding to $O_{\mathbb{Q}}(P)$, so we are done by the first case. $\square$

A partially ordered set $(\mathfrak{T}, \leq)$ is called a **tree** if it has a minimum, and for every $\lambda, \lambda', \mu \in \mathfrak{T}$ with $\lambda, \lambda' \leq \mu$ necessarily $\lambda \leq \lambda'$ or $\lambda' \leq \lambda$. Since the trivial locality is coarser than every locality on F, we deduce from Proposition 7.3.5:

COROLLARY 7.3.6. *The collection of all localities on the field F forms a tree with respect to coarsening.*

CHAPTER 8

Topologies

Every locality on a field defines on it a topology with good arithmetical properties. The coarsening relation among localities fits nicely into this topological structure: in fact, we shall see later that if one removes from the tree of localities (as discussed in §7.3) the trivial valuation (i.e., its root), then the resulting connected components correspond canonically to the distinct topologies induced by the non-trivial localities.

8.1 Basic properties

Given a locality λ on the field F we define a topology $\mathcal{T}_\lambda$ on F as follows:

- When $\lambda = v$ is a valuation, $\mathcal{T}_v$ has a basis consisting of all sets $a + bO_v$, with $a, b \in F$, $b \neq 0$. Notice that when v is non-trivial, $\mathcal{T}_v$ also has the basis consisting of all sets $a + b\mathfrak{m}_v$, with $a, b \in F$, $b \neq 0$, and in particular it is Hausdorff.
- When $\lambda = P$ is an ordering, $\mathcal{T}_P$ has a basis consisting of all non-empty sets $(a, b)_P$, where $a, b \in F$. This topology is clearly also Hausdorff.

A topology $\mathcal{T}$ on the field F is called a **ring topology** if the maps $+, \cdot : F \times F \to F$ are continuous with respect to $\mathcal{T}$. For any fixed $a \in F$ the maps

$$x \mapsto a + x, \qquad x \mapsto ax$$

are then also continuous. Consequently, for a $\mathcal{T}$-open subset U of F, the set $a + U$ is $\mathcal{T}$-open. When $a \neq 0$ so is the set aU.

PROPOSITION 8.1.1. *For every locality λ on F the topology $\mathcal{T}_\lambda$ is a ring topology on F.*

PROOF. There are two cases to consider:

CASE I: *$\lambda = v$ is a valuation.*

Consider a basic open subset $a + bO_v$ of $\mathcal{T}_v$, with $a, b \in F$, $b \neq 0$. We show that its inverse images under the maps $+, \cdot : F \times F \to F$ are $\mathcal{T}_v$-open. We may assume that $v(b) \geq 0$.

For the addition, take $x_1, x_2 \in F$ with $x_1 + x_2 \in a + bO_v$. Then

$$(x_1 + bO_v) + (x_2 + bO_v) \subseteq a + bO_v,$$

as desired.

Next take $x_1, x_2 \in F$ with $x_1 \cdot x_2 \in a + bO_v$. Choose $c_1, c_2 \in F^\times$ with

$$v(c_1) = \max\{v(b), v(b) - v(x_2)\}, \quad v(c_2) = \max\{0, v(b) - v(x_1)\}.$$

Then

$$(x_1 + c_1 O_v)(x_2 + c_2 O_v) \subseteq x_1 x_2 + x_1 c_2 O_v + x_2 c_1 O_v + c_1 c_2 O_v \subseteq a + bO_v.$$

CASE II: $\lambda = P$ *is an ordering.*

Here we consider a basic open subset $(a - \epsilon, a + \epsilon)_P$ of $\mathcal{T}_P$, where $a \in F$ and $\epsilon \in P$.

First take $x_1, x_2 \in F$ such that $x_1 + x_2 \in (a - \epsilon, a + \epsilon)_P$. Let

$$\delta = \frac{\epsilon - |x_1 + x_2 - a|_P}{2} >_P 0.$$

Whenever $y_1, y_2 \in F$ satisfy $|y_i - x_i|_P <_P \delta$, $i = 1, 2$, we have

$$|y_1 + y_2 - a|_P \leq_P |y_1 - x_1|_P + |y_2 - x_2|_P + |x_1 + x_2 - a|_P <_P \epsilon$$

by the triangle inequality (Proposition 6.1.2(a)). Thus $y_1 + y_2 \in (a - \epsilon, a + \epsilon)_P$.

Next let $x_1, x_2 \in F$ satisfy $x_1 \cdot x_2 \in (a - \epsilon, a + \epsilon)_P$. Choose $\delta \in F$ with

$$0 <_P \delta <_P \min\left\{1, \frac{\epsilon - |x_1 x_2 - a|_P}{|x_1|_P + |x_2|_P + 1}\right\}.$$

Whenever $y_1, y_2 \in F$ satisfy $|y_i - x_i|_P <_P \delta$, the triangle inequality gives:

$$\begin{aligned} |y_1 y_2 - a|_P &= |(y_1 - x_1) y_2 + x_1 (y_2 - x_2) + (x_1 x_2 - a)|_P \\ &\leq_P \delta |y_2|_P + \delta |x_1|_P + |x_1 x_2 - a|_P \\ &\leq_P \delta(|x_2|_P + 1) + \delta |x_1|_P + |x_1 x_2 - a|_P <_P \epsilon. \end{aligned}$$

Thus $y_1 y_2 \in (a - \epsilon, a + \epsilon)_P$. □

Furthermore, we have:

PROPOSITION 8.1.2. *For a non-trivial locality λ on the field F, the map $x \mapsto x^{-1}$ is $\mathcal{T}_\lambda$-continuous on $F^\times$.*

PROOF. Again, there are two cases to consider:

CASE I: $\lambda = v$ *is a non-trivial valuation.*

It suffices to prove continuity on $F \setminus cO_v$ for $c \in F^\times$. To this end take arbitrary $a, b \in F$ with $b \neq 0$, and assume that $x \in F \setminus cO_v$ and $x^{-1} \in a + bO_v$. The set $(x + bc^2 O_v) \setminus cO_v$ is open and contains x. For every y in this set we have

$$v(x^{-1} - y^{-1}) = v(x - y) - v(x) - v(y) > v(bc^2) - 2v(c) = v(b),$$

so $y^{-1} \in a + bO_v$ as well.

CASE II: $\lambda = P$ *is an ordering.* Here it suffices to prove continuity on $F \setminus (-\delta, \delta)_P$ for some $\delta \in P$. Assume that $x \in F$ satisfies $\delta \leq_P |x|_P$, and let $\epsilon \in P$. For every $y \in F$ with $\delta \leq_P |y|_P$ and $|x - y|_P < \epsilon \delta^2$ we get

$$|x^{-1} - y^{-1}|_P = \frac{|x - y|_P}{|xy|_P} < \epsilon. \qquad \square$$

LEMMA 8.1.3. *Let λ_1 and λ_2 be non-trivial localities on F, with λ_1 finer than λ_2. Then $\mathcal{T}_{\lambda_1} = \mathcal{T}_{\lambda_2}$.*

PROOF. It suffices to consider the following two cases:

CASE I: *λ_1 and λ_2 are valuations.*

Then $O_{\lambda_1} \subseteq O_{\lambda_2}$ and $\mathfrak{m}_{\lambda_2} \subseteq \mathfrak{m}_{\lambda_1}$. Hence every basic $\mathcal{T}_{v_1}$-open set contains a basic $\mathcal{T}_{v_2}$-open set, and vice versa. We conclude that $\mathcal{T}_{\lambda_1} = \mathcal{T}_{\lambda_2}$.

CASE II: *$\lambda_1 = P$ is an ordering and $\lambda_2 = v$ is a valuation.*

Then $G_v \leq P$, so $P = \bigcup_{a \in P} aG_v$. Since aG_v is $\mathcal{T}_v$-open for every $a \in P$, so is P. Now $\mathcal{T}_v$ is invariant under maps $x \mapsto c + dx$, where $c, d \in F$ and $d \neq 0$. Hence

$$(a, b)_P = (a + P) \cap (b - P)$$

is also $\mathcal{T}_v$-open. We conclude that $\mathcal{T}_P \subseteq \mathcal{T}_v$.

Likewise, to prove that $\mathcal{T}_v \subseteq \mathcal{T}_P$ it suffices to show that $\mathfrak{m}_v$ is $\mathcal{T}_P$-open. The non-triviality of v yields $a \in \mathfrak{m}_v \cap P$. Now let $x_0 \in \mathfrak{m}_v$. When $x_0 \in P$, the $\mathcal{T}_P$-open set

$$(-a - x_0, a + x_0)_P$$

contains x_0. By Lemma 7.1.1, it is contained in $(a + x_0)O_v \subseteq \mathfrak{m}_v$, as desired. When $x_0 \in -P$ we similarly obtain that the $\mathcal{T}_P$-open set

$$(-a + x_0, a - x_0)_P$$

contains x_0 and is contained in $(a - x_0)O_v \subseteq \mathfrak{m}_v$. □

We will return to the connection between the local topologies $\mathcal{T}_\lambda$ and the coarsening relation in §10.

8.2 Continuity of roots

Theorem 8.2.3 below shows that, roughly speaking, roots of polynomials in a valued field (F, v) are continuous in the valuation topology $\mathcal{T}_v$. We first need the following two estimates. Here $\bar{v}$ is the classical restricted Gauss valuation on $F(X)$ (Example 4.3.3), i.e.,

$$\bar{v}\Big(\sum_{k=0}^{n} c_k X^k\Big) = \min_{0 \leq k \leq n} v(c_k)$$

for $c_k \in F$.

LEMMA 8.2.1. *Let (F, v) be a valued field, let $b_1, \ldots, b_n \in F$, and let $g(X) = \prod_{i=1}^{n}(X - b_i)$. If $v(b_i) < 0$ for some $1 \leq i \leq n$, then $\bar{v}(g) \leq v(b_j)$ for all $1 \leq j \leq n$.*

PROOF. Write

$$g(X) = X^n + c_1 X^{n-1} + \cdots + c_{n-1}X + c_n,$$

with $c_1, \ldots, c_n \in F$. Let

$$\delta = \min\{v(b_1), \ldots, v(b_n)\},$$

and suppose that precisely k elements of the list $b_1, \ldots, b_n$ have value δ. Using Lemma 3.2.1(c) we compute:

$$\bar{v}(g) \leq v(c_k) = v\Big(\pm \sum_{i_1 < \cdots < i_k} b_{i_1} \cdots b_{i_k}\Big) = k\delta.$$

Now if $v(b_i) < 0$ for some i, then $\delta < 0$. It follows that

$$\bar{v}(g) \leq k\delta \leq \delta \leq v(b_j)$$

for all j, as desired. □

LEMMA 8.2.2. *Let $h(X) = \sum_{k=0}^{n} c_k X^k \in F[X]$ and let $b \in F$. Then*

$$\bar{v}(h) + \min\{0, nv(b)\} \leq v(h(b)) \quad .$$

PROOF. By the ultrametric inequality:

$$\begin{aligned}\bar{v}(h) + \min\{0, nv(b)\} &\leq \bar{v}(h) + \min_{1\leq k\leq n} kv(b) \\ &\leq \min_{1\leq k\leq n} (v(c_k) + kv(b)) \\ &= \min_{1\leq k\leq n} v(c_k b^k) \leq v(h(b)) \quad .\end{aligned}$$

□

We can now prove the desired result about continuity of roots in a valuation topology.

THEOREM 8.2.3. *Let (F, v) be a valued field. Let $a_1, \ldots, a_n$ be distinct elements of F and let $f(X) = \prod_{i=1}^{n}(X - a_i)$. For every $\gamma \in v(F^\times)$ there exists $\alpha \in v(F^\times)$ such that if $b_1, \ldots, b_n \in F$ and if $g(X) = \prod_{i=1}^{n}(X - b_i)$ satisfies $\bar{v}(f - g) > \alpha$, then, after renumbering, $v(a_i - b_i) > \gamma$, $i = 1, \ldots, n$.*

PROOF. We may assume that the valuation v is non-trivial.

Since $a_1, \ldots, a_n$ are distinct, we can choose $\gamma_1 \in v(F^\times)$ with $\gamma < \gamma_1$ and $v(a_i - a_{i'}) < \gamma_1$ for all $i \neq i'$. Choose $\alpha \in v(F^\times)$ such that

$$0 \leq \alpha, \quad \bar{v}(f) \leq \alpha, \quad n\gamma_1 - \min\{0, n\bar{v}(f)\} \leq \alpha.$$

Now let $b_1, \ldots, b_n \in F$, let $g(X) = \prod_{i=1}^{n}(X - b_i)$, and suppose that $\alpha < \bar{v}(f - g)$. Then $\bar{v}(f) = \bar{v}(g)$. It follows from Lemma 8.2.1 that

$$\min\{0, n\bar{v}(f)\} = \min\{0, n\bar{v}(g)\} \leq \min\{0, nv(b_j)\}$$

for every $1 \leq j \leq n$. We now compute using Lemma 8.2.2:

$$\begin{aligned}n\gamma_1 &\leq \alpha + \min\{0, n\bar{v}(f)\} \\ &< \bar{v}(f - g) + \min\{0, nv(b_j)\} \\ &\leq v((f - g)(b_j)) \\ &= v(f(b_j)) \\ &= v\Big(\prod_{i=1}^{n}(b_j - a_i)\Big) = \sum_{i=1}^{n} v(b_j - a_i).\end{aligned}$$

Consequently, there exists $1 \leq i \leq n$ such that $\gamma_1 \leq v(b_j - a_i)$. Moreover, by our choice of γ_1 and the ultrametric inequality, for each j there is precisely one such i. After renumbering we may therefore assume that

$$\gamma < \gamma_1 \leq v(b_i - a_i), \quad i = 1, \ldots, n.$$

□

8.3 Bounded sets

We conclude this chapter by some elementary comments on bounded sets in ring topologies, which will be needed in §10.1. Let $\mathcal{T}$ be a ring topology on a field F and let $\mathcal{T}^0$ be the set of all $\mathcal{T}$-open neighborhoods of 0. We say that a subset B of F is **$\mathcal{T}$-bounded** if for every $U \in \mathcal{T}^0$ there exists $W \in \mathcal{T}^0$ such that $W \cdot B \subseteq U$.

8.3.1 EXAMPLES.
(1) If v is a non-trivial valuation on F, then aO_v is $\mathcal{T}_v$-bounded for every $a \in F^\times$.
(2) If P is an ordering on F, then $(-a, a)_P$ is $\mathcal{T}_P$-bounded for every $a \in P$.

LEMMA 8.3.2.
(a) *A subset of a $\mathcal{T}$-bounded set is also $\mathcal{T}$-bounded.*
(b) *The union of finitely many $\mathcal{T}$-bounded subsets is $\mathcal{T}$-bounded.*
(c) *Every finite set in F is $\mathcal{T}$-bounded.*
(d) *If B is $\mathcal{T}$-bounded, then so is xB for every $x \in F^\times$.*
(e) *If B, C are $\mathcal{T}$-bounded subsets in F, then so is $B + C$.*
(f) *If the topology $\mathcal{T}$ is non-discrete and $B \in \mathcal{T}^0$ is $\mathcal{T}$-bounded, then the sets xB, with $x \in F^\times$, form a local neighborhood system for $\mathcal{T}$ at 0.*

PROOF. The proofs of (a)–(d) are straightforward.

(e) Let $U \in \mathcal{T}^0$. The continuity of addition gives $U', U'' \in \mathcal{T}^0$ such that $U' + U'' \subseteq U$. By assumption there exist $W', W'' \in \mathcal{T}^0$ with $W' \cdot B \subseteq U'$ and $W'' \cdot C \subseteq U''$. Then

$$(W' \cap W'')(B + C) \subseteq U' + U'' \subseteq U.$$

(f) Given $U \in \mathcal{T}^0$ take W with $W \cdot B \subseteq U$. Since $\mathcal{T}$ is non-discrete, $W \neq \{0\}$, so we can pick a non-zero $x \in W$. Then $0 \in xB \subseteq W \cdot B \subseteq U$. □

The following lemma is a special case of an argument of H. Weber [**We**] and M. Marshall [**Mar3**].

LEMMA 8.3.3. *Let $\mathcal{T}_1, \mathcal{T}_2, \ldots, \mathcal{T}_n$, where $n \geq 2$, be ring topologies on F. Assume that every set in $\mathcal{T}_1^0$ is $\mathcal{T}_i$-unbounded for $i = 2, \ldots, n$. Let $S \in \mathcal{T}_1^0$ and for each $2 \leq i \leq n$ let $B_i \in \mathcal{T}_i^0$ be $\mathcal{T}_i$-bounded. Then*

$$S \not\subseteq B_2 \cup \cdots \cup B_n.$$

PROOF. We argue by induction on n. For $n = 2$ this follows Lemma 8.3.2(a).
Suppose that $n \geq 3$ and assume that $S \subseteq B_2 \cup \cdots \cup B_n$ with $S, B_2, \ldots, B_n$ as above. Since B_n is $\mathcal{T}_n$-bounded, so is $B_n - B_n$, by Lemma 8.3.2(d)(e). By assumption, S is $\mathcal{T}_n$-unbounded, so $S \not\subseteq B_n - B_n$. Choose $x \in S$ with $x \notin B_n - B_n$. As $x + 0 = x \in S$ the $\mathcal{T}_1$-continuity of addition gives rise to $U \in \mathcal{T}_1^0$ such that

$$x + U \subseteq S \subseteq B_2 \cup \cdots \cup B_n, \quad U \subseteq S.$$

By the choice of x we therefore have

$$x + (U \cap B_n) \subseteq B_2 \cup \cdots \cup B_{n-1}.$$

Moreover, $U \setminus B_n \subseteq S \setminus B_n \subseteq \bigcup_{i=2}^{n-1} B_i$. It follows that

$$U \subseteq \bigcup_{i=2}^{n-1} [(B_i - x) \cup B_i].$$

But $(B_i - x) \cup B_i$ is also a $\mathcal{T}_i$-bounded set in $\mathcal{T}_i^0$ (by Lemma 8.3.2). This contradicts the induction hypothesis. $\square$

CHAPTER 9

Complete Fields

Generally speaking, metric space methods are inadequate for studying local topologies, in the sense of the previous chapter. However, as we will see in this chapter, if the locality is Archimedean (i.e., a valuation of rank 1, or an Archimedean ordering in the sense of §6.3), then a metric approach is nevertheless possible. In particular, one can then construct completions, which turn out to be again valued or ordered fields with very satisfactory properties.

9.1 Metrics

In light of Theorem 2.5.2, we may identify the value group of a valuation of rank 1 as a subgroup of the ordered group $(\mathbb{R}, \leq_{\mathbb{R}})$. Likewise, we may identify an Archimedean ordered field (F, P) as an ordered subfield of $(\mathbb{R}, (\mathbb{R}^{\times})^2)$ (Proposition 6.3.1). With these identifications we obtain "arithmetical" metrics on the field as follows.

PROPOSITION 9.1.1. *Let F be a field.*

(a) *If v is a valuation of rank 1 on F, then for every $c \in \mathbb{R}$ with $0 < c < 1$ the map $d_{v,c}(x, y) = c^{v(x-y)}$ is a metric on F which induces the topology $\mathcal{T}_v$.*

(b) *If P is an Archimedean ordering on F, then $d_P(x, y) = |x-y|_P$ is a metric on F which induces the topology $\mathcal{T}_P$.*

PROOF. (a) Clearly, $d = d_{v,c}$ is symmetric and non-negative. Furthermore, $d(x, y) = 0$ if and only if $x = y$ (where we use the conventions $v(0) = \infty$ and $c^{\infty} = 0$). To show that $d(x, z) \leq d(x, y) + d(y, z)$ for every $x, y, z \in F$, we may assume, without loss of generality, that $v(x - y) \leq v(y - z)$. By the ultrametric inequality for v,

$$v(x - z) = v\big((x - y) + (y - z)\big) \geq v(x - y).$$

Hence

$$d(x, z) = c^{v(x-z)} \leq c^{v(x-y)} = d(x, y) \leq d(x, y) + d(y, z),$$

as desired.

The open d-ball of radius $0 < \epsilon \in \mathbb{R}$ around $x \in F$ consists of all $y \in F$ such that $v(x - y) > \log_c \epsilon$. Therefore the metric topology induced by d is $\mathcal{T}_v$.

(b) Since $|x - y|_{\leq_{\mathbb{R}}}$ is a metric on $\mathbb{R}$, its restriction $d_P(x, y) = |x - y|_P$ is a metric on F. It is clear that it induces $\mathcal{T}_P$. □

Now suppose that v is an arbitrary valuation of rank 1 on F. When $0 < c_1, c_2 < 1$ we have $d_{v,c_1}(x, y) = d_{v,c_2}(x, y)^{\delta}$, where $\delta = \log c_1 / \log c_2 > 0$. Hence a sequence in F converges (resp., is a Cauchy sequence) with respect to the metric d_{v,c_1} if and only if it converges (resp., is a Cauchy sequence) with respect to the metric d_{v,c_2}.

Therefore, in metric space considerations we will suppress the constant c, and just write d_v for $d_{v,c}$. In particular, the metric space (F, d_{v,c_1}) is complete if and only if (F, d_{v,c_2}) is complete. In this case we will simply say that the valuation v is **complete**.

Let R be the set of all Cauchy sequences in (F, d_v), where as before, v is a valuation of rank 1 on F. Recall from the general theory of metric spaces that Cauchy sequences $\{a_n\}_{n=1}^\infty$ and $\{b_n\}_{n=1}^\infty$ are **equivalent** if $d_v(a_n, b_n) \to 0$ (in our situation this means that $v(a_n - b_n) \to \infty$) as $n \to \infty$. This is obviously an equivalence relation. Then the **completion** of (F, d_v) is the set of equivalence classes of R, with the (well-defined) metric

$$\hat{d}_v\left(\overline{\{a_n\}_{n=1}^\infty}, \overline{\{b_n\}_{n=1}^\infty}\right) = \lim_{n\to\infty} d_v(a_n, b_n).$$

The metric space $(R, \hat{d}_v)$ is complete and F is dense in it. Moreover, these two properties characterize the completion up to isometry over (F, d_v).

9.2 Examples

EXAMPLE 9.2.1: *The p-adic numbers.*

Let p be a prime number, and consider the p-adic valuation v_p on the field $\mathbb{Q}_p$ of p-adic numbers (Example 4.1.4). Recall that for $k \in \mathbb{Z}$ and $\alpha \in \mathbb{Q}_p$ one has $k \le v_p(\alpha)$ if and only if $\alpha \in p^k\mathbb{Z}_p$.

We claim that v_p is complete. Indeed, let $\{\alpha_n\}_{n=1}^\infty$ be a Cauchy sequence in the space $(\mathbb{Q}_p, d_{v_p})$. Thus for every non-negative integer k there exists a positive integer $N(k)$ such that for every $n, m \ge N(k)$ one has $v_p(\alpha_n - \alpha_m) \ge k$. After replacing $\{\alpha_n\}_{n=1}^\infty$ by $\{\alpha_n - \alpha_{N(0)}\}_{n=N(0)}^\infty$ we may assume that $\alpha_n \in \mathbb{Z}_p$ for all n.

Since $\mathbb{Z}/p^k \xrightarrow{\sim} \mathbb{Z}_p/p^k$ canonically, we can choose $\beta_k \in \mathbb{Z}$ with

$$v_p(\beta_k - \alpha_{N(k)}) \ge k.$$

For $n \ge N(k)$ the ultrametric inequality gives

$$v_p(\beta_k - \alpha_n) \ge \min\left\{v_p(\beta_k - \alpha_{N(k)}), v_p(\alpha_{N(k)} - \alpha_n)\right\} \ge k.$$

For any $n \ge N(k), N(k+1)$ the ultrametric inequality once again gives

$$v_p(\beta_{k+1} - \beta_k) \ge \min\left\{v_p(\beta_{k+1} - \alpha_n), v_p(\alpha_n - \beta_k)\right\} \ge k.$$

Therefore the sequence $\beta = (\beta_k + p^k\mathbb{Z})_{k=1}^\infty$ belongs to the inverse limit $\mathbb{Z}_p = \varprojlim \mathbb{Z}/p^k$. For $n \ge N(k)$ we obtain

$$v_p(\beta - \alpha_n) \ge \min\left\{v_p(\beta - \beta_k), v_p(\beta_k - \alpha_n)\right\} \ge k.$$

It follows that $\{\alpha_n\}_{n=1}^\infty$ converges to β, as desired.

We further observe that $\mathbb{Z}$ is dense in $\mathbb{Z}_p$. Therefore $\mathbb{Q}$ is dense in $\mathbb{Q}_p$. We conclude that $\mathbb{Q}_p$ is the completion of $\mathbb{Q}$ with respect to the p-adic valuation.

EXAMPLE 9.2.2: *Formal power series.*

Let K be a field, let $(\Gamma, \leq)$ be an ordered abelian group of rank 1, and let $F = K((\Gamma))$ be the field of formal power series over K with exponents in Γ (§2.8). We identify again $(\Gamma, \leq)$ as an ordered subgroup of $(\mathbb{R}, \leq_{\mathbb{R}})$. Let $v\colon F^\times \to \Gamma$ be the canonical valuation on F defined in Example 4.2.1 with respect to the trivial valuation on K, i.e., $v(f) = \min(\mathrm{Supp}(f))$ for $f \in F^\times$.

We claim that v is complete. Indeed, let $\{f_n\}_{n=1}^\infty$ be a Cauchy sequence in the space (F, d_v). Thus for every $\gamma \in \Gamma$ there exists a positive integer N such that whenever $n, m \geq N$ one has $v(f_n - f_m) > \gamma$, i.e., $f_n(\gamma') = f_m(\gamma')$ for all $\gamma' \leq \gamma$. Then the sequence $\{f_n(\gamma)\}_{n=1}^\infty$ stabilizes to some element of F, which we denote by $f(\gamma)$. Moreover, for every $\gamma \in \Gamma$ and for every sufficiently large n, the maps f, f_n coincide on all $\gamma' < \gamma$. Since $\mathrm{Supp}(f_n)$ does not contain an infinite decreasing sequence, this implies that neither does $\mathrm{Supp}(f)$. Therefore $\mathrm{Supp}(f)$ is well-ordered, so $f \in F = K((\Gamma))$. Furthermore, the sequence $\{f_n\}_{n=1}^\infty$ converges to f.

In the special case $\Gamma = \mathbb{Z}$ we obtain that the field $K((t))$ of Laurent series over K is complete. Moreover, the ring of all maps in $K((t))$ with finite support is dense in $K((t))$. Therefore so is its quotient field, which is the field $K(t)$ of rational functions in one variable over K. Consequently, $K((t))$ is the completion of $K(t)$ with respect to the metric induced by v.

9.3 Completions

Consider again a valuation v of rank 1 on the field F. Theorem 9.3.2 below shows that the completion of (F, d_v) can also be obtained as the metric space corresponding to a certain valued field extension $(\hat{F}, \hat{v})$ of (F, v). To prove it, we first make some technical observations. Let M be the set of all sequences $\{a_n\}_{n=1}^\infty$ which are equivalent to the constant sequence 0, i.e., $v(a_n) \to \infty$ as $n \to \infty$. Let R be as before the set of all Cauchy sequences in (F, d_v). Clearly, $M \subseteq R$.

LEMMA 9.3.1.

(a) *For every $\{a_n\}_{n=1}^\infty \in R$ there exists $\gamma \in v(F^\times)$ such that $v(a_n) \geq \gamma$ for all n.*

(b) *For every $\{a_n\}_{n=1}^\infty \in R \setminus M$ the value $v(a_n)$ is constant for n sufficiently large.*

PROOF. (a) There exists a positive integer N such that for every $n, m \geq N$ one has $v(a_n - a_m) \geq 0$. In particular, $v(a_n - a_N) \geq 0$, so by the ultrametric inequality, $v(a_n) \geq \min\{v(a_N), 0\}$ for $n \geq N$. We take

$$\gamma = \min\{v(a_1), \ldots, v(a_N), 0\}.$$

(b) The assumption that the sequence is not in M means that there exists $\gamma \in v(F^\times)$ such that $v(a_m) \leq \gamma$ for infinitely many m. Since this is a Cauchy sequence, there exists a positive integer N such that for every $n, m \geq N$ one has $v(a_n - a_m) > \gamma$. For every $n \geq N$ we take $m \geq N$ with $v(a_m) \leq \gamma$, to obtain that

$$v(a_n) = v(a_m + (a_n - a_m)) = v(a_m) \leq \gamma.$$

Now for all $n, k \geq N$ this inequality $v(a_n) \leq \gamma$ and $v(a_k - a_n) > \gamma$ give

$$v(a_m) = v(a_n + (a_k - a_n)) = v(a_n). \qquad \square$$

THEOREM 9.3.2. *Let v be a valuation of rank 1 on the field F and keep the previous notation.*

(a) *R is a ring with respect to componentwise addition and multiplication.*

(b) *M is an ideal in R.*

(c) *$\hat{F} = R/M$ is a field in which F embeds via the diagonal map $\iota: a \mapsto \{a\}_{n=1}^{\infty} + M$.*

(d) *The map $\hat{v}: \hat{F}^{\times} \to v(F^{\times})$, defined by*

$$\hat{v}(\{a_n\}_{n=1}^{\infty} + M) = v(a_n)$$

for n sufficiently large, is well defined, surjective, and is a valuation on $\hat{F}$ extending v.

(e) *$(\hat{F}, d_{\hat{v}})$ is isometric to the completion of the metric space (F, d_v).*

(f) *The valued field extension $(\hat{F}, \hat{v})/(F, v)$ is immediate.*

PROOF. (a) It is enough to check that R is closed under addition and multiplication. So let $\{a_n\}_{n=1}^{\infty}$ and $\{b_n\}_{n=1}^{\infty}$ be Cauchy sequences in (F, d_v). Lemma 9.3.1(a) yields $\gamma_0 \in v(F^{\times})$ such that $v(a_n), v(b_n) \geq \gamma_0$ for all n. For every $\gamma \in v(F^{\times})$ there exists a positive integer N such that for all $n, m \geq N$ one has

$$v(a_n - a_m), v(b_n - b_m) \geq \gamma.$$

Then

$$v((a_n + b_n) - (a_m + b_m)) \geq \min\{v(a_n - a_m), v(b_n - b_m)\} \geq \gamma,$$

and

$$\begin{aligned} v(a_n b_n - a_m b_m) &= v(a_n(b_n - b_m) + b_m(a_n - a_m)) \\ &\geq \min\{v(a_n) + \gamma, v(b_m) + \gamma\} \geq \gamma_0 + \gamma. \end{aligned}$$

This shows that $\{a_n + b_n\}_{n=1}^{\infty}$ and $\{a_n b_n\}_{n=1}^{\infty}$ are Cauchy sequences as well.

(b) Clearly, M is a group with respect to the addition of R, and contains the constant sequence $\{0\}_{n=1}^{\infty}$. Next let $\{a_n\}_{n=1}^{\infty} \in M$ and $\{b_n\}_{n=1}^{\infty} \in R$. Then $v(a_n) \to \infty$, and Lemma 9.3.1(a) yields $\gamma \in v(F^{\times})$ such that $v(b_n) \geq \gamma$ for all n. Hence $v(a_n b_n) \to \infty$ as $n \to \infty$, i.e., $\{a_n b_n\}_{n=1}^{\infty} \in M$.

(c) By (a) and (b), R/M is a ring. We need to show that every non-zero element of R/M is invertible. To this end take a sequence $\{a_n\}_{n=1}^{\infty} \in R \setminus M$. Lemma 9.3.1(b) gives γ and N such that $v(a_n) = \gamma$ for all $n \geq N$. In particular, $a_n \neq 0$ for $n \geq N$. Define a sequence $\{b_n\}_{n=1}^{\infty}$ by setting $b_n = a_n^{-1}$ for $n \geq N$, and $b_n = 0$ for $n < N$. It is a Cauchy sequence, since for $n, m \geq N$ one has

$$v(b_n - b_m) = v(a_m - a_n) - v(a_n) - v(a_m) = v(a_n - a_m) - 2\gamma \to \infty,$$

as $n, m \to \infty$. Moreover, $\{b_n\}_{n=1}^{\infty} + M$ is a multiplicative inverse of $\{a_n\}_{n=1}^{\infty} + M$ in R/M.

The injectivity of ι is immediate.

(d) The map $\hat{v}$ is well defined by Lemma 9.3.1(b). For every $a \in F^{\times}$ one has $\hat{v}(\iota(a)) = v(a)$, so $\hat{v}$ extends v and is surjective. The fact that $\hat{v}: \hat{F}^{\times} \to v(F^{\times})$ is a group homomorphism and the ultrametric inequality follow from the corresponding properties of v.

(e) Sequences $\{a_n\}_{n=1}^\infty, \{b_n\}_{n=1}^\infty \in R$ are equivalent if and only if they have the same coset in $\hat{F} = R/M$. Also, for a constant $0 < c < 1$ as above we have

$$d_{\hat{v}}(\overline{\{a_n\}_{n=1}^\infty}, \overline{\{b_n\}_{n=1}^\infty}) = c^{\hat{v}(\{a_n - b_n\}_{n=1}^\infty + M)} = \lim_{n\to\infty} c^{v(a_n - b_n)}$$
$$= \lim_{n\to\infty} d_v(a_n, b_n) = \hat{d}_v(\overline{\{a_n\}_{n=1}^\infty}, \overline{\{b_n\}_{n=1}^\infty}).$$

Hence $d_{\hat{v}} = \hat{d}_v$, as required (the notation being as at the end of §9.1).

(f) We have already seen in the proof of (d) that $v(F^\times) = \hat{v}(\hat{F}^\times)$. As in any extension of valued fields, the residue field of (F, v) is contained in the residue field of $(\hat{F}, \hat{v})$. To show that they coincide, take $\alpha \in O_{\hat{v}}$. By (e), F is dense in $\hat{F}$ with respect to the metric $d_{\hat{v}}$. Hence there exists $b \in F$ such that $d_{\hat{v}}(\alpha, \iota(b)) < 1$. Then $\hat{v}(\alpha - \iota(b)) > 0$, so $b \in O_v$, and α and $\iota(b)$ have the same residues with respect to $\hat{v}$. □

REMARK 9.3.3. The analogous construction for an Archimedean ordering P on a field F is much simpler, in the sense that the completion of (F, d_P) is always $\mathbb{R}$ with its canonical metric $d_\mathbb{R}(x, y) = |x - y|_\mathbb{R}$. The latter corresponds to the ordering $(\mathbb{R}^\times)^2$. In fact, by Proposition 6.3.1, (F, P) embeds in $(\mathbb{R}, (\mathbb{R}^\times)^2)$ as an ordered field. Hence (F, d_P) embeds as a metric space in $(\mathbb{R}, d_\mathbb{R})$. Since $(\mathbb{R}, d_\mathbb{R})$ is the completion of $\mathbb{Q}$ with its standard (Archimedean) metric, it is also the completion of (F, d_P).

CHAPTER 10

Approximation Theorems

In this chapter we prove one of deepest, and among the most useful, facts in the theory of valuations and orderings, which for historical reasons is called "the weak approximation theorem". It relates the coarsening relation among localities with the topologies they induce, by providing a way to simultaneously approximate finitely many elements in the field with respect to different local topologies, as long as these localities are independent, in the sense defined below. A variant of this fact for incomparable valuations is proved in §10.2.

10.1 Approximation by independent localities

Localities $\lambda_1, \ldots, \lambda_n$ on a field F are called **independent** if the finest common coarsening of any two of them is the trivial locality. Otherwise they are called **dependent**.

EXAMPLE 10.1.1. Distinct valuations of rank 1 on F are always independent. In particular, for any distinct prime numbers $p_1, \ldots, p_n$, the corresponding p_i-adic valuations on $\mathbb{Q}$ are discrete, whence of rank 1. Therefore they are independent.

EXAMPLE 10.1.2. Let $\mathbb{Q}_p$ be the field of p-adic numbers and let v_p be its p-adic valuation. Let $v \neq v_p$ be another non-trivial valuation on $\mathbb{Q}_p$. Then v and v_p are independent. Indeed, suppose they were dependent. Since v_p is discrete (hence again of rank 1), this would mean that v_p is coarser than v. But then the quotient v/v_p would be a non-trivial valuation on the residue field $\mathbb{F}_p$ of v_p, which is impossible (see Example 3.2.2(1)).

Our proof of the weak approximation theorem (Theorem 10.1.7 below) will proceed in several steps. The proof will have a mild K-theoretic flavor, in the sense that it will be based on the auxiliary maps

$$f(X) = \frac{X}{1+X}, \quad g(X) = \frac{1}{1+X}.$$

Note that

$$f(X) + g(X) = 1, \quad f(X) = g(X^{-1}), \quad f(0) = 0, \quad g(0) = 1.$$

We start with the following very special case.

PROPOSITION 10.1.3. *Let λ_1, λ_2 be independent localities on a field F. Let U_1 be a $\mathcal{T}_{\lambda_1}$-neighborhood of 1, and let U_2 be a $\mathcal{T}_{\lambda_2}$-neighborhood of 0. Then $U_1 \cap U_2 \neq \emptyset$.*

PROOF. We may assume that each λ_i is non-trivial – otherwise $U_i = F$, and we are done. There are four cases to consider.

CASE I: *$\lambda_1 = v_1$ and $\lambda_2 = v_2$ are valuations.*

We may assume that $U_1 = 1+b_1O_{v_1}$ and $U_2 = b_2O_{v_2}$ for some non-zero $b_i \in \mathfrak{m}_{v_i}$, $i = 1, 2$. By Proposition 7.3.3 and the independence assumption, $F = O_{v_1}O_{v_2}$. One can therefore write $b_1^{-1}b_2^{-1} = yz$ with $y \in O_{v_1}$ and $z \in O_{v_2}$. Let

$$x = y^{-1}b_1^{-1} = zb_2.$$

We have

$$v_1(x) \leq -v_1(b_1) < 0, \quad v_2(x) \geq v_2(b_2) > 0.$$

Hence,

$$v_1(1 - f(x)) = v_1(g(x)) = -v_1(x) \geq v_1(b_1), \quad v_2(f(x)) = v_2(x) \geq v_2(b_2).$$

It follows that $f(x) \in U_1 \cap U_2$.

CASE II: *$\lambda_1 = P$ is an ordering and $\lambda_2 = v$ is a valuation.*

In this case we may assume that $U_1 = (1-\epsilon, 1+\epsilon)_P$ for some $\epsilon \in P$, and $U_2 = bO_v$ with $0 <_P b \in \mathfrak{m}_v$.

The valuation corresponding to $O_{O_v}(P)$ is coarser than both P and v (Proposition 7.3.1(a)), so by assumption it is trivial. Thus $F = \bigcup_{c\in P\cap O_v}(-c, c)_P$. In particular, $b^{-1}(\epsilon^{-1} - 1) <_P c$ for some $c \in P \cap O_v$. Let $x = bc$. Then $0 <_P \epsilon^{-1} <_P 1+x$. Hence $0 <_P (1+x)^{-1} <_P \epsilon$. Furthermore, $x \in bO_v \subseteq \mathfrak{m}_v$. We obtain

$$f(x) = 1 - g(x) \in (1-\epsilon, 1)_P, \quad v(f(x)) = v(x) \geq v(b).$$

Therefore $f(x) \in U_1 \cap U_2$.

CASE III: *$\lambda_1 = v$ is a valuation and $\lambda_2 = P$ is an ordering.*

By Proposition 8.1.1, $\mathcal{T}_v$ and $\mathcal{T}_P$ are ring topologies. Hence $1 - U_1$ is a $\mathcal{T}_v$-neighborhood of 0, and $1 - U_2$ is a $\mathcal{T}_P$-neighborhood of 1. By Case II,

$$(1 - U_2) \cap (1 - U_1) \neq \emptyset.$$

Hence $U_1 \cap U_2 \neq \emptyset$ as well.

CASE IV: *$\lambda_1 = P_1$ and $\lambda_2 = P_2$ are orderings.*

If for some i the valuation ring $O_{\mathbb{Q}}(P_i)$ is non-trivial, then we may replace P_i by the corresponding valuation, by Proposition 7.3.1(b) and Lemma 8.1.3. Then one of cases I–III applies.

Consequently, we may assume without loss of generality that $F = O_{\mathbb{Q}}(P_i)$, $i = 1, 2$, i.e., that P_1, P_2 are Archimedean orderings (Proposition 7.3.1(c)). We may therefore assume that

$$U_1 = (1-\epsilon, 1+\epsilon)_{P_1}, \quad U_2 = (-\epsilon, \epsilon)_{P_2}$$

for some positive $\epsilon \in \mathbb{Q}$ (note that the restriction of P_i is the unique ordering on $\mathbb{Q}$, $i = 1, 2$).

Since P_1 and P_2 are independent, they are distinct, so we can choose $c \in P_1 \cap -P_2$. After replacing, if necessary, c by c^{-1} we may further assume that $-1 <_{P_2} c$. Choose

$$q_1 \in (1, 1+c)_{P_1} \cap \mathbb{Q}, \quad q_2 \in (1+c, 1)_{P_2} \cap \mathbb{Q}.$$

Now consider the map $f(X)$ on the positive cone of F with respect to any ordering. It is order-preserving, and is bounded from above by 1. For any sequence $\{r_m\}_{m=1}^{\infty}$ of positive rational numbers such that $r_m \to \infty$ (resp., $r_m \to 0$) one has

$f(r_m) \to 1$ (resp., $f(r_m) \to 0$). Hence, as $m \to \infty$, the rational sequence $f(q_1^m)$ increases to 1, and the rational sequence $f(q_2^m)$ decreases to 0. For a sufficiently large m we therefore have

$$f((q_1^m, \infty)_{P_1}) \subseteq U_1, \quad f((0, q_2^m)_{P_2}) \subseteq U_2.$$

Hence $f((1+c)^m) \in U_1 \cap U_2$, as required. □

COROLLARY 10.1.4. *Let λ_1, λ_2 be non-trivial localities on a field F. If $\mathcal{T}_{\lambda_1} \subseteq \mathcal{T}_{\lambda_2}$, then λ_1 and λ_2 are dependent.*

PROOF. Since λ_1 is non-trivial, $\mathcal{T}_{\lambda_1}$ is Hausdorff, and we can find disjoint $\mathcal{T}_{\lambda_1}$-open neighborhoods U_1, U_2 of $1, 0$ respectively. By the assumption, U_2 is $\mathcal{T}_{\lambda_2}$-open as well. In light of Proposition 10.1.3, λ_1 and λ_2 are necessarily dependent. □

PROPOSITION 10.1.5. *Let $\lambda_1, \ldots, \lambda_n$ be non-trivial valuations on F. Then $\mathcal{T}_{\lambda_1} = \cdots = \mathcal{T}_{\lambda_n}$ if and only if there exists a non-trivial valuation λ on F which is coarser than $\lambda_1, \ldots, \lambda_n$.*

PROOF. The "if" part follows from Lemma 8.1.3.

For the "only if" part we argue by induction on n. For $n = 1$ the assertion is trivial. When $n \geq 2$, let μ be the finest common coarsening of $\lambda_2, \ldots, \lambda_n$. By the induction hypothesis, it is non-trivial. By Lemma 8.1.3 again, $\mathcal{T}_{\lambda_1} = \mathcal{T}_{\lambda_2} = \mathcal{T}_\mu$. Let λ be the finest common coarsening of λ_1 and μ. It is also the finest common coarsening of $\lambda_1, \lambda_2, \ldots, \lambda_n$. By Corollary 10.1.4, λ_1, μ are dependent, i.e., λ is non-trivial. □

Next we prove the weak approximation theorem in a somewhat more general situation. The argument is inspired by the methods of [**Mar3**] and [**We**].

PROPOSITION 10.1.6. *Let $\lambda_1, \ldots, \lambda_n$ be independent localities on F, with $n \geq 2$. For each i let U_i be a $\mathcal{T}_{\lambda_i}$-open subset of F. Assume that $1 \in U_1$ and $0 \in U_2 \cap \cdots \cap U_n$. Then*

$$U_1 \cap U_2 \cap \cdots \cap U_n \neq \emptyset.$$

PROOF. We use the terminology of §8.3. In particular, $\mathcal{T}^0_{\lambda_i}$ denotes the collection of all $\mathcal{T}_{\lambda_i}$-open neighborhoods of 0. Since the λ_i are independent, they are non-trivial. For each $1 \leq i \leq n$ we choose a set B_i in $\mathcal{T}^0_{\lambda_i}$ as follows:

- if $\lambda_i = v$ is a valuation, we take $B_i = O_v$;
- if $\lambda_i = P$ is an ordering, we take $B_i = (-2, 2)_P$.

Note that for every $x \in F^\times$ at least one of x and x^{-1} is in B_i. Also, the set B_i is in both cases $\mathcal{T}_{\lambda_i}$-bounded (Example 8.3.1). By Lemma 8.3.2(f), the sets zB_i, with $z \in F^\times$, therefore form a local neighborhood system for $\mathcal{T}_{\lambda_i}$ at 0. As $g(0) = 1$ and $f(0) = 0$, a continuity argument yields $y_1, \ldots, y_n \in F^\times$ such that

$$g(y_1 B_1) \subseteq U_1, \quad f(y_i B_i) \subseteq U_i, \quad i = 2, \ldots, n.$$

Next let $2 \leq i \leq n$. By Lemma 8.3.2(d), $y_i^{-1} B_i$ is $\mathcal{T}_{\lambda_i}$-bounded. We claim that every set in $\mathcal{T}^0_1$ is $\mathcal{T}_i$-unbounded. Indeed, it is enough to verify this for sets of the form zB_1, $z \in F^\times$ (by Lemma 8.3.2(a)). But if such a set would be $\mathcal{T}_i$-bounded, then there would exist $t \in F^\times$ so that $tzB_1 \subseteq B_i$, whence $\mathcal{T}_i \subseteq \mathcal{T}_1$. However, this contradicts Corollary 10.1.4, and therefore proves the claim.

Now the claim and Lemma 8.3.3 imply that

$$y_1 B_1 \not\subseteq \bigcup_{i=2}^{n} y_i^{-1} B_i.$$

Therefore there exists $x \in y_1 B_1$ such that $xy_i \notin B_i$, whence $(xy_i)^{-1} \in B_i$ for all $2 \le i \le n$. Consequently,

$$g(x) \in g(y_1 B_1) \subseteq U_1, \quad g(x) = f(x^{-1}) \in f(y_i B_i) \subseteq U_i, \quad i = 2, \dots, n,$$

as required. □

After these preparations we can now state and prove our main result.

THEOREM 10.1.7 (The Weak Approximation Theorem). *Let $\lambda_1, \dots, \lambda_n$ be non-trivial localities on a field F. The following three conditions are equivalent:*

(a) *$\lambda_1, \dots, \lambda_n$ are independent;*

(b) *for every list of non-empty $\mathcal{T}_{\lambda_i}$-open subsets U_i of F, $i = 1, \dots, n$, one has $U_1 \cap \dots \cap U_n \neq \emptyset$;*

(c) *$\mathcal{T}_{\lambda_1}, \dots, \mathcal{T}_{\lambda_n}$ are distinct.*

PROOF. (a)⇒(b): For each $1 \le i \le n$ we choose $t_i \in U_i$. Consider the map $h\colon F^n \to F$ given by

$$h(x_1, \dots, x_n) = \sum_{j=1}^{n} t_j x_j.$$

By continuity, there exist $\mathcal{T}_{v_i}$-open subsets W_{ij}, $j = 1, \dots, n$, of F such that $1 \in W_{ii}$, $0 \in W_{ij}$ for $i \neq j$, and such that

$$h(W_{i1} \times \dots \times W_{in}) \subseteq U_i.$$

By Proposition 10.1.6, $W_{1j} \cap \dots \cap W_{nj} \neq \emptyset$ for every j. We choose z_j in this intersection. Then

$$h(z_1, \dots, z_n) \in U_1 \cap \dots \cap U_n.$$

(b)⇒(c): Suppose that $\mathcal{T}_{\lambda_i} = \mathcal{T}_{\lambda_j}$ for some distinct i, j. Use the Hausdorff property to choose non-empty disjoint $\mathcal{T}_{\lambda_i}$-open subsets U_i, U_j of F. For $l \neq i, j$ we take $U_l = F$. Then $U_1 \cap \dots \cap U_n = \emptyset$.

(c)⇒(a): Use Proposition 10.1.5 for any pair of localities from the list. □

We refer to [**St**] and [**Mar3**] for more general versions of this theorem (in the context of V-topologies on fields, and even on rings).

10.2 Approximation by incomparable valuations

We say that localities λ_1, λ_2 on the field F are **comparable** if one of them is coarser than the other. Otherwise they are called **incomparable**. Independent non-trivial localities are incomparable, but not conversely. Still, the following reminiscent of the weak approximation theorem holds for incomparable valuations.

THEOREM 10.2.1. *Let $v_1, \ldots, v_n$ be valuations on the field F. The following conditions are equivalent:*

(a) *$v_1, \ldots, v_n$ are incomparable;*

(b) *the natural homomorphism $\bigcap_{i=1}^n O_{v_i} \to \prod_{i=1}^n \bar{F}_{v_i}$ is surjective.*

PROOF. (a)$\Rightarrow$(b): We argue by induction on n. The assertion for $n = 1$ is trivial. For $n \geq 2$ let u be the finest common coarsening of $v_1, \ldots, v_n$ (Corollary 7.3.2). By (a), u is a valuation and $u \neq v_i$ for all v_i. Therefore the quotients $v_1/u, \ldots, v_n/u$ are non-trivial valuations. By Proposition 7.2.1(c), they are also incomparable. By Proposition 7.3.4 they have no non-trivial common coarsening.

Let $E = \bar{F}_u$, and let $\mathcal{T}_1, \ldots, \mathcal{T}_k$ be the distinct topologies induced on E by the quotient valuations $v_1/u, \ldots, v_n/u$. For each $1 \leq j \leq k$ let $I(j)$ be the set of all $1 \leq i \leq n$ such that $\mathcal{T}_{v_i/u} = \mathcal{T}_j$. It follows from Proposition 10.1.5 that $k \geq 2$. Hence $|I(j)| < n$ for all $1 \leq j \leq k$. For $1 \leq j \leq k$ let $\bar{w}_j$ be the finest common coarsening of the v_i/u, $i \in I(j)$. By Proposition 10.1.5 again, $\bar{w}_j$ is a non-trivial valuation on E inducing $\mathcal{T}_j$. Therefore $\mathfrak{m}_{\bar{w}_j}$ is $\mathcal{T}_j$-open.

Now for each $1 \leq j \leq k$ the homomorphism

$$\prod_{i \in I(j)} \pi_{v_i/u} \colon \bigcap_{i \in I(j)} O_{v_i/u} \to \prod_{i \in I(j)} \bar{E}_{v_i/u}$$

is surjective by the induction hypothesis. Moreover,

$$\mathfrak{m}_{\bar{w}_j} \leq \bigcap_{i \in I(j)} \mathfrak{m}_{v_i/u} \leq \bigcap_{i \in I(j)} O_{v_i/u},$$

so $\mathfrak{m}_{\bar{w}_j}$ is contained in the kernel of this epimorphism. Hence the maps $\pi_{v_i/u}$ induce an epimorphism

$$(10.2.1) \qquad \prod_{j=1}^{k} \Big[\Big(\bigcap_{i \in I(j)} O_{v_i/u}\Big)/\mathfrak{m}_{\bar{w}_j}\Big] \to \prod_{i=1}^{n} E_{v_i/u}.$$

Next, given $a_j \in \bigcap_{i \in I(j)} O_{v_i/u}$, $j = 1, \ldots, k$, the weak approximation theorem (Theorem 10.1.7) gives:

$$\emptyset \neq \bigcap_{j=1}^{k} (a_j + \mathfrak{m}_{\bar{w}_j}) \subseteq \bigcap_{j=1}^{k} \bigcap_{i \in I(j)} O_{v_i/u} = \bigcap_{i=1}^{n} O_{v_i/u}.$$

This shows that the natural homomorphism

$$(10.2.2) \qquad \bigcap_{i=1}^{n} O_{v_i/u} \to \prod_{j=1}^{k} \Big[\Big(\bigcap_{i \in I(j)} O_{v_i/u}\Big)/\mathfrak{m}_{\bar{w}_j}\Big]$$

is surjective.

We compose the epimorphisms (10.2.1) and (10.2.2) to conclude that the homomorphism

$$\bigcap_{i=1}^{n} O_{v_i/u} \to \prod_{i=1}^{n} \bar{E}_{v_i/u}$$

is also surjective. Furthermore, there is a commutative square with bijective lower row:

$$\begin{array}{ccc} O_{v_i} & \xrightarrow{\pi_u} & O_{v_i/u} \\ {\scriptstyle \pi_{v_i}}\big\downarrow & & \big\downarrow{\scriptstyle \pi_{v_i/u}} \\ \bar{F}_{v_i} & \xrightarrow{\sim} & \bar{E}_{v_i/u} \end{array}$$

(see the commutative diagram (5.2.1)). Consequently, the homomorphism

$$\bigcap_{i=1}^{n} O_{v_i} \to \prod_{i=1}^{n} \bar{F}_{v_i}$$

is also surjective.

(b)⇒(a): Suppose that v_i is coarser than v_j, where $i \neq j$. Then

$$\mathfrak{m}_{v_i} \subseteq \mathfrak{m}_{v_j} \subseteq O_{v_j} \subseteq O_{v_i},$$

and there is a commutative diagram of ring homomorphisms

$$\begin{array}{ccccc} O_{v_i} \cap O_{v_j} & \longrightarrow & (O_{v_i} \cap O_{v_j})/\mathfrak{m}_{v_i} & \hookrightarrow & \bar{F}_{v_i} \\ \big\downarrow & & \big\downarrow & & \\ O_{v_j} & & O_{v_j/v_i} & & \\ {\scriptstyle \pi_{v_j}}\big\downarrow & & \big\downarrow{\scriptstyle \pi_{v_j/v_i}} & & \\ \bar{F}_{v_j} & \xrightarrow{\sim} & (\bar{F}_{v_i})_{v_j/v_i} & & \end{array}$$

where the isomorphism is by Proposition 5.2.1(e). It follows that

$$\mathrm{Ker}\big(O_{v_i} \cap O_{v_j} \to \bar{F}_{v_i}\big) \leq \mathrm{Ker}\big(O_{v_i} \cap O_{v_j} \to \bar{F}_{v_j}\big).$$

We deduce that the natural map $O_{v_i} \cap O_{v_j} \to \bar{F}_{v_i} \times \bar{F}_{v_j}$ is not surjective, and therefore neither is the homomorphism in (b). □

COROLLARY 10.2.2. *Let $v_1, \dots, v_n$ be incomparable valuations on the field F. Then the projection $\bigcap_{i=1}^n O_{v_i}^\times \to \prod_{i=1}^n \bar{F}_{v_i}^\times$ is surjective.*

REMARK 10.2.3. Given incomparable valuations $v_1, \dots, v_n$ on F, one may also ask about the image of the homomorphism

$$\bigoplus_{i=1}^{n} v_i \colon F^\times \to \bigoplus_{i=1}^{n} v_i(F^\times).$$

In general, it need not be surjective. Indeed, if an n-tuple $(\gamma_1, \dots, \gamma_n)$ is in its image, then for every distinct i, j, the images of γ_i and γ_j in $u_{ij}(F^\times)$ must be equal, where u_{ij} is the finest common coarsening of v_i and v_j. In fact, by a theorem of Ribenboim [**Ri1**, p. 135, Th. 1], the latter condition characterizes the image of $\bigoplus_{i=1}^n v_i$. We refer to [**Ri1**, p. 136, Th. 3] for a more general (but also more technical) approximation theorem for incomparable valuations along the lines of Theorem 10.1.7.

10.3 Consequences

As an application of the weak approximation theorem we prove:

PROPOSITION 10.3.1. *Let λ_1, λ_2 be localities on the field F.*
(a) *If λ_1, λ_2 are non-trivial and independent, then $F^\times = U_1 \cdot U_2$ for every $\mathcal{T}_{\lambda_1}$-open subset U_1 and every $\mathcal{T}_{\lambda_2}$-open subgroup U_2 of $F^\times$.*
(b) *If λ_1, λ_2 are non-trivial and independent, then $F^\times = G_{\lambda_1} \cdot G_{\lambda_2}$.*
(c) *If λ_1, λ_2 are incomparable, then their finest common coarsening is a valuation u satisfying*

$$O_u^\times = (G_{\lambda_1} \cap O_u^\times)(G_{\lambda_2} \cap O_u^\times).$$

(d) *If λ_1, λ_2 are incomparable valuations, then their finest common coarsening u satisfies $O_u^\times = G_{\lambda_1} G_{\lambda_2}$.*

PROOF. (a) Let $a \in F^\times$. The sets U_1, aU_2 are open in $\mathcal{T}_{\lambda_1}, \mathcal{T}_{\lambda_2}$, respectively. Theorem 10.1.7 therefore yields $b \in U_1 \cap aU_2$. Then

$$a = b \cdot ab^{-1} \in U_1 \cdot U_2^{-1} = U_1 \cdot U_2,$$

as desired.

(b) This is a special case of (a).

(c) Since λ_1, λ_2 are incomparable they must be distinct, and strictly finer than u. Consequently, u is a valuation.

By Proposition 7.3.4, the localities $\lambda_1/u, \lambda_2/u$ on $\bar{F}_u$ have no non-trivial common coarsenings, i.e., they are independent. It therefore follows from (b) that $\bar{F}_u^\times = G_{\lambda_1/u} \cdot G_{\lambda_2/u}$. This means that

$$\pi_u(O_u^\times) = \pi_u(G_{\lambda_1} \cap O_u^\times) \cdot \pi_u(G_{\lambda_2} \cap O_u^\times)$$

(see Proposition 7.2.1(a)). Since the kernel G_u of $\pi_u \colon O_u^\times \to \bar{F}_u^\times$ is contained in $O_u^\times$, as well as in G_{λ_1} and G_{λ_2}, we deduce that

$$O_u^\times = (G_{\lambda_1} \cap O_u^\times)(G_{\lambda_2} \cap O_u^\times).$$

(d) As $G_{\lambda_i} \leq O_{\lambda_i}^\times \leq O_u^\times$, $i = 1, 2$, this follows from (c). □

Recall that by Proposition 7.3.3, the finest common coarsening u of valuations v_i, $i \in I$, on F is given by $O_u = \prod_{i \in I} O_{v_i}$, i.e., O_u consists of all products $\prod_{i \in I} z_i$, where $z_i \in O_{v_i}$ and $z_i = 1$ for all but finitely many i. Define $\prod_{i \in I} O_{v_i}^\times$ similarly. We can now complement this result as follows:

PROPOSITION 10.3.2. *Let v_i, $i \in I$, be a collection of valuations on F with the finest common coarsening u. Then*

$$O_u^\times = \prod_{i \in I} O_{v_i}^\times.$$

PROOF. We first verify this in the case of two valuations v_1, v_2.

If v_1, v_2 are comparable – say, v_1 is coarser than v_2 – then $u = v_1$, so $O_u^\times = O_{v_1}^\times \geq O_{v_2}^\times$, and the assertion is clear.

If v_1, v_2 are incomparable, then Proposition 10.3.1(d) gives

$$O_u^\times = G_{v_1} G_{v_2} \leq O_{v_1}^\times O_{v_2}^\times.$$

On the other hand, $O_{v_i}^\times \leq O_u^\times$ for $i = 1, 2$, so we are done once again.

Next, the case of finitely many valuations v_i follows from the case of two valuations by induction.

Now consider the general case. Since u is coarser than every v_i, we have $O_u^\times \geq \prod_{i \in I} O_{v_i}^\times$. Conversely, suppose that $x \in O_u^\times$. In light of Proposition 7.3.3 there exist $i_1, \ldots, i_n \in I$ such that $x, x^{-1} \in \prod_{j=1}^n O_{v_{i_j}}$. Let w be the finest common coarsening of $v_{i_1}, \ldots, v_{i_n}$. By Proposition 7.3.3 again, $O_w = \prod_{j=1}^n O_{v_{i_j}}$. By what we have just seen, $O_w^\times = \prod_{j=1}^n O_{v_{i_j}}^\times$. Therefore $x \in O_w^\times \leq \prod_{i \in I} O_{v_i}^\times$, as desired. $\square$

CHAPTER 11

Canonical Valuations

In this chapter we describe a line of results on the existence of valuations with special properties. The main such property is compatibility with respect to a given subgroup of the multiplicative group (see below). Furthermore, in applications one is often interested in such valuations which are "as non-trivial as possible". In §11.1 we use our previous investigation of the tree of localities to produce such canonical valuations. In §11.2 these valuations are related to the notion of the Σ-core, discussed in §2.3. An explicit construction of these canonical valuations is presented in §11.3, following works of Arason, Elman, Jacob, and Ware. Finally, in §11.4 we derive several criteria for the existence of valuations with prescribed non-triviality properties.

11.1 Compatible localities

Let F be a field and let S be a subgroup of $F^\times$. We say that a locality λ on F is S-**compatible** if $G_\lambda \leq S$. Thus when $\lambda = v$ is a valuation, this means that $1+\mathfrak{m}_v \leq S$ (and then the terminology goes back to [**War2**]). When $\lambda = P$ is an ordering, this means that S is either P or $F^\times$. In the valued field case, the notion of S-compatibility serves as a weak form of "completeness"; in fact, we will see later that it is a weak form of Hensel's lemma (to be discussed in §18), which for v of rank 1 is itself a consequence of the completeness of the metric space (F, d_v). Nevertheless, for many purposes it is enough to assume this much weaker compatibility condition (see, e.g., the exact sequence (3.2.7)).

EXAMPLE 11.1.1. An ordering P on a field F is $(F^\times)^2$-compatible if and only if $(F^\times)^2 = P$. A field F for which $(F^\times)^2$ is an ordering is called a **Euclidean** field [**Be1**]. E.g., this holds for $F = \mathbb{R}$. We will study such fields in more detail in §19.

For the rest of this section we fix subgroups $S \leq H$ of $F^\times$. We consider the following sets of valuations v on F:

$$\mathrm{Val}(S) = \{\ v \ \mid\ G_v \leq S\ \},$$
$$\mathrm{Val}(S,H) = \{\ v \ \mid\ G_v \leq S,\ O_v^\times \leq H\ \}.$$

By definition, $\mathrm{Val}(S)$ is the set of all S-compatible valuations on F. Thus $\mathrm{Val}(S,H)$ consists of all valuations which satisfy this weak form of completeness, and whose triviality is controlled by the condition $O_v^\times \leq H$. Finding such valuations is useful in Galois-theoretic applications of valuation theory (as explained in the Introduction).

The following discussion is a variant of a construction due to Endler and Engler ([**EnEg**]; see also [**P2**]). The original constructions in these papers is however in the context of Henselian algebraic extensions of valued fields.

LEMMA 11.1.2. *Let v, v' be valuations on F.*
(a) *If $v' \in \mathrm{Val}(S,H)$ is coarser than $v \in \mathrm{Val}(S)$, then $v \in \mathrm{Val}(S,H)$.*
(b) *If v, v' are in $\mathrm{Val}(S,H)$, then so is their finest common coarsening.*
(c) *If v, v' are incomparable valuations in $\mathrm{Val}(S)$, then $v, v' \in \mathrm{Val}(S,H)$.*
(d) *If $v \in \mathrm{Val}(S) \setminus \mathrm{Val}(S,H)$ and $v' \in \mathrm{Val}(S,H)$, then v is coarser than v'.*

PROOF. (a) We have $O_v^\times \leq O_{v'}^\times \leq H$.

(b) This follows from Proposition 10.3.2.

(c) Let u be again the finest common coarsening of v and v'. By Proposition 10.3.1(d),

$$O_v^\times, O_{v'}^\times \leq O_u^\times = G_v G_{v'} \leq S \leq H,$$

so $v, v' \in \mathrm{Val}(S,H)$.

(d) By (c), v and v' are comparable. By (a), v' cannot be coarser than v. Hence v is coarser than v'. □

Since the trivial valuation is always in $\mathrm{Val}(S)$, we have $\mathrm{Val}(S) = \mathrm{Val}(S,H)$ if and only if $H = F^\times$. By Lemma 11.1.2(c), $\mathrm{Val}(S) \setminus \mathrm{Val}(S,H)$ is a chain with respect to coarsening. If it is non-empty (i.e., if $H \neq F^\times$), then it has a coarsest common refinement (i.e., a supremum) $v_{(1)} = v_{(1)}(S,H)$. Thus

$$O_{v_{(1)}} = \bigcap_u O_u, \quad \mathfrak{m}_{v_{(1)}} = \bigcup_u \mathfrak{m}_u,$$

where u ranges over all valuations in $\mathrm{Val}(S) \setminus \mathrm{Val}(S,H)$.

If $\mathrm{Val}(S,H)$ is non-empty, then as always, there is a finest common coarsening (i.e., an infimum) $v_{(2)} = v_{(2)}(S,H)$ of its valuations - namely,

$$O_{v_{(2)}} = \prod_u O_u,$$

with u ranging over all valuations in $\mathrm{Val}(S,H)$ (see Corollary 7.3.2 and Proposition 7.3.3). Finding conditions guaranteeing the non-emptiness of $\mathrm{Val}(S,H)$ is a subtle problem, which will be discussed in detail in §11.3 and §11.4.

PROPOSITION 11.1.3.
(a) *If $H \neq F^\times$, then $v_{(1)} \in \mathrm{Val}(S)$.*
(b) *If $\mathrm{Val}(S,H) \neq \emptyset$, then $v_{(2)} \in \mathrm{Val}(S,H)$.*
(c) *If $H \neq F^\times$ and $\mathrm{Val}(S,H) \neq \emptyset$, then $v_{(1)}$ is coarser than $v_{(2)}$ (possibly $v_{(1)} = v_{(2)}$).*
(d) *If $H \neq F^\times$ and $\mathrm{Val}(S,H) \neq \emptyset$, then there is no valuation which is strictly finer than $v_{(1)}$ and strictly coarser than $v_{(2)}$.*
(e) *If $\mathrm{Val}(S,H) \neq \emptyset$, then $v_{(2)}(S,H)$ is comparable to every valuation in $\mathrm{Val}(S)$.*

PROOF. (a) Letting u range over the non-empty chain $\mathrm{Val}(S) \setminus \mathrm{Val}(S,H)$ we get $G_{v_{(1)}} = \bigcup_u G_u \leq S$.

(b) For any $v \in \mathrm{Val}(S,H)$ we have $G_{v_{(2)}} \leq G_v \leq S$. Letting u range over all valuations in $\mathrm{Val}(S,H)$, we obtain from Proposition 10.3.2 that $O_{v_{(2)}}^\times = \prod_u O_u^\times \leq H$.

(c) By (b) and by Lemma 11.1.2(d), $v_{(2)}$ is finer than all the valuations in $\mathrm{Val}(S)\setminus\mathrm{Val}(S,H)$. Hence it is finer than their coarsest common refinement $v_{(1)}$.

(d) Suppose that u is a valuation on F which is strictly coarser than $v_{(2)}$. Then $G_u < G_{v_{(2)}} \leq S$, so $u \in \mathrm{Val}(S)$. But $v_{(2)}$ is coarser than any valuation in $\mathrm{Val}(S,H)$, so necessarily $u \notin \mathrm{Val}(S,H)$. It follows that u is coarser than $v_{(1)}$.

(e) By construction, $v_{(2)}(S,H)$ is coarser than all valuations in $\mathrm{Val}(S,H)$. By (c) it is finer than all valuations in $\mathrm{Val}(S)\setminus\mathrm{Val}(S,H)$. □

The following picture describes $\mathrm{Val}(S)$ graphically:

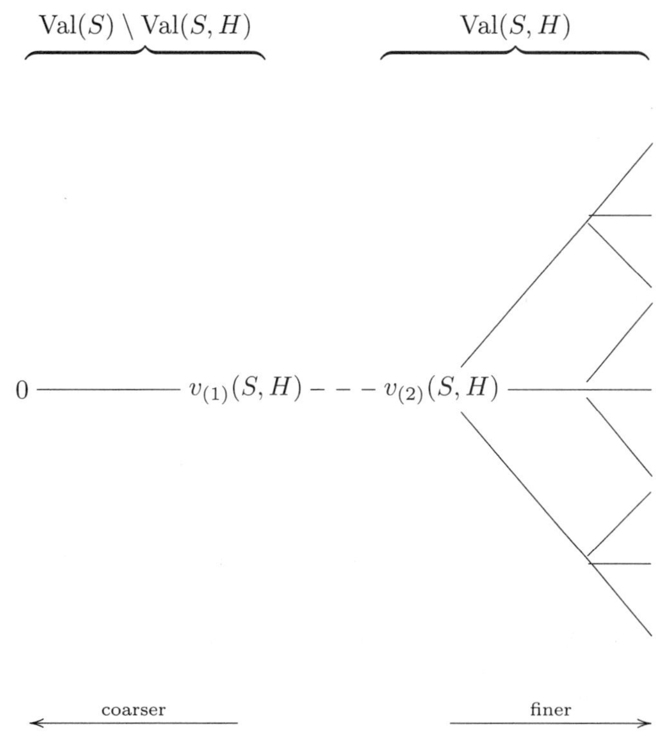

EXAMPLE 11.1.4. Let P be an ordering on F. Then $v_{(1)}(P, F^{\times})$ is the finest valuation v on F such that $G_v \leq P$. Thus it corresponds to the valuation ring $O_{\mathbb{Q}}(P)$ defined in §7.3 (see Proposition 7.3.1(b)).

REMARK 11.1.5. If $S \leq G_{\lambda}$ for some locality λ on F, then $\mathrm{Val}(S)$ is totally ordered, by Corollary 7.3.6.

PROPOSITION 11.1.6. *Assume that* $\mathrm{Val}(S,H) \neq \emptyset$.
(a) *If* $S \leq S' \leq H$, *then* $v_{(2)}(S,H) = v_{(2)}(S',H)$.
(b) *If* $S \leq H \leq H'$, *then* $v_{(2)}(S,H')$ *is coarser than* $v_{(2)}(S,H)$.

PROOF. (a) We have $\mathrm{Val}(S,H) \subseteq \mathrm{Val}(S',H)$, so $v_{(2)}(S',H)$ is well defined and is coarser than $v_{(2)}(S,H)$. It follows that

$$G_{v_{(2)}(S',H)} \leq G_{v_{(2)}(S,H)} \leq S,$$

whence $v_{(2)}(S',H) \in \mathrm{Val}(S,H)$. We conclude that $v_{(2)}(S,H)$ is coarser than $v_{(2)}(S',H)$, and therefore these valuations are in fact equal.

(b) One should just observe that $\mathrm{Val}(S,H) \subseteq \mathrm{Val}(S,H')$. □

11.2 S-cores

One important use of the construction of the previous section is to provide various canonical coarsenings of valuations with special properties. To this end we fix again the field F and a subgroup S of $F^\times$. First we need:

LEMMA 11.2.1. *Let v be a valuation on F such that $G_v \leq S$. Let u be a coarsening of v, and let $w = v/u$. The following conditions are equivalent:*
(a) $SO_u^\times = SO_v^\times$;
(b) *the canonical epimorphism*

$$v(F^\times)/v(S) \to u(F^\times)/u(S)$$

is an isomorphism;
(c) $w(\bar{F}_u^\times) = w(\bar{S}_u)$;
(d) *the kernel of the natural epimorphism $v(F^\times) \to u(F^\times)$ is contained in $v(S)$.*

PROOF. (a)⇔(b): Use the canonical isomorphisms

$$F^\times/SO_v^\times \cong v(F^\times)/v(S), \quad F^\times/SO_u^\times \cong u(F^\times)/u(S).$$

(b)⇔(c): Use Proposition 5.2.1(h).

(b)⇔(d): Let Δ be the kernel as in (d). The snake lemma (Lemma 1.1.3) gives rise to an exact sequence

$$0 \longrightarrow \Delta/(\Delta \cap v(S)) \longrightarrow v(F^\times)/v(S) \longrightarrow u(F^\times)/u(S) \longrightarrow 0.$$

This shows the desired equivalence. □

We deduce the following connection between the constructions of the previous section and the notion of the Σ-core of an ordered abelian group, as defined in §2.3.

PROPOSITION 11.2.2. *Let $S \leq H \leq F^\times$ and let $v \in \mathrm{Val}(S,H)$. Then $v = v_{(2)}(S,H)$ if and only if the $v(H)$-core of $v(F^\times)$ is trivial.*

PROOF. The coarsenings u of v correspond bijectively to the convex subgroups Δ of $v(F^\times)$ (Proposition 5.1.1). Specifically, Δ is the kernel of the epimorphism $v(F^\times) \to u(F^\times)$. By Lemma 11.2.1, $\Delta \leq v(H)$ if and only if $HO_v^\times = HO_u^\times$. Since $O_v^\times \leq H$, this means that $O_u^\times \leq H$, or equivalently, that $u \in \mathrm{Val}(S,H)$.

Now the equality $v = v_{(2)}(S,H)$ means that there are no proper coarsenings u of v in $\mathrm{Val}(S,H)$. By what we have just seen, this is equivalent to saying that there are no non-zero convex subgroups of $v(F^\times)$ which are contained in $v(H)$, i.e., that the $v(H)$-core of $v(F^\times)$ is trivial. □

Every $v \in \mathrm{Val}(S)$ belongs to $\mathrm{Val}(S, SO_v^\times)$, so this set is non-empty. Therefore we may define

$$v_S = v_{(2)}(S, SO_v^\times).$$

We call this valuation the S-**core** of v, and notice that it is coarser than v.

PROPOSITION 11.2.3. *Let $v \in \mathrm{Val}(S)$ and let u be a coarsening of v. The following conditions are equivalent:*

(a) *$u = v_S$;*
(b) *u is the coarsest coarsening of v such that $SO_v^\times = SO_u^\times$;*
(c) *u is the coarsest coarsening of v such that the canonical epimorphism*

$$v(F^\times)/v(S) \to u(F^\times)/u(S)$$

is an isomorphism;
(d) *u is the coarsest coarsening of v such that the kernel of the epimorphism $v(F^\times) \to u(F^\times)$ is contained in $v(S)$;*
(e) *the $v(S)$-core of $v(F^\times)$ is the kernel of the canonical epimorphism $v(F^\times) \to u(F^\times)$;*
(f) *the $u(S)$-core of $u(F^\times)$ is trivial.*

PROOF. (a)⇔(b): Immediate.

(b)⇔(c)⇔(d): Apply Lemma 11.2.1.

(d)⇔(e)⇔(f): This follows from Proposition 5.1.1 and Lemma 2.3.2. □

COROLLARY 11.2.4. *Let $v \in \mathrm{Val}(S)$. Then:*
(a) *$v_S = (v_S)_S$;*
(b) *v_S is comparable to any valuation in* $\mathrm{Val}(S)$.

PROOF. (a) follows from Proposition 11.2.3. (b) follows from Proposition 11.1.3(e). □

As another construction in the same spirit, one can define for $S \leq H \leq F^\times$ a valuation $v^*(S,H) \in \mathrm{Val}(S)$ as follows:

$$v^*(S,H) = \begin{cases} v_{(1)}(S,H) & \text{if } \mathrm{Val}(S,H) = \emptyset, \\ v_{(2)}(S,H) & \text{if } \mathrm{Val}(S,H) \neq \emptyset. \end{cases}$$

We notice:

PROPOSITION 11.2.5.
(a) *$v^*(S,H)$ is comparable to every valuation in* $\mathrm{Val}(S)$.
(b) *$v^*(S,H)$ is trivial if and only if $H = F^\times$ or* $\mathrm{Val}(S)$ *consists only of the trivial valuation.*

PROOF. (a) When $\mathrm{Val}(S,H)$ is empty, $v^*(S,H) = v_{(1)}(S,H)$ is finer than all the valuations in $\mathrm{Val}(S)$. When $\mathrm{Val}(S,H)$ is non-empty, $v^*(S,H) = v_{(2)}(S,H)$ is comparable to all the valuations in $\mathrm{Val}(S)$, by Proposition 11.1.3(e).

(b) If $H = F^\times$, then $\mathrm{Val}(S,H) = \mathrm{Val}(S) \neq \emptyset$, so $v^*(S,H) = v_{(2)}(S,H)$ is coarser than all the valuations in $\mathrm{Val}(S)$, i.e., it is the trivial valuation.

If $\mathrm{Val}(S)$ consists only of the trivial valuation, then obviously $v^*(S,H) \in \mathrm{Val}(S)$ is trivial.

Conversely, suppose that $v^*(S,H)$ is trivial. If $\mathrm{Val}(S,H)\neq\emptyset$, then $v^*(S,H)=v_{(2)}(S,H)$ is trivial and belongs to $\mathrm{Val}(S,H)$ (by Proposition 11.1.3(b)), whence $H=F^\times$. If $\mathrm{Val}(S,H)=\emptyset$, then every valuation in $\mathrm{Val}(S)$ is coarser than $v^*(S,H)=v_{(1)}(S,H)$, whence is also trivial. □

11.3 Explicit constructions

Let F be a field and let $S\leq H$ be subgroups of $F^\times$. The coarsest valuation $v_{(2)}(S,H)$ of $\mathrm{Val}(S,H)$ was constructed in §11.1 using the general properties of the tree $\mathrm{Val}(S)$. In this section we give an alternative construction of this valuation, which is more explicit and elementary. This is achieved in Theorem 11.3.6 below, which at the same time characterizes the non-emptiness of $\mathrm{Val}(S,H)$. This elementary construction is due to Arason, Elman, Jacob [**AEJ**], following earlier works of Jacob [**J1**] and Ware [**War2**]; see also the related works [**Ef6**], [**Eg2**], [**HwJ**], and [**Koe1**].

We first set

$$\begin{aligned}A(S)&=\{x\in F^\times \mid 1-x\notin S\cup -xS\}\\&=\{x\in F^\times\mid 1-x,1-x^{-1}\notin S\}.\end{aligned}$$

LEMMA 11.3.1.
(a) *For $x\in F^\times$ one has $x\in A(S)$ if and only if $x^{-1}\in A(S)$.*
(b) *If $x\in A(S)\setminus S$, then $1-x\in A(S)$.*
(c) *If $x\in F^\times$ and $xS\cap A(S)=\emptyset$, then $S-xS\subseteq S\cup -xS$.*
(d) *If $A(S)\subseteq S$ and $-1\notin S$, then $S+S\subseteq S$.*
(e) *If v an S-compatible valuation on F, then $A(S)\subseteq O_v^\times$.*

PROOF. (a), (c) Straightforward.

(b) Since $1-x\notin -xS$, the cosets $(1-x)S$, $-xS$ in $F^\times/S$ are disjoint. Hence $x\notin -(1-x)S$.

(d) The assumptions imply that $(-S)\cap A(S)=\emptyset$, so by (c), $S+S\subseteq S$.

(e) Let $x\in F^\times\setminus O_v^\times$. If $v(x)>0$, then $1-x\in G_v\leq S$. If $v(x)<0$, then

$$1-x=-x(1-x^{-1})\in -xG_v\subseteq -xS.$$

Thus in both cases $x\notin A(S)$. □

Now define

$$\begin{aligned}O^-(S,H)&=(1-S)\setminus H,\\O^+(S,H)&=\{x\in H\mid xO^-(S,H)\subseteq O^-(S,H)\},\\O(S,H)&=O^-(S,H)\cup O^+(S,H).\end{aligned}$$

When the context is clear, we abbreviate

$$O^-=O^-(S,H),\quad O^+=O^+(S,H),\quad O=O(S,H).$$

Clearly, $O^+O^-\subseteq O^-$ and $O^+O^+\subseteq O^+$.

11.3.2 EXAMPLES. For every valuation v on F, a direct computation shows that

$$\mathfrak{m}_v = O^-(G_v, O_v^\times), \quad O_v^\times = O^+(G_v, O_v^\times), \quad O_v = O(G_v, O_v^\times).$$

Similarly,

$$\mathfrak{m}_v = O^-(O_v^\times, O_v^\times), \quad O_v^\times = O^+(O_v^\times, O_v^\times), \quad O_v = O(O_v^\times, O_v^\times).$$

The next two facts motivate these somewhat technical definitions.

LEMMA 11.3.3. *For every valuation v in* $\mathrm{Val}(S, H)$ *one has* $O^- = O_v \setminus H$.

PROOF. Suppose that $x \in O_v \setminus H$. As $O_v^\times \leq H$ we have $x \in \mathfrak{m}_v$, so $1 - x \in G_v \leq S$ and $x \in O^-$.

Conversely, suppose that $x \in O^-$. Then $1-x \in S \leq H$, as well as $-1 \in O_v^\times \leq H$ and $x \notin H$. Therefore

$$1 - x^{-1} = -x^{-1}(1-x) \notin H,$$

so $x^{-1} \notin \mathfrak{m}_v$. Hence $x \in O_v \setminus H$. □

LEMMA 11.3.4. *Suppose that* $\mathrm{Val}(S, H) \neq \emptyset$ *and let* $v = v_{(2)}(S, H)$. *Then* $O^+ = O_v \cap H$.

PROOF. By Proposition 11.1.3(b), $v \in \mathrm{Val}(S, H)$. Lemma 11.3.3 therefore gives $O^- = O_v \setminus H$. Consequently, O^- is the preimage under v of

$$\Gamma_0 = \left\{\gamma \in v(F^\times) \mid 0 \leq \gamma \notin v(H)\right\}.$$

By Proposition 11.2.2, the $v(H)$-core of $v(F^\times)$ is trivial. We apply Lemma 2.3.3 with respect to $\Gamma = v(F^\times)$ and $\Sigma = v(H)$ to obtain that for every $x \in H$ one has $v(x) + \Gamma_0 \subseteq \Gamma_0$ if and only if $0 \leq v(x)$. In other words, $xO^- \subseteq O^-$ if and only if $x \in O_v$. The assertion follows. □

PROPOSITION 11.3.5. *Suppose that* $-1 \in H$, $A(S) \subseteq H$ *and* $1 - O^-O^- \subseteq S$. *Then:*

(a) $1 - O^- \subseteq O^+$.
(b) $O^-O^- \subseteq 1 - O^+$.
(c) $(1 - O^+) \cap H \subseteq O^+$.
(d) $(1 - O^+) \setminus H \subseteq O^-$.

PROOF. For $1 \neq y \in F$ let $\tilde{y} = y/(y-1)$. Then $y \mapsto \tilde{y}$ maps $F \setminus \{0, 1\}$ onto itself. Moreover, $y \in O^-$ if and only if $\tilde{y} \in O^-$. We use the identity

$$1 - xy = (1 - (1-x)\tilde{y})(1-y) \tag{11.3.1}$$

for $y \neq 1$.

(a) Take $x \in 1 - O^-$ and $y \in O^-$. By the identity (11.3.1) and the assumptions,

$$1 - xy \in (1 - O^-O^-)S \subseteq SS = S.$$

Since $x \in S \leq H$ and $y \notin H$ this implies $xy \in O^-$. We conclude that $x \in O^+$.

(b) Let $x, y \in O^-$. By (11.3.1) and (a),

$$\begin{aligned} 1 - xy \in (1 - (1 - O^-)O^-)(1 - O^-) &\subseteq (1 - O^+O^-)O^+ \\ &\subseteq (1 - O^-)O^+ \subseteq O^+O^+ \subseteq O^+. \end{aligned}$$

(c) Suppose that $x \in (1-O^+)\cap H$ and $y \in O^-$. By (11.3.1),

$$1-xy \in (1-O^+O^-)(1-O^-) \subseteq (1-O^-)(1-O^-) \subseteq SS = S.$$

As $xy \notin H$, this shows that $xy \in O^-$, whence $x \in O^+$.

(d) If $x \in (1-O^+)\setminus H$, then in particular $x \notin A(S)$, so

$$1-x \in H \cap (S \cup -xS) = S$$

(since $-1 \in H$). We conclude that $x \in O^-$. □

We now come to the explicit description of $v_{(2)}(S,H)$.

THEOREM 11.3.6. *Let $S \leq H$ be subgroups of $F^\times$. The following conditions are equivalent:*

(a) $\mathrm{Val}(S,H) \neq \emptyset$;

(b) *$O(S,H)$ is a valuation ring which corresponds to a valuation in* $\mathrm{Val}(S,H)$*;*

(c) *$O(S,H)$ is a valuation ring which corresponds to the coarsest valuation $v_{(2)}(S,H)$ in* $\mathrm{Val}(S,H)$*;*

(d) *$-1 \in H$, $A(S) \subseteq H$ and $1-O^-(S,H)O^-(S,H) \subseteq S$.*

PROOF. We abbreviate again $O^- = O^-(S,H)$, $O^+ = O^+(S,H)$ and $O = O(S,H) = O^- \cup O^+$.

(c)⇒(b)⇒(a): Trivial.

(a)⇒(d): Let $v \in \mathrm{Val}(S,H)$. Obviously, $-1 \in O_v^\times \leq H$. By Lemma 11.3.1(e), $A(S) \subseteq O_v^\times \leq H$. By Lemma 11.3.3, and using again the fact that $O_v^\times \leq H$, we obtain $O^- = O_v \setminus H \subseteq \mathfrak{m}_v$. Hence

$$1-O^-O^- \subseteq 1-\mathfrak{m}_v\mathfrak{m}_v \subseteq 1+\mathfrak{m}_v \leq S.$$

(d)⇒(b): We have $0 \in O^- \subseteq O$.

As has already been observed, $O^+O^- \subseteq O^-$ and $O^+O^+ \subseteq O^+$. Also

$$O^-O^- \setminus H \subseteq (1-S)\setminus H = O^-.$$

Finally, by Proposition 11.3.5(b)(c),

$$O^-O^- \cap H \subseteq (1-O^+)\cap H \subseteq O^+.$$

We conclude that $OO \subseteq O$.

We now show that for every $x \in F^\times$ either $x \in O$ or $x^{-1} \in O$. Indeed, if $x \notin H$, then $x \notin A(S)$, so either $1-x \in S$ or $1-x^{-1} \in S$. Thus either $x \in O^-$ or $x^{-1} \in O^-$ in this case. If $x \in H \setminus O = H \setminus O^+$, then there exists $y \in O^-$ such that $xy \notin O^-$. As $xy \notin H$, this implies that $xy \notin O$. By what we have just seen, $(xy)^{-1} \in O^-$. Consequently,

$$x^{-1} = (xy)^{-1}y \in O^-O^- \subseteq OO \subseteq O,$$

as desired. In particular, $\pm 1 \in O$.

Next we prove that for $0 \neq x,y \in O$ also $x+y \in O$. By symmetry we may assume that $-x^{-1}y \in O$. Then $1+x^{-1}y \in 1-O \subseteq O$ (by Proposition 11.3.5). Therefore

$$x+y = x(1+x^{-1}y) \in OO \subseteq O.$$

This completes the proof that O is a valuation ring on F. It remains to show that the corresponding valuation v belongs to $\mathrm{Val}(S, H)$.

If $x \in O^\times \setminus H$, then $x, x^{-1} \in O^-$, so $1-x, 1-x^{-1} \in S$. As $-1 \in H$, this gives

$$x = -\frac{1-x}{1-x^{-1}} \in H,$$

a contradiction. Consequently, $O^\times \leq H$.

Finally, we verify that $G_v \leq S$. By definition, $1-O^- \subseteq S$. So let $x \in O^+ \cap \mathfrak{m}_v$; we show that $1-x \in S$. As $x^{-1} \in H \setminus O = H \setminus O^+$ we have $x^{-1}y \notin O^-$ for some $y \in O^-$. Since $x^{-1}y \notin H$, this implies $x^{-1}y \notin O$. Hence

$$xy^{-1} = (x^{-1}y)^{-1} \in O \setminus H = O^-.$$

Therefore

$$1-x = 1-(xy^{-1})y \in 1 - O^-O^- \subseteq S,$$

as desired.

(a)$\Rightarrow$(c): As $\mathrm{Val}(S, H) \neq \emptyset$ we may define $v = v_{(2)}(S, H)$. By Proposition 11.1.3(b), $v \in \mathrm{Val}(S, H)$. By Lemma 11.3.3 and Lemma 11.3.4, $O^- = O_v \setminus H$ and $O^+ = O_v \cap H$, respectively. Altogether,

$$O = O^- \cup O^+ = (O_v \setminus H) \cup (O_v \cap H) = O_v,$$

proving (c). □

We can now express the S-core $v_S = v_{(2)}(S, SO_v^\times)$ of an S-compatible valuation v explicitly.

COROLLARY 11.3.7. *Let v be an S-compatible valuation. Then*

$$O_{v_S} = O(S, SO_v^\times).$$

PROOF. As $\mathrm{Val}(S, SO_v^\times)$ contains v, it is non-empty. By condition (c) of Theorem 11.3.6, $O(S, SO_v^\times)$ corresponds to v_S. □

In fact, the examples in 11.3.2 are special cases of Corollary 11.3.7, when one takes $S = G_v$ or $S = O_v^\times$, respectively. Indeed, in both cases $v = v_S$.

11.4 Existence of valuations

In §11.3 we characterized the pairs (S, H) for which $\mathrm{Val}(S, H) \neq \emptyset$, and described the canonical valuation $v_{(2)}(S, H)$ in this set (where as before $S \leq H \leq F^\times$). In practical applications the subgroup S is given, and one looks for valuations in $\mathrm{Val}(S)$ which are as non-trivial as possible. Thus for a given S we look for small subgroups H of $F^\times$ with $S \leq H$ and $\mathrm{Val}(S, H) \neq \emptyset$. Obviously, the trivial valuation is contained in $\mathrm{Val}(S, F^\times)$, and the smaller H is, the smaller $\mathrm{Val}(S, H)$ becomes. Therefore the best one can hope for is to have such $H = H_S$ which is *minimal*, in the sense that $\mathrm{Val}(S, H_S) \neq \emptyset$ and H_S is contained in any other subgroup H for which $\mathrm{Val}(S, H) \neq \emptyset$. In this section we prove that, apart from an exceptional case, there indeed exists such a canonical minimal subgroup H_S.

In light of Theorem 11.3.6, the condition

$$O^-(S, H)O^-(S, H) \subseteq 1 - S$$

has vital importance in proving existence of non-trivial valuations. When it holds, we say – following [**AEJ**] – that the pair (S, H) is **pre-additive**. Proposition 11.4.2 below (which is based on [**AEJ**] and [**HwJ**]) studies the consequences of the *failure* of this condition. In particular, it shows that after slightly enlarging H, one can always obtain a pre-additive pair. First we need the following lemma. As before, we set

$$A(S) = \{x \in F \mid 1 - x \notin S \cup -xS\}.$$

LEMMA 11.4.1. *Suppose that H contains -1 and $A(S)$. If $z, w \in O^-(S, H)$, then $zw \in 1 - S$ or $1 - zw \in zS = wS$.*

PROOF. One has $1 - z, 1 - w \in S$. Furthermore, as $z, w \notin H$, the assumptions imply that $-zS, -wS$ do not intersect $A(S)$. By Lemma 11.3.1(c),

$$1 - zw = (1 - z) + z(1 - w) \in S + zS \subseteq S \cup zS,$$

$$1 - zw = (1 - w) + w(1 - z) \in S + wS \subseteq S \cup wS.$$

The assertion follows. □

PROPOSITION 11.4.2. *Suppose that H contains -1 and $A(S)$. Suppose that $a, b \in O^-(S, H)$ and $1 - ab \notin S$. Then:*

(a) $a^2 \in S$;
(b) $O^-(S, \langle H, a\rangle)O^-(S, \langle H, a\rangle) \subseteq 1 - S$;
(c) $(A(S) \setminus S)a \subseteq O^-(S, H)$;
(d) $A(S) \subseteq S$.

PROOF. (a) Lemma 11.4.1 implies that $1 - ab \in aS = bS$. Since $a \notin H$, the assumptions imply that $aS \cap A(S) = \emptyset$. It follows from Lemma 11.3.1(c) that

$$ab \in 1 - aS \subseteq S - aS \subseteq S \cup -aS.$$

But $-aS \leq aH$ and $ab \notin aH$, so necessarily $ab \in S$. We conclude that

$$a^2 \in aS \cdot aS = aS \cdot bS = abS = S.$$

(b) Let $H^* = \langle H, a\rangle$. As already noted, $1 - ab \in aS = bS$, whence $b \in \langle S, a\rangle \leq H^*$. Suppose that $0 \neq x, y \in O^-(S, H^*)$ but $xy \notin 1 - S$. As $a, b \in H^*$, $x, y \notin H^*$, and $S \leq H^*$, we have $aS \neq xS$ and $bS \neq yS$. Lemma 11.4.1 therefore implies that $ax, by \in 1 - S$. Furthermore, $ax, by \notin H^*$, so $ax, by \in O^-(S, H^*)$. Also, $ay^{-1} \notin H^*$, whence $ay^{-1} \notin A(S)$, so $1 - ay^{-1} \in S \cup -ay^{-1}S$. Therefore one of the following cases holds:

CASE I: $ay^{-1} \in 1 - S$. Then $ay^{-1} \in O^-(S, H^*)$ and $(ay^{-1})(by) = ab \notin 1 - S$. By Lemma 11.4.1, $1 - ab \in ay^{-1}S$, contrary to $1 - ab \in aS \subseteq H^*$, $a \in H^*$, and $y \notin H^*$.

CASE II: $a^{-1}y \in 1 - S$. Then $a^{-1}y \in O^-(S, H^*)$ and $xy = (ax)(a^{-1}y) \notin 1 - S$. By applying Lemma 11.4.1 twice we obtain that $1 - xy \in xS \cap axS$, contrary to $a \notin H$ and $S \leq H$.

(c) Assume that $x \in A(S) \setminus S$. Then $1 - x \notin S \cup -xS$. By Lemma 11.3.1(b), $1 - x \in A(S) \subseteq H$. We have

$$x \in A(S) \leq H, \quad -1 \in H, \quad 1 - a \in S \leq H, \quad a \notin H.$$

Hence $xa, -(1-x)a \notin H$. In particular, $xa \notin A(S)$, and $-(1-x)aS$ does not intersect $A(S)$. We obtain that

$$1 - xa \in S \cup -xaS,$$

and by Lemma 11.3.1(c),

$$1 - xa = (1-a) + (1-x)a \in S + (1-x)aS \subseteq S \cup (1-x)aS.$$

As $x \in A(S)$ we have $1 - x \notin -xS$, so $-xaS$ and $(1-x)aS$ are disjoint. Consequently, $1 - xa \in S$, whence $xa \in O^-(S, H)$.

(d) Let $x \in A(S) \setminus S$. By Lemma 11.3.1(a) also $x^{-1} \in A(S) \setminus S$. It follows from (c) that $xa \in O^-(S,H)$ and $x^{-1}b \in O^-(S,H)$. As $ab = (xa)(x^{-1}b) \notin 1 - S$, Lemma 11.4.1 implies that $1 - ab \in aS$, as well as $1 - ab \in xaS$. Therefore $x \in S$, a contradiction. □

REMARK 11.4.3. More concretely, the condition $A(S) \subseteq S$ of Proposition 11.4.2(d) means that for every $x \in F^\times$ at least one of $x, 1-x, 1-x^{-1}$ lies in S.

Now let

$$H_S = \langle -1, S, A(S) \rangle.$$

We obtain the first main result of this section, showing that H_S almost always has the desired minimality property:

THEOREM 11.4.4.
(a) *If* $\mathrm{Val}(S,H) \neq \emptyset$, *then* $H_S \leq H$.
(b) *If* $\mathrm{Val}(S, H_S) = \emptyset$, *then* $A(S) \subseteq S$, *and for some* $H_S < H \leq F^\times$ *with* $(H : H_S) = 2$ *one has* $\mathrm{Val}(S,H) \neq \emptyset$.
(c) *If every element of* $F^\times/S$ *has a finite odd order, then* $\mathrm{Val}(S, H_S) \neq \emptyset$.

PROOF. (a) This is an immediate consequence of Theorem 11.3.6.

(b) By Theorem 11.3.6, (S, H_S) is not pre-additive. Proposition 11.4.2(d) therefore implies that $A(S) \subseteq S$. Moreover, Proposition 11.4.2(a)(b) (for the pair (S, H_S)) give rise to a subgroup H of $F^\times$ containing H_S with $(H : H_S) = 2$ and such that (S, H) is pre-additive. By Theorem 11.3.6 again, $\mathrm{Val}(S,H) \neq \emptyset$.

(c) Otherwise (b) would yield a subgroup $H \leq F^\times$ containing H_S as a subgroup of index 2. Then $F^\times/H_S$ would contain elements of order 2. However, being a quotient of $F^\times/S$, all its elements have finite odd orders, a contradiction. □

We will return to the question whether $\mathrm{Val}(S, H_S) \neq \emptyset$ in §§26.5–26.6, and answer it in K-theoretic terms.

As another application of these computations we now obtain criteria for the existence of valuations satisfying some non-triviality conditions, but without the S-compatibility requirement.

THEOREM 11.4.5. *Let* S *be a subgroup of* $F^\times$ *such that every element of* $F^\times/S$ *has finite odd order. The following conditions are equivalent:*
(a) *there exists a valuation* v *on* F *with* $v(F^\times) \neq v(S)$;

(b) *there exists a proper subgroup T of $F^\times$ containing S such that for every $x \in F^\times$ at least one of*

$$x, 1-x, 1-x^{-1}$$

lies in T;

(c) *there exists a proper subgroup T of $F^\times$ containing S such that $H_T = T$;*

(d) *there exists a proper subgroup T of $F^\times$ containing S such that* $\mathrm{Val}(T,T) \neq \emptyset$;

(e) *there exists a proper subgroup T of $F^\times$ containing S such that $O(T,T)$ is a valuation ring in* $\mathrm{Val}(T,T)$.

PROOF. (a)$\Rightarrow$(b): Let $T = SO_v^\times$. By assumption, it is a proper subgroup of $F^\times$. Now consider $x \in F^\times$. When $v(x) = 0$ (resp., $v(x) > 0$, $v(x) < 0$), we have $x \in O_v^\times \leq T$ (resp., $1 - x \in O_v^\times \leq T$, $1 - x^{-1} \in O_v^\times \leq T$).

(b)$\Rightarrow$(c): Take T as in (b). Since $(-1)^2 = 1 \in S$, the assumption on S implies that $-1 \in S \leq T$. In light of Remark 11.4.3, $A(T) \subseteq T$. Hence $H_T = T$.

(c)$\Rightarrow$(d): Use Theorem 11.4.4(c).

(d)$\Leftrightarrow$(e): This follows from Theorem 11.3.6.

(e)$\Rightarrow$(a): Take T as in (e). The valuation v corresponding to the valuation ring $O(T,T)$ satisfies $O_v^\times \leq T$. Therefore

$$(v(F^\times) : v(S)) \geq (v(F^\times) : v(T)) = (F^\times : TO_v^\times) = (F^\times : T) > 1. \qquad \square$$

EXAMPLE 11.4.6. The assumption in Theorem 11.4.5 about the orders in $F^\times/S$ holds in particular when S contains $(F^\times)^q$ for q odd. For $S = (F^\times)^p$ with p prime, it gives necessary and sufficient conditions for the existence of a valuation v whose value group is not p-divisible (i.e., $v(F^\times) \neq pv(F^\times)$).

CHAPTER 12

Valuations of Mixed Characteristics

Valued fields (F, v) of mixed characteristics, i.e., for which $\operatorname{char} F = 0$ and $\operatorname{char} \bar{F}_v = p > 0$ are in general relatively difficult to investigate. Still, for complete discrete valuations, there is a very satisfactory structure theory for such fields, which goes back to O. Teichmüller and E. Witt, and which we now describe. §12.1 constructs the so-called Teichmüller character. It allows one to expand elements in such fields "λ-adically" (§12.2). In §§12.3–12.4 we use these expansions to construct the ring of Witt vectors over a perfect field of characteristic p. Its quotient field serves as a mixed characteristic analog of the field of Laurent series over a given residue field. §12.5 studies a different situation of general mixed characteristic valuations, but under the assumption that the index of $(F^\times)^p$ in $F^\times$ is finite.

12.1 Multiplicative representatives

In the generalized formal power series field $F = K((\Gamma))$ with its canonical valuation (Example 4.2.1), we had a natural identification of the residue field K as a subfield of F. For general valued fields this need not be true anymore. However, as we show in this section, for discrete complete valued fields (F, v) with perfect residue field of characteristic $p > 0$ one can still find such canonical *multiplicative representatives*, i.e., a homomorphic section of the residue homomorphism $\pi_v \colon O_v^\times \to \bar{F}_v^\times$. Moreover, in the positive equal characteristic case ($\operatorname{char} F = \operatorname{char} \bar{F}_v = p > 0$) this embeds $\bar{F}_v$ canonically as a subfield of F.

Throughout this section p will be a fixed prime number. For a subset B of a ring A let B^{p^∞} be the set of all elements of B which are p^n-powers of elements of B for every $n \geq 0$.

PROPOSITION 12.1.1. *Let A be a ring and let I be an ideal in A containing p. Suppose that the exponentiation by p is injective on A/I and that $A \cong \varprojlim A/I^{r+1}$ canonically. Then the projection $\pi \colon A \to A/I$ induces a bijection between A^{p^∞} and $(A/I)^{p^\infty}$. This bijection is multiplicative, and when $pA = 0$, also additive.*

PROOF. The fact that π induces a map $A^{p^\infty} \to (A/I)^{p^\infty}$ is clear.

For $1 \leq i \leq p-1$ we have $p \mid \binom{p}{i}$, so $\binom{p}{i} \in I$. If $x, y \in A$ and $x - y \in I^r$, then the binomial formula gives

$$x^p - y^p = (x-y)^p + \sum_{i=1}^{p-1} \binom{p}{i} y^{p-i}(x-y)^i \in (x-y)^p + pI^r \subseteq I^{rp} + pI^r \subseteq I^{r+1}.$$

Using induction we deduce that $x^{p^n} - y^{p^n} \in I^{n+r}$ for all $n \geq 0$.

Now to prove the injectivity, suppose that $a, b \in A^{p^\infty}$ and $\pi(a) = \pi(b)$. For each $n \geq 0$ pick $a_n, b_n \in A$ with $a_n^{p^n} = a$ and $b_n^{p^n} = b$. Then $\pi(a_n)^{p^n} = \pi(b_n)^{p^n}$. Since exponentiation by p is injective on A/I, this implies that $\pi(a_n) = \pi(b_n)$, i.e.,

$a_n - b_n \in I$. It follows from what we have just seen that $a - b = a_n^{p^n} - b_n^{p^n} \in I^{n+1}$. By assumption, $\bigcap_{n=0}^{\infty} I^{n+1} = \{0\}$, so we obtain $a = b$, as desired.

To prove the surjectivity, let $\bar{a} \in (A/I)^{p^\infty}$. For each n we can choose $c_n \in A$ with $\pi(c_n)^{p^n} = \bar{a}$. Then

$$(\pi(c_{n+1}^p - c_n))^{p^n} = \pi(c_{n+1})^{p^{n+1}} - \pi(c_n)^{p^n} = 0.$$

Therefore $\pi(c_{n+1}^p - c_n) = 0$, i.e., $c_{n+1}^p - c_n \in I$. As we have seen, this implies that $c_{n+1}^{p^{r+1}} - c_n^{p^r} \in I^{r+1}$ for all $n, r \geq 0$. It follows that

$$(c_{m+r}^{p^r} + I^{r+1})_{r=0}^{\infty} \in \varprojlim A/I^{r+1}$$

for every m. Since the canonical map $A \to \varprojlim A/I^{r+1}$ is an isomorphism, for every m there exists a unique $a_m \in A$ such that

$$a_m - c_{m+r}^{p^r} \in I^{r+1}, \quad r = 0, 1, 2, \ldots . \tag{12.1.1}$$

In particular, for $m = r = 0$ (12.1.1) gives $a_0 - c_0 \in I$, so $\pi(a_0) = \bar{a}$. Substituting in (12.1.1) 0 for m and $m + r$ for r, we obtain $a_0 - c_{m+r}^{p^{m+r}} \in I^{m+r+1}$. Furthermore, as we have seen above, (12.1.1) implies that $a_m^{p^m} - c_{m+r}^{p^{m+r}} \in I^{m+r+1}$. Hence $a_0 - a_m^{p^m} \in I^{m+r+1}$ for all m, r. As $\bigcap_{r=0}^{\infty} I^{m+r+1} = \{0\}$, this gives $a_0 = a_m^{p^m}$ for all m. Thus $a_0 \in A^{p^\infty}$ is a pre-image of $\bar{a}$ as required.

Finally, let $\bar{a}, \bar{b} \in (A/I)^{p^\infty}$. As we have just seen, there exist $a_0, b_0 \in A^{p^\infty}$ such that $\pi(a_0) = \bar{a}$ and $\pi(b_0) = \bar{b}$. Then $a_0 b_0 \in A^{p^\infty}$ and $\pi(a_0 b_0) = \bar{a}\bar{b}$. When $pA = 0$ the binomial formula again shows that $a_0 + b_0 \in A^{p^\infty}$, and we obviously have $\pi(a_0 + b_0) = \bar{a} + \bar{b}$. □

COROLLARY 12.1.2 (Teichmüller [**T**]). *Let v be a complete discrete valuation on a field F such that* $\operatorname{char} \bar{F}_v = p > 0$. *Then:*

(a) *π_v induces a group isomorphism $(O_v^\times)^{p^\infty} \xrightarrow{\sim} (\bar{F}_v^\times)^{p^\infty}$;*
(b) *when $\bar{F}_v$ is perfect, π_v induces an isomorphism $(O_v^\times)^{p^\infty} \xrightarrow{\sim} \bar{F}_v^\times$;*
(c) *when $\bar{F}_v$ is perfect and* $\operatorname{char} F = p$, *the ring epimorphism $\pi_v \colon O_v \to \bar{F}_v$ has a homomorphic section.*

PROOF. (a) Consider the canonical ring homomorphism $O_v \to \varprojlim O_v/\mathfrak{m}_v^{r+1}$. Since v is discrete, this map is injective. Since (F, v) is complete, it is also surjective. As $\operatorname{char} \bar{F}_v = p$, exponentiation by p is injective on $\bar{F}_v$. By Proposition 12.1.1, $\pi_v \colon O_v^{p^\infty} \to \bar{F}_v^{p^\infty}$ is bijective and multiplicative. It induces a group isomorphism as required.

(b) This is a special case of (a).

(c) Use the last assertion of Proposition 12.1.1. □

In the situation of Corollary 12.1.2(a), the inverse group isomorphism

$$\rho_v \colon (\bar{F}_v^\times)^{p^\infty} \to (O_v^\times)^{p^\infty}$$

is called the **Teichmüller character** corresponding to v. When $\bar{F}_v$ is perfect (of characteristic p), it is a group isomorphism $\rho_v \colon \bar{F}_v^\times \xrightarrow{\sim} (O_v^\times)^{p^\infty}$. It is convenient to define $\rho_v(0) = 0$. Then ρ_v is a *set-theoretic* section of the projection $\pi_v \colon O_v \to \bar{F}_v$.

EXAMPLE 12.1.3. Let $F = \mathbb{Q}_p$ and let v_p be its canonical (p-adic) valuation. Then $\bar{F}_{v_p} = \mathbb{F}_p$. The group $(\mathbb{Z}/p^{r+1})^\times$ has order $p^r(p-1)$. Therefore, for every $1 \le k \le p-1$ we have

$$k^{p^{r+1}} = k^{p^r(p-1)} k^{p^r} \equiv k^{p^r} \pmod{p^{r+1}}.$$

Thus the sequence $\alpha = (k^{p^r} + p^{r+1}\mathbb{Z})_{r=0}^\infty$ is in $O_{v_p} = \mathbb{Z}_p = \varprojlim \mathbb{Z}/p^{r+1}$. Its projection to $\mathbb{F}_p^\times$ is $\bar{k}$. Moreover, $\alpha^p = \alpha$, so $\alpha \in O_{v_p}^{p^\infty}$. We conclude that the map $\mathbb{F}_p^\times \to O_{v_p}^\times$, $\bar{k} \mapsto \alpha$ is the Teichmüler character of v_p.

12.2 λ-adic expansions

Using the Teichmüller character, one can expand elements of complete discrete valuation rings in a way which resembles the usual expansions of p-adic numbers and power series. To show this, we first need a technical lemma.

LEMMA 12.2.1. *Let I be an ideal of a ring A, and let ρ be a set-theoretic section of the projection $\pi\colon A \to A/I$. Suppose that $0 \ne \lambda_0, \lambda_1, \dots, \lambda_n \in A$ are not zero divisors, and consider the map*

$$\Lambda_n\colon (A/I)^{n+1} \to \prod_{j=0}^n (A/\lambda_j I), \quad (\bar{a}_i)_{i=0}^n \mapsto \Big(\sum_{i=0}^j \rho(\bar{a}_i)\lambda_i + \lambda_j I\Big)_{j=0}^n.$$

(a) *Λ_n is injective.*
(b) *Assume in addition that $\lambda_0 = 1$ and $I^j = \lambda_j A$ for all j. For every $b \in A$ the sequence $(b + I^{j+1})_{j=0}^n$ is in the image of Λ_n.*

PROOF. (a) Suppose that $\bar{a}_i, \bar{b}_i \in A/I$, $i = 0, 1, 2, \dots, n$, and that

$$\sum_{i=0}^j \rho(\bar{a}_i)\lambda_i + \lambda_j I = \sum_{i=0}^j \rho(\bar{b}_i)\lambda_i + \lambda_j I, \quad j = 0, 1, \dots, n.$$

We show by induction on $0 \le j \le n$ that $\bar{a}_j = \bar{b}_j$. Indeed, suppose that $\bar{a}_i = \bar{b}_i$ for all $0 \le i < j$. Then $\rho(\bar{a}_j)\lambda_j - \rho(\bar{b}_j)\lambda_j \in \lambda_j I$. Cancelling λ_j and projecting into A/I, we deduce that $\bar{a}_j = \bar{b}_j$ as well.

(b) We need to construct $\bar{a}_0, \dots, \bar{a}_n$ such that

$$b - \sum_{i=0}^j \rho(\bar{a}_i)\lambda_i \in \lambda_j I = I^{j+1} = \lambda_{j+1} A, \quad j = 0, 1, , \dots, n.$$

We proceed by induction on j. First let $\bar{a}_0 = \pi(b)$. Then

$$b - \rho(\bar{a}_0)\lambda_0 = b - \rho(\bar{a}_0) \in \mathrm{Ker}(\pi) = I = \lambda_0 I.$$

Next suppose that $\bar{a}_0, \dots, \bar{a}_{j-1}$ have already been defined. Thus

$$b - \sum_{i=0}^{j-1} \rho(\bar{a}_i)\lambda_i = \lambda_j c$$

for some $c \in A$. We take $\bar{a}_j = \pi(c)$. Again, $c - \rho(\bar{a}_j) \in I$. It follows that

$$b - \sum_{i=0}^j \rho(\bar{a}_i)\lambda_i = \lambda_j c - \rho(\bar{a}_j)\lambda_j \in \lambda_j I,$$

completing the inductive construction. □

COROLLARY 12.2.2. *Let I be an ideal of a ring A and let ρ be a set-theoretic section of the projection $A \to A/I$. Suppose that $0 \neq \lambda_0, \lambda_1, \lambda_2, \ldots \in A$ are not zero divisors, $\lambda_0 = 1$, and $I^j = \lambda_j A$ for all j. Then there is an injection*

$$(A/I)^{\mathbb{N}} \to \varprojlim A/I^{j+1}, \quad (\bar{a}_i)_{i=0}^{\infty} \mapsto \Big(\sum_{i=0}^{j} \rho(\bar{a}_i)\lambda_i + I^{j+1}\Big)_{j=0}^{\infty}$$

whose image contains the image of the diagonal map $A \to \varprojlim A/I^{j+1}$.

PROOF. The maps

$$(A/I)^{\mathbb{N}} \longrightarrow (A/I^{j+1})_{j=0}^{n}, \quad (\bar{a}_i)_{i=0}^{\infty} \mapsto \Lambda_n(\bar{a}_0, \ldots, \bar{a}_n),$$

of Lemma 12.2.1 are compatible in the natural sense. Hence they combine to a map $(A/I)^{\mathbb{N}} \to \varprojlim A/I^{j+1}$ as desired. □

It is convenient to write the image of $(\bar{a}_i)_{i=0}^{\infty}$ above formally as

$$\sum_{i=0}^{\infty} \rho(\bar{a}_i)\lambda_i.$$

This is especially useful when $\lambda_i = \lambda^i$ for some $\lambda \in I$. Then it is called a **λ-adic expansion**.

We obtain the following characterization of the complete discrete valued fields with perfect residue field in the positive characteristic case, which is due to H. Hasse, F.K. Schmidt, and S. MacLane (see [**HasS**], [**Mac1**], also for the non-perfect residue field case).

THEOREM 12.2.3. *Let (F, v) be a complete discrete valued field of characteristic $p > 0$ with $\bar{F}_v$ perfect. Then $F \cong \bar{F}_v((t))$ as valued fields.*

PROOF. Choose a uniformizer λ for (F, v). The assumptions imply that $O_v \cong \varprojlim O_v/\lambda^j O_v$ canonically. It follows from Corollary 12.1.2(c) that the Teichmüller character $\rho_v \colon \bar{F}_v \to O_v$ is a ring homomorphism. By Corollary 12.2.2, there is a well-defined bijection

$$\sum_{i=0}^{\infty} \bar{a}_i t^i \mapsto \sum_{i=0}^{\infty} \rho_v(\bar{a}_i)\lambda^i$$

from the discrete valuation ring $\bar{F}_v[[t]]$ on $\bar{F}_v((t))$ to O_v (the notation being as in §2.8). This map is continuous with respect to the valuation topologies, and is clearly additive. Moreover, it is multiplicative on polynomials. Since the polynomials are dense in $\bar{F}_v[[t]]$, it is multiplicative on the whole ring, i.e., is a ring homomorphism. Taking quotient fields, we obtain the desired isomorphism. □

12.3 p-perfect structures

Throughout this and the next section we fix a prime number p. We give a construction due to E. Witt [**Wit**] of a complete discretely valued field of characteristic 0 with a prescribed perfect residue field of characteristic p. This extends the construction of the field $\mathbb{Q}_p$ of p-adic numbers, and is a mixed-characteristic analog of the field of Laurent series over a given residue field (Example 4.2.2). Our approach differs from Witt's original one (which was based on what are now called "Witt vectors"), and roughly follows the approach of Lazard ([**Lz1**], [**Lz2**]), as presented in [**Se2**]. Another approach to the construction of this ring was developed by H. Lenstra [**Len**].

Recall that the **characteristic** of a ring A is the minimal positive integer n – if such an integer exists – such that $nA = 0$. When there is no such n, we define the characteristic to be 0. We say that a ring $\bar{A}$ is **p-perfect** if it has characteristic p and the map $\bar{A} \to \bar{A}$, $\bar{a} \mapsto \bar{a}^p$, is bijective. A **$p$-perfect structure** will be a triple $\mathcal{A} = (A, \bar{A}, \lambda)$ consisting of a ring A of characteristic 0, a ring $\bar{A}$, and a non-zero element λ of A, such that:

(i) λ is not a zero divisor in A;
(ii) $\bar{A} \cong A/\lambda A$;
(iii) $\bar{A}$ is p-perfect; and
(iv) $A \cong \varprojlim A/\lambda^{n+1}A$ via the canonical projection.

We call $\bar{A}$ the **residue ring** of the structure $\mathcal{A}$, and λ its **uniformizer**.

EXAMPLE 12.3.1. If O is a complete discrete valuation ring of characteristic 0 with perfect residue field $\bar{F}$ of characteristic p and with uniformizer λ, then $(O, \bar{F}, \lambda)$ is a p-perfect structure (see Proposition 3.4.3).

Now let $\mathcal{A} = (A, \bar{A}, \lambda)$ be a p-perfect structure and let $\pi_{\mathcal{A}}: A \to \bar{A}$ be the induced epimorphism with kernel λA. We note that $p \in \lambda A$. Proposition 12.1.1 gives a *canonical* bijection $\rho_{\mathcal{A}}: \bar{A} \xrightarrow{\sim} A^{p^\infty}$ which is a (set-theoretic) section of $\pi_{\mathcal{A}}$. Corollary 12.2.2 shows that every element of A has a unique λ-adic expansion $\sum_{i=0}^{\infty} \rho_{\mathcal{A}}(\bar{a}_i)\lambda^i$ in $\varprojlim A/\lambda^n A$, with $\bar{a}_i \in \bar{A}$. Conversely, for every $\mu \in \lambda A$ and every sequence $(\bar{a}_i)_{i=0}^{\infty}$ in $\bar{A}$ there is a well-defined element $\sum_{i=0}^{\infty} \rho_{\mathcal{A}}(\bar{a}_i)\mu^i$ of A.

Let

$$\mathcal{A} = (A, \bar{A}, \lambda_A), \quad \mathcal{B} = (B, \bar{B}, \lambda_B)$$

be p-perfect structures. Then any ring homomorphism $\theta: A \to B$ such that $\theta(\lambda_A) \in \lambda_B B$ induces a ring homomorphism $\bar{\theta}: \bar{A} \to \bar{B}$ such that the following square commutes:

$$\begin{array}{ccc} A & \xrightarrow{\pi_{\mathcal{A}}} & \bar{A} \\ {\scriptstyle\theta}\downarrow & & \downarrow{\scriptstyle\bar{\theta}} \\ B & \xrightarrow{\pi_{\mathcal{B}}} & \bar{B}. \end{array}$$

LEMMA 12.3.2. *For $\mathcal{A}, \mathcal{B}, \theta$ as above:*
(a) $\theta \circ \rho_{\mathcal{A}} = \rho_{\mathcal{B}} \circ \bar{\theta}$;
(b) $\theta(\sum_{i=0}^{\infty} \rho_{\mathcal{A}}(\bar{a}_i)\lambda_A^i) = \sum_{i=0}^{\infty} \rho_{\mathcal{B}}(\bar{\theta}(\bar{a}_i))\theta(\lambda_A)^i$ *for every* $\bar{a}_i \in \bar{A}$.

PROOF. (a) We have

$$\pi_{\mathcal{B}} \circ \theta \circ \rho_{\mathcal{A}} = \bar{\theta} \circ \pi_{\mathcal{A}} \circ \rho_{\mathcal{A}} = \bar{\theta} = \pi_{\mathcal{B}} \circ \rho_{\mathcal{B}} \circ \bar{\theta}.$$

Furthermore, θ maps A^{p^∞} to B^{p^∞}. Hence the images of both $\theta \circ \rho_{\mathcal{A}}$ and $\rho_{\mathcal{B}} \circ \bar{\theta}$ are contained in B^{p^∞}. Since $\pi_{\mathcal{B}}$ is injective on this set, we deduce that $\theta \circ \rho_{\mathcal{A}} = \rho_{\mathcal{B}} \circ \bar{\theta}$.

(b) For each $n \geq 0$, the definition of the λ-adic expansions and (a) give

$$\begin{aligned}\theta\Big(\sum_{i=0}^{\infty}\rho_{\mathcal{A}}(\bar{a}_i)\lambda_A^i\Big)+\theta(\lambda_A)^nB &= \theta\Big(\sum_{i=0}^{n-1}\rho_{\mathcal{A}}(\bar{a}_i)\lambda_A^i\Big)+\theta(\lambda_A)^nB\\ &= \sum_{i=0}^{n-1}\rho_{\mathcal{B}}(\bar{\theta}(\bar{a}_i))\theta(\lambda_A)^i+\theta(\lambda_A)^nB\\ &= \sum_{i=0}^{\infty}\rho_{\mathcal{B}}(\bar{\theta}(\bar{a}_i))\theta(\lambda_A)^i+\theta(\lambda_A)^nB.\end{aligned}$$

The assertion follows. □

Next we investigate the behavior of the coefficients in λ-adic expansions under addition, subtraction, and multiplication of the "infinite sums". We will first do this generically as follows.

Given a ring A and a collection $\{X_i\}_{i\in I}$ of formal variables, let

$$A\big[X_i^{p^{-n}} \mid i\in I, n\in\mathbb{N}\big]$$

be the quotient of the polynomial ring $A[X_{i,n}\,|\,i\in I, n\in\mathbb{N}]$ by the ideal generated by the polynomials $X_{i,n+1}^p - X_{i,n}$ for $i,n\geq 0$. Here X_i stands for the image of $X_{i,0}$ in this quotient. Note that if A is a p-perfect ring, then so is $A[X_i^{p^{-n}}\,|\,i\in I, n\in\mathbb{N}]$. Also, if $\bar{B}$ is a p-perfect ring, then for every ring homomorphism $\theta_0\colon A\to\bar{B}$ and elements $\bar{b}_i\in\bar{B}$, $i\in I$, there is a unique ring homomorphism

$$\theta\colon A[X_i^{p^{-n}}\,|\,i\in I, n\in\mathbb{N}]\to\bar{B}$$

extending θ_0 and mapping X_i to $\bar{b}_i$ for each i. We denote the image of $f\in A[X_i^{p^{-n}}\,|\,i\in I, n\in\mathbb{N}]$ under θ by $f(\bar{b})$. This construction is functorial in the sense that for any other p-perfect ring $\bar{C}$ and any ring homomorphism $\theta'\colon\bar{B}\to\bar{C}$ one has

$$\theta'(f(\bar{b})) = f(\theta'(\bar{b})).$$

Define similarly

$$\begin{aligned}C &= \mathbb{Z}\big[X_i^{p^{-n}}, Y_i^{p^{-n}} \mid i\in I, n\in\mathbb{N}\big],\\ \bar{C} &= \mathbb{F}_p\big[X_i^{p^{-n}}, Y_i^{p^{-n}} \mid i\in I, n\in\mathbb{N}\big],\\ \hat{C} &= \varprojlim C/p^nC.\end{aligned}$$

Then $\mathcal{C} = (\hat{C},\bar{C},p)$ is a p-perfect structure. As before, let $\pi_{\mathcal{C}}\colon\hat{C}\to\bar{C}$ be the canonical projection and $\rho_{\mathcal{C}}\colon\bar{C}\to\hat{C}^{p^\infty}$ its canonical section.

Now assume that $I=\mathbb{N}$. Consider the elements

$$x=\sum_{i=0}^{\infty}\rho_{\mathcal{C}}(X_i)p^i,\quad y=\sum_{i=0}^{\infty}\rho_{\mathcal{C}}(Y_i)p^i$$

of $\hat{C}$. If $*$ is one of the binary ring operations $+,-,\times$, then $x*y\in\hat{C}$ has a p-adic expansion

$$x*y=\sum_{i=0}^{\infty}\rho_{\mathcal{C}}(f_i^*)p^i,$$

for some $f_i^* \in \bar{C}$, $i = 0, 1, 2, \dots$. These generic elements of C provide rules for carrying arithmetical operations in p-perfect structures, as the following Lemma shows.

LEMMA 12.3.3. *Let $\mathcal{A} = (A, \bar{A}, \lambda)$ be a p-perfect structure. Given sequences $\bar{a}_i, \bar{b}_i \in \bar{A}$, $i = 0, 1, 2, \dots$, let $\alpha = \sum_{i=0}^{\infty} \rho_{\mathcal{A}}(\bar{a}_i)p^i$ and $\beta = \sum_{i=0}^{\infty} \rho_{\mathcal{A}}(\bar{b}_i)p^i$. Then for any operation $*$ as above,*

$$\alpha * \beta = \sum_{i=0}^{\infty} \rho_{\mathcal{A}}(f_i^*(\bar{a}, \bar{b}))p^i.$$

PROOF. For each $i, n \geq 0$ let $\bar{a}_{i,n}$ and $\bar{b}_{i,n}$ be the unique elements on $\bar{A}$ satisfying $\bar{a}_{i,n}^{p^n} = \bar{a}_i$ and $\bar{b}_{i,n}^{p^n} = \bar{b}_i$. Then the map

$$X_{i,n} \mapsto \rho_{\mathcal{A}}(\bar{a}_{i,n}), \quad Y_{i,n} \mapsto \rho_{\mathcal{A}}(\bar{b}_{i,n})$$

induces a ring homomorphism $\theta\colon C \to A$ mapping X_i, Y_i to $\rho_{\mathcal{A}}(\bar{a}_i), \rho_{\mathcal{A}}(\bar{b}_i)$, respectively. Furthermore, θ induces a ring homomorphism $\hat{\theta}\colon \hat{C} \to A$ with the same property. The induced homomorphism $\bar{\theta}\colon \bar{C} \to \bar{A}$ is the unique ring homomorphism such that $\bar{\theta}(X_i) = \bar{a}_i$, $\bar{\theta}(Y_i) = \bar{b}_i$ for each i. By Lemma 12.3.2(b),

$$\hat{\theta}(\sum_{i=0}^{\infty} \rho_C(\bar{c}_i)p^i) = \sum_{i=0}^{\infty} \rho_{\mathcal{A}}(\bar{\theta}(\bar{c}_i))p^i$$

for any $\bar{c}_i \in \bar{C}$. Consequently,

$$\begin{aligned}
\alpha * \beta &= \hat{\theta}(x) * \hat{\theta}(y) = \hat{\theta}(x * y) \\
&= \hat{\theta}\Big(\sum_{i=0}^{\infty} \rho_C(f_i^*)p^i\Big) \\
&= \sum_{i=0}^{\infty} \rho_{\mathcal{A}}(\bar{\theta}(f_i^*))p^i \\
&= \sum_{i=0}^{\infty} \rho_{\mathcal{A}}(f_i^*(\bar{a}, \bar{b}))p^i. \quad \square
\end{aligned}$$

PROPOSITION 12.3.4. *Let $\mathcal{A} = (A, \bar{A}, p), \mathcal{B} = (B, \bar{B}, \lambda)$ be p-perfect structures, and let $\pi_{\mathcal{A}}, \pi_{\mathcal{B}}$ be as before. For every ring homomorphism $\bar{\theta}\colon \bar{A} \to \bar{B}$ there exists a unique ring homomorphism $\theta\colon A \to B$ such that $\pi_{\mathcal{B}} \circ \theta = \bar{\theta} \circ \pi_{\mathcal{A}}$. Moreover, if $\bar{\theta}$ is injective and $\lambda = p$, then θ is also injective.*

PROOF. Necessarily $\theta(p) = p \in \lambda B$, so the uniqueness of θ follows from Lemma 12.3.2(b).

We prove the existence of θ. Given $\bar{b}_i \in \bar{B}$ we have a well-defined sum $\sum_{i=0}^{\infty} \rho_{\mathcal{B}}(\bar{b}_i)p^i$ in $B \cong \varprojlim B/\lambda^i B$. Therefore we may define

$$\theta(\sum_{i=0}^{\infty} \rho_{\mathcal{A}}(\bar{a}_i)p^i) = \sum_{i=0}^{\infty} \rho_{\mathcal{B}}(\bar{\theta}(\bar{a}_i))p^i$$

for $\bar{a}_i \in \bar{A}$. To show that this is a ring homomorphism, let $*$ be one of the operations $+$, $-$, $\times$, and let $\bar{a}_i, \bar{a}'_i \in \bar{A}$. From Lemma 12.3.3 and Lemma 12.3.2(b) we get:

$$\begin{aligned}
\theta\Big(\Big(\sum_{i=0}^{\infty} \rho_{\mathcal{A}}(\bar{a}_i)p^i\Big) * \Big(\sum_{i=0}^{\infty} \rho_{\mathcal{A}}(\bar{a}'_i)p^i\Big)\Big) & \\
&= \theta\Big(\sum_{i=0}^{\infty} \rho_{\mathcal{A}}(f_i^*(\bar{a}, \bar{a}'))p^i\Big) \\
&= \sum_{i=0}^{\infty} \rho_{\mathcal{B}}(\bar{\theta}(f_i^*(\bar{a}, \bar{a}'))p^i \\
&= \sum_{i=0}^{\infty} \rho_{\mathcal{B}}(f_i^*(\bar{\theta}(\bar{a}), \bar{\theta}(\bar{a}'))p^i \\
&= \Big(\sum_{i=0}^{\infty} \rho_{\mathcal{B}}(\bar{\theta}(\bar{a}_i))p^i\Big) * \Big(\sum_{i=0}^{\infty} \rho_{\mathcal{B}}(\bar{\theta}(\bar{a}'_i))p^i\Big) \\
&= \theta\Big(\sum_{i=0}^{\infty} \rho_{\mathcal{A}}(\bar{a}_i)p^i\Big) * \theta\Big(\sum_{i=0}^{\infty} \rho_{\mathcal{A}}(\bar{a}'_i)p^i\Big),
\end{aligned}$$

as required.

Finally, as $p \in \lambda_B B$,

$$(\pi_{\mathcal{B}} \circ \theta)\Big(\sum_{i=0}^{\infty} \rho_{\mathcal{A}}(\bar{a}_i)p^i\Big) = \pi_{\mathcal{B}}\Big(\sum_{i=0}^{\infty} \rho_{\mathcal{B}}(\bar{\theta}(\bar{a}_i))p^i\Big) = \bar{\theta}(\bar{a}_0) = (\bar{\theta} \circ \pi_{\mathcal{A}})\Big(\sum_{i=0}^{\infty} \rho_{\mathcal{A}}(\bar{a}_i)p^i\Big)$$

for any $\bar{a}_i \in \bar{A}$. Thus $\pi_{\mathcal{B}} \circ \theta = \bar{\theta} \circ \pi_{\mathcal{A}}$ on A.

For the last assertion assume that $\bar{\theta}$ is injective, $\lambda = p$, and $0 \neq \alpha \in \mathrm{Ker}(\theta)$. As $\bigcap_{n=0}^{\infty} p^n A = \{0\}$, we may write $\alpha = p^n \alpha'$ for some non-negative integer n and some $\alpha' \in A \setminus pA$. We have $p^n \theta(\alpha') = \theta(\alpha) = 0$. Since $\lambda = p$ is not a zero divisor in B, necessarily $\theta(\alpha') = 0$. Now expand $\alpha' = \sum_{i=0}^{\infty} \rho_{\mathcal{A}}(\bar{a}_i)p^i$ with $\bar{a}_i \in \bar{A}$. Then

$$\bar{\theta}(\bar{a}_0) = (\bar{\theta} \circ \pi_{\mathcal{A}})(\alpha') = (\pi_{\mathcal{B}} \circ \theta)(\alpha') = 0.$$

The injectivity of $\bar{\theta}$ implies that $\bar{a}_0 = 0$, and we get the contradiction $\alpha' \in pA$. $\square$

PROPOSITION 12.3.5. *Let $\mathcal{A} = (A, \bar{A}, p)$ be a p-perfect structure. Let $\bar{B}$ be a p-perfect ring and let $\bar{\theta}\colon \bar{A} \to \bar{B}$ be a ring epimorphism. Then $\bar{B}$ can be completed into a p-perfect structure $\mathcal{B} = (B, \bar{B}, p)$ such that there is a ring epimorphism $\theta\colon A \to B$ satisfying $\pi_{\mathcal{B}} \circ \theta = \bar{\theta} \circ \pi_{\mathcal{A}}$.*

PROOF. We define an equivalence relation on A by letting $\alpha = \sum_{i=0}^{\infty} \rho_{\mathcal{A}}(\bar{a}_i)p^i$ and $\alpha' = \sum_{i=0}^{\infty} \rho_{\mathcal{A}}(\bar{a}'_i)p^i$ be equivalent if $\bar{\theta}(a_i) = \bar{\theta}(a'_i)$ for all i. Let B be the set of all resulting equivalence classes, and let $\theta\colon A \to B$ be the canonical projection. If $\beta = \sum_{i=0}^{\infty} \rho_{\mathcal{A}}(\bar{b}_i)p^i$ is another element of A and $*$ is one of $+, -, \times$, then for every i,

$$\bar{\theta}(f_i^*(\bar{a}, \bar{b})) = f_i^*(\bar{\theta}(\bar{a}), \bar{\theta}(\bar{b})) = f_i^*(\bar{\theta}(\bar{a}'), \bar{\theta}(\bar{b})) = \bar{\theta}(f_i^*(\bar{a}', \bar{b})).$$

Hence the elements

$$\alpha * \beta = \sum_{i=0}^{\infty} f_i^*(\bar{a}, \bar{b})p^i, \quad \alpha' * \beta = \sum_{i=0}^{\infty} f_i^*(\bar{a}', \bar{b})p^i$$

of A are equivalent. Consequently, A induces on B a ring structure, with respect to which θ is an epimorphism.

Observe that $\alpha \in A$ is equivalent to 0 if and only if $p\alpha$ is equivalent to 0. Thus, $\theta(\alpha) = 0$ if and only if $p\theta(\alpha) = \theta(p\alpha) = 0$. This proves that p is not a zero divisor in B.

Next, suppose that $\alpha, \alpha' \in A$ are equivalent. Then $(\bar{\theta} \circ \pi_{\mathcal{A}})(\alpha) = (\bar{\theta} \circ \pi_{\mathcal{A}})(\alpha')$. Hence the epimorphism $\bar{\theta} \circ \pi_{\mathcal{A}}: A \to \bar{B}$ induces an epimorphism $\pi_{\mathcal{B}}: B \to \bar{B}$ such that $\pi_{\mathcal{B}} \circ \theta = \bar{\theta} \circ \pi_{\mathcal{A}}$. Then for every $\alpha = \sum_{i=0}^{\infty} \rho_{\mathcal{A}}(\bar{a}_i)p^i \in A$ we have $\pi_{\mathcal{B}}(\theta(\alpha)) = 0$ if and only if $\bar{\theta}(\bar{a}_0) = 0$. But the latter equality means that α is equivalent to an element of pA, i.e., that $\theta(\alpha) \in \theta(pA) = pB$. Consequently, $\mathrm{Ker}(\pi_{\mathcal{B}}) = pB$, whence $B/pB \cong \bar{B}$.

It remains to show that the canonical map $B \to \varprojlim B/p^nB$ is an isomorphism. For its injectivity, let $\alpha = \sum_{i=0}^{\infty} \rho_{\mathcal{A}}(\bar{a}_i)p^i \in A$ and suppose that $\theta(\alpha) \in \bigcap_{n=0}^{\infty} p^nB$. Then for each $n \geq 0$ there exists $\alpha_n \in A$ such that α and $p^n\alpha_n$ are equivalent. It follows that $\bar{\theta}(\bar{a}_i) = 0$ for $i = 0, \ldots, n-1$. Since n is arbitrary, we conclude that $\bar{\theta}(\bar{a}_i) = 0$ for all i, so α is equivalent to 0, i.e., $\theta(\alpha) = 0$, as required.

To prove the surjectivity of this map, take $\beta_n \in B$ such that

$$(\beta_n + p^nB)_{n=0}^{\infty} \in \varprojlim B/p^nB.$$

For each n choose $\alpha_n = \sum_{i=0}^{\infty} \rho_{\mathcal{A}}(\bar{a}_{in})p^i \in A$ with $\theta(\alpha_n) = \beta_n$. For $i < n$ the compatibility of the sequence $(\beta_n + p^nB)_{n=0}^{\infty}$ implies that

$$\theta(\alpha_i) - \theta(\alpha_n) = \beta_i - \beta_n \in p^nB.$$

Therefore $\bar{\theta}(\bar{a}_{in})$ is independent of n when $i < n$. We conclude that for each i there exists $\bar{a}_i \in A$ such that $\bar{\theta}(\bar{a}_i) = \bar{\theta}(\bar{a}_{in})$ when $i < n$. Now let $\alpha = \sum_{i=0}^{\infty} \rho_{\mathcal{A}}(\bar{a}_i)p^i$ and let $\beta = \theta(\alpha) \in B$. Then

$$\beta - \beta_n = \theta(\alpha - \alpha_n) \in \theta(p^nA) = p^nB$$

for all $n \geq 0$, as required. □

COROLLARY 12.3.6. *Let $\bar{A}$ be a p-perfect ring. Then $\bar{A}$ can be completed into a p-perfect structure $\mathcal{A} = (A, \bar{A}, p)$. The ring A is unique up to an isomorphism.*

PROOF. Take a set $\{X_i\}_{i \in I}$ of variables. Let

$$C = \mathbb{Z}[X_i^{p^{-n}} \mid i \in I, n \in \mathbb{N}], \quad \hat{C} = \varprojlim C/p^nC,$$

$$\bar{C} = \mathbb{F}_p[X_i^{p^{-n}} \mid i \in I, n \in \mathbb{N}], \quad \mathcal{C} = (\hat{C}, \bar{C}, p)$$

be as before. Assuming that the cardinality of I is sufficiently large, there is a ring epimorphism $\bar{\theta}: \bar{C} \to \bar{A}$. Proposition 12.3.5 (with $\mathcal{A}, \bar{B}$ replaced by $\mathcal{C}, \bar{A}$, respectively) gives rise to a structure $\mathcal{A} = (A, \bar{A}, p)$ as required.

The uniqueness follows from Proposition 12.3.4 (for the identity map $\bar{A} \to \bar{A}$) in a standard way. □

We denote the ring A obtained in Corollary 12.3.6 by $W(\bar{A})$. It comes along with a ring epimorphism $\pi: W(\bar{A}) \to \bar{A}$ which has kernel $pW(\bar{A})$. It follows from Proposition 12.3.4 that the map $\bar{A} \to W(\bar{A})$ is a covariant functor from the category of p-perfect rings to the category of all rings. Furthermore, if $\bar{A} \subseteq \bar{B}$ are p-perfect rings, then the induced homomorphism $W(\bar{A}) \to W(\bar{B})$ is injective.

EXAMPLE 12.3.7. For $\bar{A} = \mathbb{F}_p$ the ring $\mathbb{Z}_p$ satisfies the defining properties of $W(\mathbb{F}_p)$, i.e., $(\mathbb{Z}_p, \mathbb{F}_p, p)$ is a p-perfect structure. By the uniqueness part of Corollary 12.3.6, $\mathbb{Z}_p \cong W(\mathbb{F}_p)$.

THEOREM 12.3.8. *Let $\bar{A}, \bar{B}$ be p-perfect rings such that $\bar{B}$ is a free module of finite rank d over $\bar{A}$. Then $W(\bar{B})$ is a free module of rank d over $W(\bar{A})$.*

PROOF. Let $\pi_{\mathcal{A}}: W(\bar{A}) \to \bar{A}$, $\pi_{\mathcal{B}}: W(\bar{B}) \to \bar{B}$ be the canonical projections. Let $\rho_{\mathcal{A}}: \bar{A} \to W(\bar{A})$ be an arbitrary set-theoretic section of $\pi_{\mathcal{A}}$. We may view the functorial homomorphism $W(\bar{A}) \to W(\bar{B})$ as the inclusion map. It makes the following square commutative:

$$\begin{array}{ccc} W(\bar{A}) & \hookrightarrow & W(\bar{B}) \\ {\scriptstyle \pi_{\mathcal{A}}} \downarrow & & \downarrow {\scriptstyle \pi_{\mathcal{B}}} \\ \bar{A} & \longrightarrow & \bar{B}. \end{array}$$

Next we take an $\bar{A}$-basis $\bar{b}_1, \ldots, \bar{b}_f$ of $\bar{B}$, and lift each $\bar{b}_i$ to an element b_i of $W(\bar{B})$ with $\pi_{\mathcal{B}}(b_i) = \bar{b}_i$. We define a map $\rho_{\mathcal{B}}: \bar{B} \to W(\bar{B})$ by setting

$$\rho_{\mathcal{B}}\Big(\sum_{l=1}^{f} \bar{b}_l \bar{a}_l\Big) = \sum_{l=1}^{f} b_l \rho_{\mathcal{A}}(\bar{a}_l)$$

for $\bar{a}_1, \ldots, \bar{a}_f \in \bar{A}$. Then $\rho_{\mathcal{B}}$ is a section of the ring epimorphism $\pi_{\mathcal{B}}$.

Now every element of $W(\bar{B})$ has a p-adic expansion $\sum_{i=0}^{\infty} \rho_{\mathcal{B}}(\bar{c}_i)p^i$, with $\bar{c}_i \in \bar{B}$. For each i write $\bar{c}_i = \sum_{l=1}^{f} \bar{b}_l \bar{a}_{il}$, with $\bar{a}_{il} \in \bar{A}$. Then

$$\sum_{i=0}^{\infty} \rho_{\mathcal{B}}(\bar{c}_i)p^i = \sum_{i=0}^{\infty} \rho_{\mathcal{B}}\Big(\sum_{l=1}^{f} \bar{b}_l \bar{a}_{il}\Big)p^i = \sum_{i=0}^{\infty}\sum_{l=1}^{f} b_l \rho_{\mathcal{A}}(\bar{a}_{il})p^i = \sum_{l=1}^{f} b_l \Big(\sum_{i=0}^{\infty} \rho_{\mathcal{A}}(\bar{a}_{il})p^i\Big).$$

We conclude that

$$W(\bar{B}) = b_1 W(\bar{A}) + \cdots + b_f W(\bar{A}).$$

To show that $b_1, \ldots, b_f$ is a $W(\bar{A})$-basis of $W(\bar{B})$, suppose that $\sum_{l=1}^{f} b_l a_l = 0$ with $a_1, \ldots, a_f \in W(\bar{A})$. We need to show that $a_1 = \cdots = a_f = 0$. If not, then there would exist a maximal positive integer k such that $a_1, \ldots, a_f \in p^k W(\bar{A})$. By the definition of a p-perfect structure, p is not a zero divisor in $W(\bar{B})$, so we may assume that $k = 0$. Thus $\pi_{\mathcal{A}}(a_l) \neq 0$ for some l. On the other hand,

$$\sum_{l=1}^{d} \bar{b}_l \pi_{\mathcal{A}}(a_l) = \pi_{\mathcal{B}}\Big(\sum_{l=1}^{f} b_l a_l\Big) = 0,$$

contrary to the assumption that $\bar{b}_1, \ldots, \bar{b}_f$ are linearly dependent over $\bar{A}$. $\square$

12.4 Rings of Witt vectors

Let $\bar{F}$ be a perfect field of characteristic p. It is then a p-perfect ring in the sense of the preceding section, so we can construct the ring $W(\bar{F})$. We call it the **ring of Witt vectors over $\bar{F}$**. This notion should not be confused with the "Witt ring" which is studied in the context of quadratic form theory ([**Lam3**], [**Sch2**]). For the actual meaning and description of the objects known as "Witt vectors" we refer, e.g., to [**FV**, Ch. I, §8] or to [**Se2**, Ch. II, §6].

By definition, p is not a zero divisor in $W(\bar{F})$. Since $\bar{F}$ is a field, the kernel $pW(\bar{F})$ of the ring epimorphism $W(\bar{F}) \to \bar{F}$ is a maximal ideal. Also, the bijectivity of the canonical homomorphism $W(\bar{F}) \to \varprojlim W(\bar{F})/p^n W(\bar{F})$ implies that $\bigcap_{n=0}^{\infty} p^n W(\bar{F}) = \{0\}$. Proposition 3.4.3 therefore implies:

THEOREM 12.4.1. *Let $\bar{F}$ be a perfect field of characteristic p. Then $W(\bar{F})$ is a complete discrete valuation ring with residue field $\bar{F}$. Its quotient field F has characteristic 0. The valuation v corresponding to $W(\bar{F})$ is given for a non-negative $\alpha \in W(\bar{F})$ by $v(\alpha) = k$, where k is the maximal non-negative integer such that $\alpha \in p^k W(\bar{F})$.*

The ring $W(\bar{F})$ has the following universal property:

THEOREM 12.4.2. *Let F be a field of characteristic 0 and let v be a complete discrete valuation on F such that $\bar{F}_v$ is perfect of characteristic p. Then there exists a unique ring homomorphism $\iota\colon W(\bar{F}_v) \to O_v$ which commutes with the canonical projection $W(\bar{F}_v) \to \bar{F}_v$ and the residue map $\pi_v\colon O_v \to \bar{F}_v$.*

Moreover, ι is injective and makes O_v a free $W(\bar{F}_v)$-module of rank $(v(F^\times) : \mathbb{Z}v(p))$.

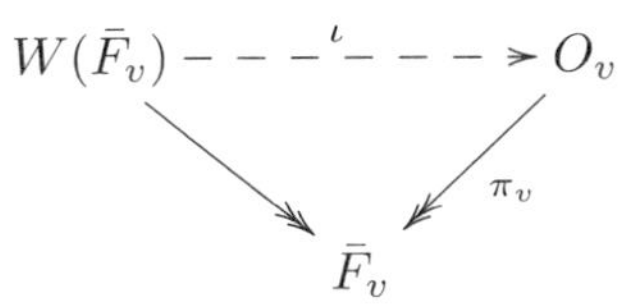

PROOF. Let λ be a uniformizer for v. We have p-perfect structures $\mathcal{A} = (W(\bar{F}_v), \bar{F}_v, p)$ and $\mathcal{B} = (O_v, \bar{F}_v, \lambda)$ (see Example 12.3.1). By Proposition 12.3.4 (with $\bar{\theta}$ being the identity map on $\bar{F}_v$) there is a unique homomorphism $\iota\colon W(\bar{F}_v) \to O_v$ which commutes with the canonical projection $W(\bar{F}_v) \to \bar{F}_v$ and with π_v. Since $p^n \neq 0$ in F, the homomorphism ι maps the ideals $p^n W(\bar{F}_v)$, $n = 1, 2, \ldots$, non-trivially. Since these are the only non-zero ideals in $W(\bar{F}_v)$, we must have $\mathrm{Ker}(\iota) = \{0\}$, proving the injectivity of ι.

Now set $e = (v(F^\times) : \mathbb{Z}v(p))$, i.e., $\mathfrak{m}_v^e = pO_v$. For integers $0 \le i < e$ and $k \ge 0$ we have $\mathfrak{m}_v^{ke+i} = p^k \lambda^i O_v$. By Corollary 12.2.2 (for the ideal $\mathfrak{m}_v = \lambda O_v$ of O_v), every element of O_v can be uniquely written as

$$\sum_{i=0}^{e-1} \sum_{k=0}^{\infty} \rho_{\mathcal{B}}(\bar{a}_{ki}) p^k \lambda^i$$

with $\bar{a}_{ki} \in \bar{F}_v$. Furthermore, by Lemma 12.3.2(b),

$$\sum_{k=0}^{\infty} \rho_{\mathcal{B}}(\bar{a}_{ki}) p^k = \iota\Big(\sum_{k=0}^{\infty} \rho_{\mathcal{A}}(\bar{a}_{ki}) p^k\Big) \in \iota(W(\bar{F}_v)).$$

Thus λ^i, $i = 0, \ldots, e-1$, form a basis of O_v as a free $\iota(W(\bar{F}_v))$-module. $\square$

THEOREM 12.4.3. *Let $\bar{E}/\bar{F}$ be an extension of perfect fields of characteristic p. Let F and E be the quotient fields of $W(\bar{F})$ and $W(\bar{E})$, respectively. Then:*

(a) *E extends F;*

(b) *the valuation ring $W(\bar{E})$ extends the valuation ring $W(\bar{F})$.*

PROOF. (a) As before, the functorial map embeds $W(\bar{F})$ inside $W(\bar{E})$. Hence F embeds inside E.

(b) The valuation ring $W(\bar{F})$ is discrete, and hence has no non-trivial proper coarsenings. Also, $W(\bar{F}) \subseteq W(\bar{E})$ and $1/p \notin W(\bar{E})$. Conclude that $W(\bar{F}) = W(\bar{E}) \cap F$. □

12.5 Mixed valuations under a finiteness assumption

In this section we study a different family of valued fields (F, v) with mixed characteristics, namely, the valuation is no longer assumed to be discrete, but we assume that $F^\times/p = F^\times/(F^\times)^p$ is finite. This leads to a quite satisfactory structure theory, which we will also treat from the K-theoretic viewpoint in §27. The results here generalize arguments of Pop [**P1**] and follow [**Ef9**]. First we need two elementary field-theoretic facts.

LEMMA 12.5.1. *If E/F is an extension of infinite fields and $(E^\times : F^\times) < \infty$, then $E = F$.*

PROOF. For $a \in E$ the set $a - F$ is infinite, so we can find distinct elements $x_1, x_2 \in F$ with $x_1, x_2 \neq a$ and $(a - x_1)F^\times = (a - x_2)F^\times$. But this can happen only when $a \in F$. □

LEMMA 12.5.2. *If E is a field of characteristic p, then $E^\times/p$ is either trivial or infinite.*

PROOF. When E is finite, it is perfect, so $E^\times/p = 1$. Suppose that E is infinite. Then the group-isomorphism $E^\times \xrightarrow{\sim} (E^\times)^p$, $x \mapsto x^p$, shows that the subfield $F = E^p$ of E is also infinite. Now apply Lemma 12.5.1. □

For a valued field (F, v) and for $\alpha \in v(F^\times)$ with $\alpha > 0$ let

$$W_\alpha = \left\{x \in F^\times \mid v(1 - x) \geq \alpha\right\}.$$

Observe that W_α is a subgroup of $G_v = 1 + \mathfrak{m}_v$.

LEMMA 12.5.3. *Let (F, v) be a valued field and let S be a subgroup of $F^\times$ such that $G_v/(S \cap G_v)$ is a finitely generated group. Then there exists a positive $\alpha \in v(F^\times)$ such that:*

(i) *$SW_\alpha = SG_v$;*

(ii) *if* char $\bar{F}_v = p$, *then $1 + pO_v \leq W_\alpha$.*

PROOF. We choose $a_1, \dots, a_n \in \mathfrak{m}_v$ such that the cosets of $1 - a_i$, $i = 1, \dots, n$, generate $G_v/(S \cap G_v)$. Hence $(1 - a_i)S$, $i = 1, \dots, n$, generate SG_v/S. Take any α with

$$0 < \alpha \leq \min\{v(a_1), \dots, v(a_n)\}.$$

Then for each i we have $1 - a_i \in W_\alpha$. Combined with $W_\alpha \leq G_v$, this shows that $SW_\alpha/S = SG_v/S$, so (i) holds.

To prove (ii), suppose that char $\bar{F}_v = p$, i.e., $v(p) > 0$. We take

$$\alpha = \min\{v(p), v(a_1), \dots, v(a_n)\}.$$

□

THEOREM 12.5.4. *Let v be a valuation of rank* 1 *on a field F such that* $\operatorname{char} F = 0$ *and* $\operatorname{char} \bar{F}_v = p > 0$. *Suppose that $F^\times/(F^\times)^p(1+p\mathfrak{m}_v)$ is finite. Then either:*
(a) *$v(F^\times)$ is discrete and $\bar{F}_v$ is finite; or*
(b) *$v(F^\times)$ is p-divisible and $\bar{F}_v$ is perfect.*

PROOF. Let $S = (F^\times)^p(1+p\mathfrak{m}_v)$. We break the argument into five steps.

PART I: *$\bar{F}_v$ is perfect.* Indeed, $\bar{S}_v = (\bar{F}_v^\times)^p$. Since $F^\times/S$ is by assumption finite, the exact sequences (3.2.5)–(3.2.6) imply that $\bar{F}_v^\times/(\bar{F}_v^\times)^p$ is also finite. Since $\operatorname{char} \bar{F}_v = p$ and by Lemma 12.5.2, this quotient must be trivial, as desired.

PART II: $S \cap G_v = G_v^p(1+p\mathfrak{m}_v)$. Since $\operatorname{char} \bar{F}_v = p$, the p-torsion subgroup of $\bar{F}_v^\times$ is trivial. Therefore Lemma 3.2.4 implies that $(F^\times)^p \cap G_v = G_v^p$. Since $1+p\mathfrak{m}_v \leq G_v$, we obtain

$$S \cap G_v = ((F^\times)^p \cap G_v)(1+p\mathfrak{m}_v) = G_v^p(1+p\mathfrak{m}_v).$$

PART III: $S \cap G_v \subseteq (1-S)(1+p\mathfrak{m}_v)$. Recall that $p | \binom{p}{i}$ for $i = 1, \dots, p-1$. Hence for every $a \in \mathfrak{m}_v$ we have

$$(1-a)^p \in 1 - a^p + p\mathfrak{m}_v = (1-a^p)(1+p\mathfrak{m}_v) \subseteq (1-S)(1+p\mathfrak{m}_v).$$

Thus $G_v^p \subseteq (1-S)(1+p\mathfrak{m}_v)$. Now use Part II.

PART IV: *$v(F^\times)$ is either discrete or p-divisible.* By Corollary 2.5.4, it suffices to find a positive $\alpha \in v(F^\times)$ such that for every $b \in F^\times$ with $0 < v(b) < \alpha$ one has $v(b) \in pv(F^\times)$. Since $F^\times/S$ is finite, the sequences (3.2.5)–(3.2.6) imply that $G_v/(S \cap G_v)$ is also finite. Hence we may take α as in Lemma 12.5.3. By property (i) of W_α, and since $W_\alpha \leq G_v$, we have

$$1 - b \in G_v = (SW_\alpha) \cap G_v = (S \cap G_v)W_\alpha.$$

It therefore follows from part III and from property (ii) of W_α that $1-b \in (1-S)W_\alpha$. So choose $s \in S$ with $1-b \in (1-s)W_\alpha$. As $W_\alpha \leq G_v$ we get $1-s \in G_v$. Hence

$$v(b-s) = v\left(\frac{b-s}{1-s}\right) = v\left(1 - \frac{1-b}{1-s}\right) \geq \alpha.$$

Since $v(b) < \alpha$, necessarily $v(b) = v(s) \in v(S) = pv(F^\times)$, as desired.

PART V: *When $v(F^\times)$ is discrete, $\bar{F}_v$ is finite.* In this case Proposition 3.4.4 gives

$$G_v/(1+\mathfrak{m}_v^2) = (1+\mathfrak{m}_v)/(1+\mathfrak{m}_v^2) \cong \bar{F}_v.$$

Using again that $p \left| \binom{p}{i} \right.$ for $1 \leq i \leq p-1$, we get

$$G_v^p(1+p\mathfrak{m}_v) \leq 1 + \mathfrak{m}_v^2.$$

In light of Part II, this gives rise to a group epimorphism $G_v/(S \cap G_v) \to \bar{F}_v$. We have already noted that $G_v/(S \cap G_v)$ is finite. Consequently, $\bar{F}_v$ is also finite. $\square$

COROLLARY 12.5.5. *Let (F, v) be a valued field with* $\mathrm{char}\, F = 0$, $\mathrm{char}\, \bar{F}_v = p > 0$, *and with $F^\times/(F^\times)^p(1 + p\mathfrak{m}_v)$ finite. Then there is a coarsening u^+ of v with residue characteristic p and a coarsening u^- of u^+ with residue characteristic 0, such that u^+/u^- has rank 1 and either:*

(a) *v/u^- is discrete, $\bar{F}_v$ is finite, and $v = u^+$; or*

(b) *the value group of u^+/u^- is p-divisible.*

PROOF. The slicing lemma (Lemma 5.3.1) yields a coarsening u^+ of v and a coarsening u^- of u^+ such that

$$\mathrm{char}\, \bar{F}_{u^-} = 0, \quad \mathrm{char}\, \bar{F}_{u^+} = \mathrm{char}\, (\bar{F}_{u^-})_{u^+/u^-} = p$$

and u^+/u^- has rank 1. We have $\mathfrak{m}_v \subseteq O_v \subseteq O_{u^-}$, whence $1+p\mathfrak{m}_v \leq O_{u^-}^\times$. Therefore the push-down of $S = (F^\times)^p(1 + p\mathfrak{m}_v)$ under π_{u^-} is $\bar{S} = (\bar{F}_{u^-}^\times)^p(1 + p\mathfrak{m}_{v/u^-})$ (see Proposition 5.2.1(b)). As $F^\times/S$ is finite, the exact sequences (3.2.5)–(3.2.6) imply that so is $\bar{F}_{u^-}^\times/\bar{S}$. We can now apply Theorem 12.5.4 to the valued field $(\bar{F}_{u^-}, u^+/u^-)$ and the group $\bar{S}$. It shows that one of the following cases holds:

(i) u^+/u^- is discrete and $\bar{F}_{u^+}$ is finite. In particular, the valuation v/u^+ on $\bar{F}_{u^+}$ is trivial. Hence $v = u^+$, so (a) holds.

(ii) The value group of u^+/u^- is p-divisible, i.e., (b) holds. □

Part III

GALOIS THEORY

In part II we studied valuations and orderings on a given field. We now let the field vary, and study the behavior of valuations and orderings in the resulting extensions. In the valued field case this theory has its roots in algebraic number theory, especially in the investigation by Dedekind and Hilbert of ramification and splitting of primes in number field extensions. For this reason, it is sometimes called the *"Hilbert ramification theory"* or *"Dedekind–Hilbert theory"* [**D**], [**Ro**], although most of its main ingredients in the herein presented form were developed later by W. Krull, M. Deuring, I. Kaplansky, M. Krasner, A. Ostrowski, F.K. Schmidt and others. The ordered field analog of this theory is somewhat degenerate, since the relevant Galois groups are of order ≤ 2. Therefore most of this part (§§14–18, 21–22) will be devoted to valuations, and only in §19 will we describe the situation for orderings. §20 integrates these two cases by studying the reflection of the coarsening relation among localities in Galois groups. Naturally, we will use here intensively the language of infinite Galois theory, which is reviewed in §13.

CHAPTER 13

Infinite Galois Theory

For the reader's convenience, we review briefly the main facts of infinite Galois theory which will be needed in the sequel. We refer to standard textbooks on Galois theory such as [**FJr**], [**Ko3**], [**NSW**], [**Ris**], [**Se3**] for more details and proofs.

A **supernatural number** (also called a **Steintz number**) is a formal expression $\prod_p p^{\alpha_p}$, where p ranges over all prime numbers, and $0 \leq \alpha_p \leq \infty$ for each p. There is an obvious divisibility relation for such numbers. Given a family of supernatural numbers, one defines its product, greatest common divisor, and least common multiple in the natural way.

We define the supernatural order $|\Gamma|$ of a torsion abelian (discrete) group Γ to be the least common multiple of the orders of its finite subgroups, considered as a supernatural number.

Next suppose that G is a profinite group. Thus $G = \varprojlim G_i$, for some inverse system G_i, $i \in I$, of finite groups. When each G_i is in addition a p-group (resp., solvable, abelian, etc.) we say that G is **pro-p** (resp., **pro-solvable**, **pro-abelian**, etc.). Considering each G_i as a discrete topological group, the inverse limit topology on G is Hausdorff, compact, and totally disconnected. One defines the **order** $|G|$ of G to be the (supernatural) least common multiple of the $|G_i|$, $i \in I$ (this is independent of the choice of the inverse system). Thus G is pro-p for a prime number p if and only if $|G| \mid p^\infty$.

If H is a closed subgroup of the profinite group G, then $H \cong \varprojlim H_i$, where H_i is the projection of H in G_i. Therefore H is also profinite and $|H| \mid |G|$. We define the **index** $(G : H)$ of H in G to be the least common multiple of the finite indices $(G_i : H_i)$. In particular, $|G| = (G : 1)$. If K is a closed subgroup of H, then the following tower property holds:

$$(G : K) = (G : H)(H : K).$$

Hence $|G| = (G : H)|H|$. The usual homomorphism theorems for finite groups extend to the profinite context once we restrict ourselves to continuous homomorphisms and closed subgroups.

A **p-Sylow subgroup** of a profinite group $G = \varprojlim G_i$, where p is a prime number, is a closed subgroup H of G such that $|H| \mid p^\infty$ and $p \nmid (G : H)$. The classical Sylow theorems for the finite quotients G_i of G imply (using standard compactness arguments) that a p-Sylow subgroup of G exists and is unique up to conjugation by an element of G.

The **degree** of an algebraic extension E/F of fields is the least common multiple of the degrees $[E_1 : F]$, where E_1 ranges over all finite extensions of F contained in E. If $F \subseteq M \subseteq E$ is an intermediate field, then one has the tower property:

$$[E : F] = [E : M][M : F].$$

Now suppose that the algebraic field extension E/F is normal. Let $\mathrm{Aut}(E/F)$ be its automorphism group. When E/F is in addition separable we say that it is Galois, and as usual, call $\mathrm{Gal}(E/F) = \mathrm{Aut}(E/F)$ its **Galois group**. For a general normal extension, $\mathrm{Aut}(E/F) \cong \varprojlim \mathrm{Aut}(E_1/F)$ as abstract groups, with E_1 ranging over all finite normal extensions of F contained in E. By standard Galois theory, $\mathrm{Aut}(E_1/F)$ is a finite group, and we equip it with the discrete topology. This makes $\mathrm{Aut}(E/F)$ a profinite group with respect to the inverse limit topology, which is called the **Krull topology**. It has a local base at $\sigma \in \mathrm{Aut}(E/F)$ consisting of the subgroups $\sigma\mathrm{Aut}(E/F_1)$, where F_1 ranges over all finite extensions of F inside E. If E_1 is another normal extension of F contained in E, then the restriction map $\mathrm{Res}\colon \mathrm{Aut}(E/F) \to \mathrm{Aut}(E_1/F)$ is a continuous epimorphism with respect to the Krull topologies.

The main theorem of finite Galois theory extends to the infinite context once we restrict ourselves to closed subgroups of $\mathrm{Aut}(E/F)$:

THEOREM 13.1.1. *Let E/F be a normal extension of fields. The fixed field map*

$$H \mapsto E^H$$

gives a bijective order-reversing correspondence between the closed subgroups of $\mathrm{Aut}(E/F)$ and the intermediate fields $F \subseteq M \subseteq E$ with E/M Galois. The inverse map is given by

$$M \mapsto \mathrm{Aut}(E/M).$$

For E/F normal, $|\mathrm{Aut}(E/F)| \mid [E : F]$ as supernatural numbers, with equality holding if and only if E/F is Galois. Recall that in any normal extension E/F there is a maximal subextension F_s which is separable over F, as well as a maximal extension F_i which is purely inseparable over F. The extensions F_s/F and E/F_i are Galois, and the extensions E/F_s and F_i/F are purely inseparable (whence normal). One has $F = F_s \cap F_i$ and $E = F_sF_i$. We conclude that

$$|\mathrm{Aut}(E/F)| = [F_s : F] = [E : F_i].$$

CHAPTER 14

Valuations in Field Extensions

In this chapter we begin our investigation of the behavior of valuations in algebraic extensions of fields. We prove a fundamental result of Chevalley, guaranteeing that such extensions always exists, and study the connections among the extensions of a given valuation.

14.1 Chevalley's theorem

A given valuation on a field always extends to a valuation on any extension of the field. The proof of this fundamental fact is based on the following lemma.

LEMMA 14.1.1. *Let R be a subring of a field E, let Ω be an algebraically closed field, and let $h: R \to \Omega$ be a ring homomorphism. Also, suppose that $0 \neq x \in E$. Then h extends to a homomorphism $h': R[x] \to \Omega$ or to a homomorphism $h': R[x^{-1}] \to \Omega$.*

PROOF. We may assume that $x, x^{-1} \notin R$. Let $\mathfrak{p} = \mathrm{Ker}(h)\,(\neq R)$. Since h extends in the obvious way to the localization $R_{\mathfrak{p}}$, we may replace R by $R_{\mathfrak{p}}$ to assume, without loss of generality, that R is a local ring and that $\mathfrak{p}$ is its unique maximal ideal. Let $\bar{h}: R/\mathfrak{p} \hookrightarrow \Omega$ be the embedding induced by h, and let $\mathfrak{p}R[x]$ be the ideal in $R[x]$ generated by $\mathfrak{p}$.

Suppose that $1 \notin \mathfrak{p}R[x]$. Then $\mathfrak{p}R[x]$ is a proper ideal in $R[x]$, and is therefore contained in a maximal ideal $\mathfrak{m}$ of $R[x]$. One has $\mathfrak{p} = R \cap \mathfrak{m}$, so the field $R/\mathfrak{p}$ embeds in the field $R[x]/\mathfrak{m}$. Let $\bar{x}$ be the image of x modulo $\mathfrak{m}$. Hence, $R[x]/\mathfrak{m} = (R/\mathfrak{p})[\bar{x}]$ is a simple extension of $R/\mathfrak{p}$. Since Ω is algebraically closed, the embedding $\bar{h}$ therefore extends to a homomorphism $\hat{h}: R[x]/\mathfrak{m} \to \Omega$. From the commutative diagram below we see that $\hat{h}$ induces a homomorphism $h': R[x] \to \Omega$ extending h:

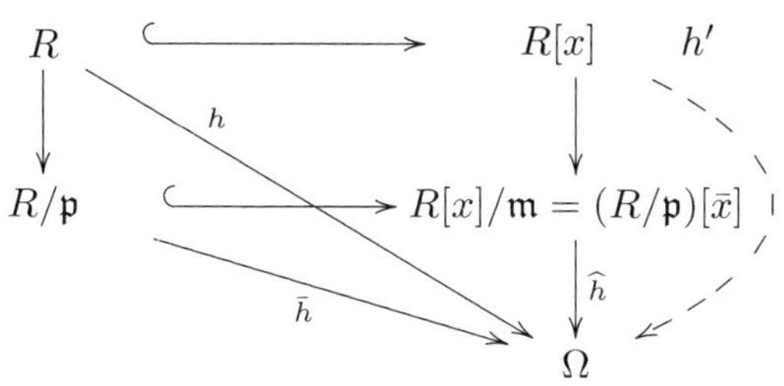

as required.

Similarly, if $1 \notin \mathfrak{p}R[x^{-1}]$, then h extends to a homomorphism $h': R[x^{-1}] \to \Omega$.

Finally, suppose that $1 \in \mathfrak{p}R[x] \cap \mathfrak{p}R[x^{-1}]$. Then we may write

$$1 = a_0 + a_1 x + \cdots + a_n x^n = b_0 + b_1 x^{-1} + \cdots + b_m x^{-m},$$

with $a_i, b_i \in \mathfrak{p}$ and $n, m \geq 1$ minimal. Without loss of generality, $n \geq m$. But then

$$\begin{aligned}1 &= a_1x + \cdots + a_{n-1}x^{n-1} + a_nx^n(1-b_0)^{-1}(b_1x^{-1} + \cdots + b_mx^{-m}) \\ &= a_1x + \cdots + a_{n-1}x^{n-1} + a_n(1-b_0)^{-1}(b_1x^{n-1} + \cdots + b_mx^{n-m}),\end{aligned}$$

with $(1-b_0)^{-1} \in R$. This contradicts the minimality of n. □

COROLLARY 14.1.2 (Chevalley [**Ch**]). *Let O be a valuation ring on a field F and let E be a field extending F. Then there exists a valuation ring O' on E such that $O' \cap F = O$.*

PROOF. Let $\mathfrak{m}$ be the maximal ideal of O, let Ω be an algebraic closure of the residue field $O/\mathfrak{m}$, and let $h: O \to \Omega$ be the natural ring homomorphism. Consider the collection of all pairs (O_1, h_1) consisting of a subring O_1 of E containing O and a homomorphism $h_1: O_1 \to \Omega$ extending h. We partially order this collection by setting $(O_1, h_1) \leq (O_2, h_2)$ if $O_1 \subseteq O_2$ and h_2 extends h_1. Zorn's lemma yields a maximal pair (O', h') in this collection.

Now consider a non-zero $x \in E$. By Lemma 14.1.1, h' extends to a homomorphism $h'': O'[x] \to \Omega$ or to a homomorphism $h'': O'[x^{-1}] \to \Omega$. The maximality of (O', h') implies that $x \in O'$ or $x^{-1} \in O'$. Thus O' is a valuation ring on E.

By construction, $O' \cap F \supseteq O$. Conversely, suppose that $y \in O' \cap F$ but $y \notin O$. Then $y \neq 0$ and $y^{-1} \in \mathfrak{m} = \mathrm{Ker}(h) \leq \mathrm{Ker}(h')$, so we get the contradiction $1 = h'(1) = h'(y \cdot y^{-1}) = h'(y) \cdot 0 = 0$. We conclude that $O' \cap F \subseteq O$. □

In the situation of Corollary 14.1.2 we say that O' **extends** O. Denoting the corresponding valuations by v' and v, respectively, this means that (E', v') extends (E, v) as valued fields (see §3.2).

EXAMPLE 14.1.3. We can now prove the converse of Example 3.2.2(1), namely, that every field which is not algebraic over a finite field carries a non-trivial valuation. Indeed, such a field contains a copy of either $\mathbb{Q}$ or $\mathbb{F}_p(t)$, where t is transcendental and p is prime. This copy carries non-trivial valuations (see Examples 4.1.2 and 4.1.3), that we can extend to E using Corollary 14.1.2.

14.2 Valuations in algebraic extensions

Let $(E, u)/(F, v)$ be a valued field extension. Recall from §3.2 that then $v(F^\times) \cong F^\times/O_v^\times$ embeds canonically in $u(E^\times) \cong E^\times/O_u^\times$ as ordered abelian groups, and $\bar{F}_v \cong O_v/\mathfrak{m}_v$ embeds canonically as a subfield of $\bar{E}_u = O_u/\mathfrak{m}_u$. One calls

$$e(u/v) = (u(E^\times) : v(F^\times))$$

the **ramification index** of the extension, and

$$f(u/v) = [\bar{E}_u : \bar{F}_v]$$

its **inertia degree**. Of course, here $e(u/v)$ and $f(u/v)$ can be infinite.

The following theorem is a special case of the so-called "fundamental inequality of valuation theory", which will be proved in its full generality in §17.1.

THEOREM 14.2.1. *Let $(E, u)/(F, v)$ be a valued field extension. Then*

$$e(u/v)f(u/v) \leq [E : F].$$

PROOF. Choose $x_1, \dots, x_e \in E^\times$ such that the cosets of $u(x_1), \dots, u(x_e)$ in $u(E^\times)/v(F^\times)$ are distinct. Choose $y_1, \dots, y_f \in O_u^\times$ whose residues $\bar{y}_1, \dots, \bar{y}_f$ in $\bar{E}_u$ are linearly independent over $\bar{F}_v$. It suffices to show that the ef elements

$$x_i y_j, \quad i = 1, \dots, e, \quad j = 1, \dots, f,$$

of E are linearly independent over F.

Suppose that this is not the case, i.e., $\sum_{i=1}^{e} \sum_{j=1}^{f} c_{ij} x_i y_j = 0$ for some $c_{ij} \in F$ which are not all zero. Without loss of generality, $u(c_{11}x_1) \leq u(c_{ij}x_i)$ for all $1 \leq i \leq e$, $1 \leq j \leq f$. It follows from the choice of $x_1, \dots, x_e$ that

$$u(c_{11}x_1) < u(c_{ij}x_i), \quad i = 2, \dots, e, \quad j = 1, \dots, f.$$

In particular, $c_{11}x_1 \neq 0$. Putting $z_{ij} = c_{ij}x_i/c_{11}x_1$, we obtain that

$$\sum_{i=1}^{e} \sum_{j=1}^{f} z_{ij} y_j = 0.$$

Now $z_{1j} \in F$ for $1 \leq j \leq f$. Furthermore, $z_{ij} \in O_u$ for all i, j. Hence $z_{ij} y_j \in O_u$, and when $2 \leq i$ one even has $z_{ij} y_j \in \mathfrak{m}_u$. For the residues in $\bar{E}_u$ we therefore have $\sum_{j=1}^{f} \overline{z_{1j}}\, \overline{y_j} = 0$. The choice of $y_1, \dots, y_f$ now forces that $\overline{z_{1j}} = 0$ for all j. However, $z_{11} = 1$, and we got a contradiction. □

COROLLARY 14.2.2. *Let $(E, u)/(F, v)$ be a finite extension of valued fields. Then v is discrete if and only if u is discrete.*

PROOF. By Theorem 14.2.1, $e(u/v) < \infty$. Also, by Lemma 2.1.1(c), $u(E^\times)$ is torsion-free. Now apply Proposition 2.4.1. □

COROLLARY 14.2.3. *Let $(E, u)/(F, v)$ be a valued field extension with E algebraic over F. Then:*

(a) *$u(E^\times)/v(F^\times)$ is a torsion group;*
(b) *$u(E^\times)$ embeds over $v(F^\times)$ in the divisible hull $v(F^\times)_{\mathrm{div}}$ (see §1.2);*
(c) *$\mathrm{rank}(u) = \mathrm{rank}(v)$;*
(d) *$\bar{E}_u$ is an algebraic extension of $\bar{F}_v$;*
(e) *$v(F^\times)$ is cofinal in $u(E^\times)$.*

PROOF. (a), (d) Write $E = \bigcup_{i \in I} E_i$, for some family of finite extensions E_i of F. For every $i \in I$ let u_i be the restriction of u to E_i. Theorem 14.2.1 implies that $u_i(E_i^\times)/v(F^\times)$ is a finite group, and $\overline{(E_i)}_{u_i}/\bar{F}_v$ is a finite field extension. Now $u(E^\times) = \varinjlim u_i(E_i^\times)$ and $\bar{E}_u = \varinjlim \overline{(E_i)}_{u_i}$. This gives both assertions.

(b) Since $u(E^\times)$ is torsion-free, this follows from (a) and from the first universal property of the divisible hull, as in Theorem 1.2.6(b).

(c) Use (a) and Proposition 2.2.4(b).

(e) Use (a) and Lemma 2.4.4. □

In light of Corollary 14.2.3(a)(d), we may refine the definition of the ramification index and the inertia degree for an *algebraic* extension $(E, u)/(F, v)$ of valued fields by setting $e(u/v)$ to be the supernatural order of the torsion group $u(E^\times)/v(F^\times)$, and by setting $f(u/v)$ to be the supernatural degree $[\bar{E}_u : \bar{F}_v]$ (see

§13). From the tower properties for indices of subgroups and for degrees of field extensions we deduce:

PROPOSITION 14.2.4. *Let* $(F,v) \subseteq (K,w) \subseteq (E,u)$ *be a tower of algebraic extensions of valued fields. Then:*
(a) $e(u/v) = e(u/w) \cdot e(w/v)$;
(b) $f(u/v) = f(u/w) \cdot f(w/v)$.

The case of purely inseparable extensions of valued fields is relatively simple:

PROPOSITION 14.2.5. *Suppose that* E *is a purely inseparable algebraic extension of a field* F *of characteristic* $p > 0$ *and let* v *be a valuation on* F. *Then:*
(a) v *has a unique extension* u *to* E;
(b) $u(E^\times)/v(F^\times)$ *is a p-primary group;*
(c) $\bar{E}_u/\bar{F}_v$ *is a purely inseparable extension.*

PROOF. (a), (b) The existence of u was shown in general in Corollary 14.1.2.
For the uniqueness, let u be any extension of v to E. We may identify $u(E^\times)$ as a subgroup of $v(F^\times)_{\mathrm{div}}$ (Corollary 14.2.3(b)). Given $a \in E^\times$, there exists a positive integer m such that $a^{p^m} \in F$. Then $p^m u(a) = v(a^{p^m}) \in v(F^\times)$. Since $v(F^\times)_{\mathrm{div}}$ is torsion-free (Proposition 1.2.2(d)), this determines $u(a)$ uniquely, and also proves (b).

(c) Let $\bar{a} \in \bar{E}_u$. Choose $a \in O_u$ with residue $\bar{a}$. Again, $a^{p^m} \in F$ for some $m > 0$, so $\bar{a}^{p^m} \in \bar{F}_v$. □

Given a field extension E/F, the restriction of valuations from E to F is clearly a monotonic map with respect to the coarsening partial order; namely, if u_1, u_2 are valuations on E with $O_{u_1} \subseteq O_{u_2}$, then their restrictions v_1, v_2, respectively, to F satisfy $O_{v_1} \subseteq O_{v_2}$. When E/F is algebraic, this map is even *strictly* monotonic on the equivalence classes of valuations, as our next proposition shows.

PROPOSITION 14.2.6. *Let* E/F *be an algebraic extension. Let* u_1, u_2 *be comparable valuations on* E *whose restrictions* v_1, v_2, *respectively, to* F *are equivalent. Then* u_1 *and* u_2 *are also equivalent.*

PROOF. Suppose for instance that u_1 is finer than u_2, i.e., $O_{u_1}^\times \leq O_{u_2}^\times$. By assumption, $O_{v_1}^\times = O_{v_2}^\times$. Equip the abelian groups

$$\Gamma = F^\times/O_{v_1}^\times = F^\times/O_{v_2}^\times, \quad \Gamma_1 = E^\times/O_{u_1}^\times, \quad \Gamma_2 = E^\times/O_{u_2}^\times$$

with the natural orders, as in Proposition 3.2.3. For $i = 1, 2$, Γ_i is torsion-free (Lemma 2.1.1(c)) and Γ_i/Γ is a torsion group (Corollary 14.2.3(a)). The projection $\Gamma_1 \to \Gamma_2$ is the identity on Γ. Lemma 1.1.2(a) therefore implies that it is also injective. Hence $O_{u_1}^\times = O_{u_2}^\times$, and therefore $O_{u_1} = O_{u_2}$ (see §3.1). Thus u_1, u_2 are equivalent. □

14.3 The Galois action

Suppose that E/F is a normal extension and that v is a valuation on F. Then the automorphism group $\mathrm{Aut}(E/F)$ acts from the right on the set of all extensions u of v to E by

$$\sigma \in \mathrm{Aut}(E/F) \quad \mapsto \quad u \circ \sigma.$$

PROPOSITION 14.3.1. *Let $(E, u)/(F, v)$ be a normal extension of valued fields and let $\sigma \in \mathrm{Aut}(E/F)$. The following conditions are equivalent:*
(a) *u and $u \circ \sigma$ are comparable;*
(b) *u and $u \circ \sigma$ are equivalent;*
(c) *$u = u \circ \sigma$.*

PROOF. (c)⇒(a): Trivial.

(a)⇒(b): This follows from Proposition 14.2.6.

(b)⇒(c): Consider $\Gamma = F^\times/O_v^\times$ and $\Gamma' = E^\times/O_u^\times$ as ordered abelian groups in the sense of Proposition 3.2.3. By (b), $O_u = O_{u\circ\sigma} = \sigma^{-1}(O_u)$. Hence σ induces a Γ-automorphism of Γ' as an ordered abelian group. Now Γ'/Γ is a torsion group (Corollary 14.2.3(a)) and Γ' is torsion-free (Lemma 2.1.1(c)). Lemma 1.1.2(c) therefore implies that the above-mentioned automorphism is the identity. Thus $aO_u^\times = \sigma(a)O_u^\times$ for all $a \in E^\times$. This gives (c). □

We can now show that the action described above is transitive.

THEOREM 14.3.2. *Let E/F be a normal algebraic extension of fields. Let v be a valuation on F and let u_1, u_2 be valuations on E extending v. Then there exists $\sigma \in \mathrm{Aut}(E/F)$ such that u_1 and $u_2 \circ \sigma$ are equivalent.*

PROOF. Let F' be the maximal purely inseparable extension of F inside E. By Proposition 14.2.5(a), v has a unique extension v' to F', so u_1, u_2 extend v'. We may therefore replace (F, v) by (F', v') to assume that E/F is separable, whence Galois. Also, a standard inverse limit argument reduces the theorem to the case where the extension E/F is finite.

So assume this, let $G = \mathrm{Gal}(E/F)$, and suppose that there is no σ as in the assertion. Then for every $\sigma, \tau \in G$, the valuations $u_1 \circ \sigma$, $u_2 \circ \tau$ are not equivalent. By Proposition 14.2.6 they are even incomparable. The approximation theorem for incomparable valuations (Theorem 10.2.1) therefore yields $a \in F$ such that $(u_1 \circ \sigma)(a) > 0$ for every $\sigma \in G$, and $(u_2 \circ \tau)(a - 1) > 0$ for every $\tau \in G$.

Now consider the norm $N_{E/F}(a) = \prod_{\sigma\in G} \sigma(a) \in F$. On one hand,

$$v(N_{E/F}(a)) = u_1(N_{E/F}(a)) = \sum_{\sigma\in G} (u_1 \circ \sigma)(a) > 0,$$

while on the other hand

$$v(N_{E/F}(a)) = u_2(N_{E/F}(a)) = \sum_{\tau\in G} (u_2 \circ \tau)(a) = 0,$$

a contradiction. □

We further observe:

PROPOSITION 14.3.3. *Suppose that E/F is a normal extension of fields, v is a valuation on F, and u, u' are extensions of v to E. Then:*
(a) *$u(E^\times)$ and $u'(E^\times)$ are $v(F^\times)$-isomorphic;*
(b) *$\bar{E}_u$ and $\bar{E}_{u'}$ are isomorphic over $\bar{F}_v$;*
(c) *$e(u/v) = e(u'/v)$;*
(d) *$f(u/v) = f(u'/v)$.*

PROOF. Theorem 14.3.2 yields $\sigma \in \mathrm{Aut}(E/F)$ such that u' and $u \circ \sigma$ are equivalent. Then

$$O_{u'} = O_{u\circ\sigma} = \sigma^{-1}(O_u),\ \mathfrak{m}_{u'} = \mathfrak{m}_{u\circ\sigma} = \sigma^{-1}(\mathfrak{m}_u),\ O_{u'}^\times = O_{u\circ\sigma}^\times = \sigma^{-1}(O_u^\times).$$

The map σ^{-1} induces an isomorphism

$$E^\times/O_u^\times \xrightarrow{\sim} E^\times/O_{u'}^\times, \quad aO_u^\times \mapsto \sigma^{-1}(a)O_{u'}^\times$$

which is the identity on $F^\times/O_v^\times$, proving (a).

Similarly, σ^{-1} induces an isomorphism

$$\bar{E}_u \xrightarrow{\sim} \bar{E}_{u'}, \quad \pi_u(a) \mapsto \pi_{u'}(\sigma^{-1}(a)),$$

which is the identity on $\bar{F}_v$. This gives (b).

Parts (c) and (d) follow immediately from parts (a) and (b), respectively. □

CHAPTER 15

Decomposition Groups

For a fixed normal extension E/F and to a given valuation on F we associate a subgroup of $\mathrm{Aut}(E/F)$ called the decomposition group (defined up to conjugation). When the extension is Galois, the fixed field of this group in E is an immediate extension of F, called the relative Henselization. It plays a role similar to that of the completion in the rank 1 case. However, its algebraic nature makes it very useful in the Galois-theoretic context.

15.1 Definition and basic properties

Throughout this section we fix a normal extension E/F. Consider a valuation v on F and a valuation u on E extending F. Let $Z(u/v)$ be the stabilizer of u under the action of $\mathrm{Aut}(E/F)$, i.e.,

$$Z(u/v) = \{\sigma \in \mathrm{Aut}(E/F) \mid u \circ \sigma = u\}.$$

The group $Z(u/v)$ is a closed subgroup of $\mathrm{Aut}(E/F)$ with respect to the Krull topology. Indeed, take $\sigma \in \mathrm{Aut}(E/F) \setminus Z(u/v)$. Thus there exists $a \in E$ such that $u(\sigma(a)) \neq u(a)$. Let E_1 be a finite normal extension of F inside E that contains a. Then the open neighborhood $\sigma\mathrm{Aut}(E/E_1)$ of σ is contained in $\mathrm{Aut}(E/F)\setminus Z(u/v)$, showing that the latter set is open, as required.

We call $Z(u/v)$ the **decomposition group** of the extension $(E,u)/(F,v)$. We call the fixed field E_Z of $Z(u/v)$ in E the **decomposition field** of this extension. Then E is a Galois extension of E_Z. Let u_Z be the restriction of u to E_Z.

15.1.1 REMARKS.

(a) Let (F_1, v_1) be a valued field with $F \subseteq F_1 \subseteq E$ and where v_1 is the restriction of u to F_1. Then

$$Z(u/v_1) = \mathrm{Aut}(E/F_1) \cap Z(u/v).$$

(b) For $\tau \in \mathrm{Aut}(E/F)$ one has

$$Z(u\circ\tau/v) = \tau^{-1}Z(u/v)\tau.$$

(c) Letting E_1 range over all finite normal extensions of F inside E and denoting the restriction of u to E_1 by u_1, we have $Z(u/v) = \varprojlim Z(u_1/v)$ as profinite groups.

(d) If u, u' are equivalent valuations on E and v, v', respectively, are their restrictions to F, then $Z(u/v) = Z(u'/v')$.

(e) If E_1/F is a normal subextension of E/F with $E_1 \subseteq E$ and if u_1 is the restriction of u to E_1, then $\mathrm{Res}_{E_1} Z(u/v) = Z(u_1/v)$. Indeed, the left-hand side is clearly contained in the right-hand side. Conversely, given $\bar\sigma \in Z(u_1/v)$ we lift it to some $\sigma \in \mathrm{Aut}(E/F)$. Then $u, u\circ\sigma$ have the same restriction u_1 to E_1, so by Theorem 14.3.2 and Proposition 14.3.1, $u = u \circ \sigma \circ \tau$ for some $\tau \in \mathrm{Aut}(E/E_1)$. Thus $\sigma \circ \tau \in Z(u/v)$ is a lifting of $\bar{\sigma}$ as required.

PROPOSITION 15.1.2. *Let E_Z and u_Z be as above. Then:*

(a) *the distinct extensions of v to E (up to equivalence) are $u \circ \sigma$, where σ ranges over a system of representatives for the right cosets of $Z(u/v)$ in $\mathrm{Aut}(E/F)$;*

(b) *u is the unique extension of u_Z to E (up to equivalence).*

PROOF. (a) Take a system of representatives Σ for the right cosets of $Z(u/v)$ in $\mathrm{Aut}(E/F)$. Let u' be an extension of v to E. Theorem 14.3.2 yields $\tau \in \mathrm{Gal}(E/F)$ such that u' is equivalent to $u \circ \tau$. Choose $\sigma \in \Sigma$ for which $\tau \in Z(u/v)\sigma$. Therefore $u \circ \tau = u \circ \sigma$.

Next suppose that $u \circ \sigma$ and $u \circ \sigma'$ are equivalent for some $\sigma, \sigma' \in \Sigma$. By Proposition 14.3.1 they are equal. Thus $u \circ \sigma \circ (\sigma')^{-1} = u$, so $\sigma \circ (\sigma')^{-1} \in Z(u/v)$. Hence $\sigma = \sigma'$.

(b) The extension $(E, u)/(E_Z, u_Z)$ is Galois, and its decomposition group is $Z(u/v) = \mathrm{Gal}(E/E_Z)$ (by Remark 15.1.1(a)). Now apply (a) with respect to this extension. □

15.2 Immediateness of decomposition fields

As in the previous section, we assume that $(E, u)/(F, v)$ is a normal extension of valued fields, with decomposition field (E_Z, u_Z). To study the valuation u_Z we first need the following technical lemma. It will also play a crucial role at several points in the sequel.

LEMMA 15.2.1. *For every $a_1 \in O_{u_Z}$ and $a_2 \in O_v$ there exists $b \in O_{u_Z}$ such that:*

(i) $u(a_1 - b) > 0$;

(ii) $(u \circ \sigma)(a_2 - b) > 0$ *whenever* $\sigma \in \mathrm{Aut}(E/F) \setminus Z(u/v)$.

PROOF. Let $w_1, \ldots, w_k$ be a system of representatives for the equivalence classes of the extensions of v to E_Z. By Proposition 14.2.6, they are incomparable. We may assume that $u_Z = w_1$. The approximation theorem for incomparable valuations (Theorem 10.2.1) yields $b \in O_{w_1} \cap \cdots \cap O_{w_k} \subseteq E_Z$ such that:

(1) a_1 and b have the same residue with respect to w_1;

(2) a_2 and b have the same residue with respect to $w_2, \ldots, w_k$.

Then (i) is clear.

For (ii), let $\sigma \in \mathrm{Aut}(E/F) \setminus Z(u/v)$. By Proposition 14.3.1, $O_{u\circ\sigma} \neq O_u$. By Proposition 15.1.2(b), O_u is the only valuation ring on E extending O_{w_1}. In particular, $O_{u\circ\sigma} \cap E_Z \neq O_{w_1}$, so $O_{u\circ\sigma} \cap E_Z = O_{w_i}$ for some $i \geq 2$. Since a_2 and b have the same residue under w_i, they have the same residue also under $u \circ \sigma$. Thus $(u \circ \sigma)(a_2 - b) > 0$. □

THEOREM 15.2.2. *Suppose E/F is Galois. Then the extension $(E_Z, u_Z)/(F, v)$ is immediate.*

PROOF. We follow an argument of Ax [**Ax**]. In view of Remark 15.1.1(c), we may assume that the extension E/F is finite.

PART I: *The residue fields of v and u_Z coincide.*

To prove this we take $a \in O_{u_Z}$ and show that its residue under u is already in $\bar{F}_v$. Indeed, Lemma 15.2.1 yields $b \in O_{u_Z}$ such that:

(i) $u(a-b) > 0$;

(ii) $(u \circ \sigma)(1-b) > 0$ whenever $\sigma \in \mathrm{Gal}(E/F) \setminus Z(u/v)$.

By (i), a, b have the same residue in $\bar{E}_u$. Also, if $\sigma \in \mathrm{Gal}(E/F)$ and $\sigma(b) \neq b$, then $\sigma \notin Z(u/v)$, so by (ii), $\sigma(b)$ is in O_u and its residue in $\bar{E}_u$ is 1. Consequently, the norm $N_{F(b)/F}(b)$ is in $F \cap O_u = O_v$ and has the same residue as a, as desired.

PART II: *The value groups of v and u_Z coincide.*

To prove this we take $a \in E_Z^\times$ and show that $u(a) \in v(F^\times)$. Lemma 15.2.1 again yields $b \in E_Z^\times$ with $u(b) > 0$ and $(u \circ \sigma)(1-b) > 0$ whenever $\sigma \in \mathrm{Gal}(E/F) \setminus Z(u/v)$. Then $u(a) = u((1-b)^n a)$ for all n, while for $\sigma \in \mathrm{Gal}(E/F) \setminus Z(u/v)$ the infinite sequence

$$(u \circ \sigma)((1-b)^n a), \quad n = 1, 2, \ldots$$

is strictly increasing. Since there are only finitely many such σ, there exists a positive integer n such that

$$u((1-b)^n a) \neq (u \circ \sigma)((1-b)^n a)$$

for all $\sigma \in \mathrm{Gal}(E/F) \setminus Z(u/v)$. Set

$$c = (1-b)^n a.$$

Next we choose representatives $\sigma_1, \ldots, \sigma_m$ for the left cosets of $Z(u/v)$ in $\mathrm{Gal}(E/F)$. These are the distinct F-embeddings of E_Z into E. After re-ordering, we may assume that the sequence $(u \circ \sigma_1)(c), \ldots, (u \circ \sigma_m)(c)$ is non-decreasing, and that $\sigma_k \in Z(u/v)$. For every $i \neq k$ one has $\sigma_i \notin Z(u/v)$, so by construction, $u(c) \neq (u \circ \sigma_i)(c)$. Thus

$$(u \circ \sigma_1)(c) \leq \cdots \leq (u \circ \sigma_{k-1})(c) < u(c) < (u \circ \sigma_{k+1})(c) \leq \cdots \leq (u \circ \sigma_m)(c).$$

Let

$$d = \sum_{1 \leq i_1 < \cdots < i_k \leq m} \sigma_{i_1}(c) \cdots \sigma_{i_k}(c), \quad e = \sum_{1 \leq i_1 < \cdots < i_{k-1} \leq m} \sigma_{i_1}(c) \cdots \sigma_{i_{k-1}}(c).$$

Using Lemma 3.2.1(c) we get

$$u(d) = u(\sigma_1(c) \cdots \sigma_k(c)), \quad u(e) = u(\sigma_1(c) \cdots \sigma_{k-1}(c)).$$

Moreover $d, e \in F$. We conclude that

$$u(a) = u(c) = u(\sigma_k(c)) = u(d/e) \in v(F^\times). \qquad \square$$

More generally, we may drop the separability assumption in Theorem 15.2.2 to obtain the following slightly weaker statement:

COROLLARY 15.2.3. *Let $(E,u)/(F,v)$ be a normal extension of valued fields and assume that $p = \mathrm{char}\, F > 0$. Then:*

(a) *$u_Z(E_Z^\times)/v(F^\times)$ is p-primary;*

(b) *the field extension $\overline{(E_Z)}_{u_Z}/\bar{F}_v$ is purely inseparable.*

PROOF. Let F' be the maximal purely inseparable extension of F inside E and let v' be the restriction of u to F'. Then E/F' is Galois and $F' \subseteq E_Z$. Moreover, the decomposition field of the extension $(E,u)/(F',v')$ is also (E_Z,u_Z), by Remark 15.1.1(a). By Theorem 15.2.2, $(E_Z,u_Z)/(F',v')$ is an immediate extension. Moreover, by Proposition 14.2.5 $v'((F')^\times)/v(F^\times)$ is a p-primary group, and the extension $\bar{F}'_{v'}/\bar{F}_v$ is purely inseparable. This gives both assertions. □

As an application of Corollary 15.2.3 we prove that the normality of a valued field extension implies the normality of the residue field extension. The proof will be based also on the following lemma.

LEMMA 15.2.4. *Let $(E,u)/(F,v)$ be a normal extension of valued fields. Let $a \in O_u$, let $\bar{a}$ be the residue of a in $\bar{E}_u$, and let $f(X) = \mathrm{irr}(a, E_Z)$ be the irreducible polynomial of a over E_Z. Then:*

(a) *the roots of $f(X)$ are in O_u;*
(b) *$f(X) \in O_{u_Z}[X]$;*
(c) *$\mathrm{irr}(\bar{a},\bar{F}_v)$ divides $\bar{f}(X)^q$ in $\bar{E}_u[X]$, where $\bar{f}(X) \in \bar{E}_u[X]$ is the residue polynomial of $f(X)$ and q is a p-power if $p = \mathrm{char}\,\bar{F}_v > 0$ and is 1 otherwise.*

PROOF. (a) Since E/E_Z is a normal extension, $f(X)$ splits as

$$f(X) = \prod_{i=1}^{d} (X - \sigma_i(a))$$

for some $\sigma_i \in \mathrm{Gal}(E/E_Z) = Z(u/v)$. Then $(u \circ \sigma_i)(a) = u(a) \geq 0$, i.e., $\sigma_i(a) \in O_u$, $i = 1, \ldots, d$.

(b) follows from (a).

(c) By Corollary 15.2.3(b), $\overline{(E_Z)}_{u_Z}/\bar{F}_v$ is a purely inseparable extension. By (b), $\bar{f}(X) \in \overline{(E_Z)}_{u_Z}[X]$, so $\bar{f}(X)^q \in \bar{F}_v[X]$, with q as above. Since $f(a) = 0$, also $\bar{f}(\bar{a})^q = 0$, and therefore $\mathrm{irr}(\bar{a},\bar{F}_v)|\bar{f}(X)^q$. □

COROLLARY 15.2.5. *If $(E,u)/(F,v)$ is a normal extension of valued fields, then the residue field extension $\bar{E}_u/\bar{F}_v$ is also normal.*

PROOF. Let $\bar{a} \in \bar{E}_u$. Choose $a \in O_u$ with residue $\bar{a}$ and let $f(X) = \mathrm{irr}(a, E_Z)$. Let $\bar{f}(X)$ and q be as in Lemma 15.2.4. Since $f(X)$ splits into linear factors in $O_u[X]$, the polynomial $\bar{f}(X)$ splits into linear factors in $\bar{E}_u[X]$. Hence $\mathrm{irr}(\bar{a},\bar{F}_v)$ also splits in $\bar{E}_u[X]$, as required. □

15.3 Relatively Henselian fields

As before, we consider a normal extension E/F of fields.

PROPOSITION 15.3.1. *The following conditions on a valuation v on F are equivalent:*

(a) *v has a unique extension to a valuation on E (up to equivalence);*
(b) *$\mathrm{Aut}(E/F) = Z(u/v)$ for every extension u of v to E;*
(c) *$\mathrm{Aut}(E/F) = Z(u/v)$ for some extension u of v to E.*

PROOF. (a)⇒(b): Let u be the unique extension of v to E (up to equivalence). If $\sigma \in \mathrm{Aut}(E/F)$, then $u \circ \sigma$ must be equivalent to u. By Proposition 14.3.1, $u = u \circ \sigma$, i.e., $\sigma \in Z(u/v)$. This proves that $\mathrm{Aut}(E/F) = Z(u/v)$.

(b)⇒(c): Just note that v extends to E, by Corollary 14.1.2.

(c)⇒(a): Take u as in (c). By Theorem 14.3.2, every other extension of v to E is equivalent to $u \circ \sigma$ for some $\sigma \in \mathrm{Aut}(E/F)$. Then, by (c), $u = u \circ \sigma$. □

When the equivalent conditions of Proposition 15.3.1 are satisfied we say that the valuation v on F is **Henselian relative to** E (or that v is **indecomposed** in E). The origin of this terminology is *Hensel's lemma*, which will be discussed in §18. Since valuations always extend in field extensions, we obtain from condition (a):

COROLLARY 15.3.2. *Let $(E, u)/(F, v)$ be a normal extension of valued fields. Let $F \subseteq F_1 \subseteq E_1 \subseteq E$ be intermediate fields with E_1/F_1 normal, and let v_1 be the restriction of u to F_1. If v is Henselian relative to E, then v_1 is Henselian relative to E_1.*

Another easy consequence of the definition is:

PROPOSITION 15.3.3. *Let $(E, u)/(F, v)$ and $(E', u')/(F', v')$ be normal extensions of valued fields such that (E', u') extends (E, u) and (F', v') extends (F, v). Assume that $E \cap F'$ is a purely inseparable extension of F. If v' is Henselian relative to E', then v is Henselian relative to E.*

PROOF. By Galois theory, the restriction homomorphism

$$\mathrm{Aut}(E'/F') \to \mathrm{Aut}(E/F)$$

is surjective. Clearly, it maps $Z(u'/v')$ into $Z(u/v)$. Since $\mathrm{Aut}(E'/F') = Z(u'/v')$ this gives $\mathrm{Aut}(E/F) = Z(u/v)$. □

REMARK 15.3.4. Suppose that $(E, u)/(F, v)$ is a normal extension of valued fields and that v is Henselian relative to E. Then E_Z is the fixed field of $\mathrm{Aut}(E/F)$ in E, i.e., it is the maximal inseparable extension of F inside E.

Our next goal is to construct *minimal* Henselian extensions of valued fields relative to a given normal extension.

Let (L_1, w_1) and (L_2, w_2) be extensions of the valued field (F, v). An **embedding** of (L_1, w_1) into (L_2, w_2) over (F, v) is a field embedding $\sigma\colon L_1 \to L_2$ such that $\sigma(O_{w_1}) = O_{w_2}$ and σ is the identity on F.

THEOREM 15.3.5. *Let E/F be a normal extension of fields and let v be a valuation on F. There is a separable valued field extension $(\hat{F}, \hat{v})/(F, v)$ contained in E such that:*

(i) *$(\hat{F}, \hat{v})$ is Henselian relative to E;*
(ii) *for every normal extension E'/F' with $F \subseteq F'$ and $E \subseteq E'$, and for every valuation v' on F' extending v and which is Henselian relative to E' there exists an embedding of $(\hat{F}, \hat{v})$ into (F', v') over (F, v).*

Furthermore, $(\hat{F}, \hat{v})$ is the decomposition field of some extension of v to the maximal separable extension F_s of F inside E.

PROOF. Since valuations extend uniquely in purely inseparable extensions (by Proposition 14.2.5), we may replace E by F_s, to assume that E/F is Galois.

We choose an extension u of v to E (Corollary 14.1.2), and let $(\hat{F}, \hat{v}) = (E_Z, u_Z)$. By definition, it is Henselian relative to E, i.e., (i) holds.

Now let $(E', u')/(F', v')$ be a normal extension as in (ii) (where u' is the unique extension of v' to E'). Let $F_1 = E \cap F'$, let v_1 be the restriction of v' to F_1, and let u_1 be the restriction of u' to E (see diagram below). By Proposition 15.3.3, v_1 is Henselian relative to E.

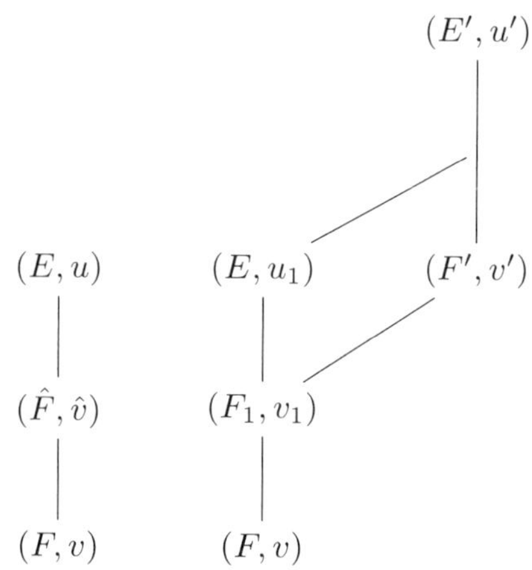

Since both u and u_1 extend v, Theorem 14.3.2 yields $\sigma \in \mathrm{Gal}(E/F)$ such that u and $u_1 \circ \sigma$ are equivalent. Thus $O_{u_1} = O_{u \circ \sigma^{-1}} = \sigma(O_u)$. By Remarks 15.1.1(a)(b)(d),

$$\sigma Z(u/v)\sigma^{-1} = Z(u \circ \sigma^{-1}/v) = Z(u_1/v) \geq Z(u_1/v_1) = \mathrm{Aut}(E/F_1).$$

Taking fixed fields in E, we obtain that $\sigma(\hat{F}) \subseteq F_1$. Furthermore, σ maps $O_{\hat{v}} = O_u \cap \hat{F}$ into

$$O_{v_1} = O_{u_1} \cap F_1 \subseteq O_{u'} \cap F' = O_{v'}.$$

We conclude that it embeds $(\hat{F}, \hat{v})$ in (F', v') over (F, v). □

An extension $(\hat{F}, \hat{v})$ of (F, v) as in Theorem 15.3.5 is called an **E-Henselization** of (F, v). From the Theorem we obtain in a standard way:

COROLLARY 15.3.6. *Let (F, v) be a valued field and let E be a normal extension of F. Any two E-Henselizations of (F, v) are isomorphic over (F, v).*

From the last sentence of Theorem 15.3.5 and from Theorem 15.2.2 we deduce:

PROPOSITION 15.3.7. *In the above setup, the E-Henselization $(\hat{F}, \hat{v})$ is an immediate extension of (F, v).*

The notions of Henselity relative to E and of E-Henselizations are especially useful when E is a *canonical* normal extension of F. We list a few important cases.

15.3.8 EXAMPLES.

(1) Suppose that E is the algebraic closure of F. Then an E-Henselization (resp., a Henselian valuation relative to E) is simply called a **Henselization** of F (resp., a **Henselian** valuation). Alternatively, here we could have taken E to be the separable closure F_{sep} of F (by Proposition 14.2.5 again).

(2) Let $E = F_{\mathrm{sol}}$ be the solvable closure of F, i.e., the compositum of all finite Galois extensions of F with solvable Galois group. Then an E-Henselization (resp., a Henselian valuation relative to E) is called a **solvable Henselization** of F (resp., a **solvably Henselian** valuation).

(3) Let p be a prime number and let $E = F(p)$ be the maximal pro-p Galois extension of F, i.e., the compositum of all finite Galois extensions of F of p-power order. Then an E-Henselization of a valuation on F (resp., a Henselian valuation relative to E) is called a **p-Henselization** of F (resp., a **p-Henselian** valuation).

CHAPTER 16

Ramification Theory

We now go deeper into the structure of decomposition groups, and investigate two arithmetically significant subgroups of such groups: the inertia group, and the (first) ramification group.

16.1 Inertia groups

Assume that $(E,u)/(F,v)$ is a fixed normal extension of valued fields. We define

$$T(u/v) = \Big\{ \sigma \in \mathrm{Aut}(E/F) \quad \Big| \quad \forall a \in O_u \ : \quad u(\sigma(a) - a) > 0 \Big\}.$$

We consider again $\mathrm{Aut}(E/F)$ and $\mathrm{Aut}(\bar{E}_u/\bar{F}_v)$ as profinite groups with respect to the Krull topologies.

THEOREM 16.1.1. *$T(u/v)$ is a normal closed subgroup of $Z(u/v)$, and there is a canonical short exact sequence of continuous homomorphisms of profinite groups*

$$1 \ \to \ T(u/v) \ \hookrightarrow \ Z(u/v) \ \xrightarrow{\Phi} \ \mathrm{Aut}(\bar{E}_u/\bar{F}_v) \ \to \ 1.$$

More specifically, Φ is given by $(\Phi(\sigma))(\bar{a}) = \overline{\sigma(a)}$ for $\sigma \in Z(u/v)$ and $a \in O_u$ with residue $\bar{a}$.

PROOF. Let $\sigma \in T(u/v)$. For every $a \in O_u$ the definition of $T(u/v)$ and the ultrametric inequality imply that $(u \circ \sigma)(a) \geq 0$. Thus u is finer than $u \circ \sigma$. We conclude from Proposition 14.3.1 that these valuations are in fact equal. This proves that $T(u/v) \leq Z(u/v)$.

Now define the map $\Phi\colon Z(u/v) \to \mathrm{Aut}(\bar{E}_u/\bar{F}_v)$ as above. To see that it is well defined, suppose that $\sigma \in Z(u/v)$ and that $a, a' \in O_u$ have the same residue $\bar{a}$ in $\bar{E}_u$. Since $u \circ \sigma = u$ we have

$$u(\sigma(a) - \sigma(a')) = (u \circ \sigma)(a - a') = u(a - a') > 0.$$

Thus $\overline{\sigma(a)} = \overline{\sigma(a')}$.

It is straightforward to see that Φ is a continuous group homomorphism and that its kernel is $T(u/v)$. In particular, $T(u/v)$ is a normal closed subgroup of $Z(u/v)$.

Finally, we show that Φ is surjective. An inverse limit argument reduces this to the case where E/F is finite. Let $\bar{a}$ be a primitive element for the maximal separable extension $\bar{F}'$ of $\bar{F}_v$ inside $\bar{E}_u$, and lift it to some $a \in O_u$. Let $f(X) = \mathrm{irr}(a, E_Z)$. By Lemma 15.2.4, $f(X) \in O_{u_Z}[X]$ and $\mathrm{irr}(\bar{a}, \bar{F}_v)$ divides $\bar{f}(X)^q$, where $\bar{f}(X) \in \bar{E}_u[X]$ is the residue polynomial of $f(X)$ and q is a positive integer. Hence, for every $\bar{\tau} \in \mathrm{Aut}(\bar{E}_u/\bar{F}_v)$ we have

$$X - \bar{\tau}(\bar{a}) \mid \mathrm{irr}(\bar{a}, \bar{F}_v) \mid \bar{f}(X)^q$$

in $\bar{E}_u[X]$. We conclude that $\bar{f}(\bar{\tau}(\bar{a})) = 0$, so

$$\bar{\tau}(\bar{a}) = \overline{\sigma(a)} = (\Phi(\sigma))(\bar{a})$$

for some $\sigma \in Z(u/v)$. Therefore $\bar{\tau}$ and $\Phi(\sigma)$ coincide on $\bar{F}'$. Since the restriction homomorphism $\mathrm{Aut}(\bar{E}_u/\bar{F}_v) \to \mathrm{Gal}(\bar{F}'/\bar{F}_v)$ is an isomorphism, they coincide on all of $\bar{E}_u$. □

We call $T(u/v)$ the **inertia group** of the extension $(E,u)/(F,v)$. The fixed field E_T of $T(u/v)$ in E is called the **inertia field** of this extension. Notice that the extensions E/E_T and E_T/E_Z are Galois. We denote the restriction of u to E_T by u_T.

16.1.2 REMARKS.
(a) Let (F_1, v_1) be a valued field with $F \subseteq F_1 \subseteq E$ and where v_1 is the restriction of u to F_1. Then

$$T(u/v_1) = \mathrm{Aut}(E/F_1) \cap T(u/v).$$

(b) For $\tau \in \mathrm{Aut}(E/F)$ one has

$$T(u \circ \tau/v) = \tau^{-1} T(u/v)\tau.$$

(c) Letting E_1 range over all finite normal extensions of F inside E and denoting the restriction of u to E_1 by u_1, we have $T(u/v) = \varprojlim T(u_1/v)$ as profinite groups.
(d) If u, u' are equivalent valuation on E and v, v' are their restrictions to F, respectively, then $T(u/v) = T(u'/v')$.

PROPOSITION 16.1.3. *In the above notation:*
(a) $\bar{E}_u/\overline{(E_T)}_{u_T}$ *is a purely inseparable extension;*
(b) $\overline{(E_T)}_{u_T}/\overline{(E_Z)}_{u_Z}$ *is a Galois extension;*
(c) $\mathrm{Gal}(E_T/E_Z) \cong \mathrm{Gal}(\overline{(E_T)}_{u_T}/\overline{(E_Z)}_{u_Z})$ *canonically;*
(d) $[E_T : E_Z] = f(u_T/u_Z)$*;*
(e) $u_T(E_T^\times) = u_Z(E_Z^\times)$*.*

PROOF. In light of Remarks 15.1.1(c) and 16.1.2(c), we may assume that the extension E/F is finite.

By Remark 16.1.2(a), $T(u/u_T) = T(u/v)$. Furthermore,

$$Z(u/u_T) \leq \mathrm{Aut}(E/E_T) = T(u/v),$$

so $T(u/u_T) = Z(u/u_T)$. Now by Theorem 16.1.1, $\mathrm{Aut}(\bar{E}_u/\overline{(E_T)}_{u_T}) = 1$. Since the extension $\bar{E}_u/\overline{(E_T)}_{u_T}$ is normal (Corollary 15.2.5), it is therefore purely inseparable, proving (a).

Since the extension E_T/E_Z is normal, so is the extension $\overline{(E_T)}_{u_T}/\overline{(E_Z)}_{u_Z}$, by Corollary 15.2.5 again. Also, since the extensions $\bar{E}_u/\overline{(E_T)}_{u_T}$ and $\overline{(E_Z)}_{u_Z}/\bar{F}_v$ are purely inseparable (by (a) and by Corollary 15.2.3(b), respectively), the groups $\mathrm{Aut}(\bar{E}_u/\bar{F}_v)$ and $\mathrm{Aut}(\overline{(E_T)}_{u_T}/\overline{(E_Z)}_{u_Z})$ are isomorphic. Using Theorem 16.1.1 and

Theorem 14.2.1 we now compute:

$$\begin{aligned}[E_T : E_Z] &= (Z(u/v) : T(u/v)) \\ &= |\mathrm{Aut}(\bar{E}_u/\bar{F}_v)| \\ &= |\mathrm{Aut}(\overline{(E_T)}_{u_T}/\overline{(E_Z)}_{u_Z})| \\ &\le [\overline{(E_T)}_{u_T} : \overline{(E_Z)}_{u_Z}] \\ &= f(u_T/u_Z) \\ &\le e(u_T/u_Z)f(u_T/u_Z) \le [E_T : E_Z].\end{aligned}$$

Consequently, these are all equalities.

As

$$|\mathrm{Aut}(\overline{(E_T)}_{u_T}/\overline{(E_Z)}_{u_Z})| = [\overline{(E_T)}_{u_T} : \overline{(E_Z)}_{u_Z}],$$

the extension $\overline{(E_T)}_{u_T}/\overline{(E_Z)}_{u_Z}$ is separable, whence (b).

For (c) recall that $\mathrm{Gal}(E_T/E_Z) \cong \mathrm{Aut}(\bar{E}_u/\bar{F}_v)$ by Theorem 16.1.1, and that the extensions $\overline{(E_Z)}_{u_Z}/\bar{F}_v$ and $\bar{E}_u/\overline{(E_T)}_{u_T}$ are purely inseparable.

Part (d) is clear.

Part (e) follows from $e(u_T/u_Z) = 1$. □

COROLLARY 16.1.4. *The residue field map gives an order-preserving bijection between the intermediate fields $E_Z \subseteq K \subseteq E_T$ and the intermediate fields*

$$\overline{(E_Z)}_{u_Z} \subseteq \bar{K} \subseteq \overline{(E_T)}_{u_T}.$$

PROOF. Use Proposition 16.1.3(c) and the Galois correspondence. □

16.2 Ramification groups

In this section we investigate in more detail the internal structure of the inertia group of a fixed normal extension $(E, u)/(F, v)$ of valued fields. We apply the techniques of §1.4, with $\bar{\mu}$ taken to be the torsion subgroup of $\bar{E}_u^\times$, i.e., the group of all roots of unity in $\bar{E}_u$. Note that it embeds as a subgroup of $\mathbb{Q}/\mathbb{Z}$.

PROPOSITION 16.2.1.

(a) *There is a well-defined pairing*

$$\langle\,\cdot\,,\,\cdot\,\rangle : T(u/v) \;\times\; u(E^\times)/v(F^\times) \to \bar{\mu},$$

given by

$$(\tau, u(a) + v(F^\times)) \mapsto \overline{\tau(a)/a}$$

for $\tau \in T(u/v)$ and $a \in E^\times$ (where the bar denotes residue in $\bar{E}_u$).

(b) *For $\sigma \in Z(u/v)$ and for τ, a as above one has*

$$\langle \sigma\tau\sigma^{-1}, u(a) + v(F^\times)\rangle = \bar{\sigma}(\langle \tau, u(a) + v(F^\times)\rangle),$$

where $\bar{\sigma} = \Phi(\sigma)$ is the image of σ in $\mathrm{Aut}(\bar{E}_u/\bar{F}_v)$ *in the sense of Theorem 16.1.1.*

PROOF. (a) Since $T(u/v) \le Z(u/v)$, for $\tau \in T(u/v)$ one has $\tau(a)/a \in O_u^\times$. Therefore we may indeed take the residue in $\bar{E}_u^\times$.

Next we verify that this residue does not depend on the choice of a. To this end take $a' \in E^\times$ with

$$u(a) + v(F^\times) = u(a') + v(F^\times).$$

Then $a' = abc$, where $b \in O_u^\times$, and $c \in F^\times$. For every $\sigma \in T(u/v)$ we have $u(\sigma(b) - b) > 0$, i.e., $\overline{\sigma(b)} = \bar{b}$. Hence

$$\overline{\frac{\sigma(a')}{a'}} = \overline{\frac{\sigma(abc)}{abc}} = \overline{\frac{\sigma(a)}{a}} \cdot \overline{\frac{\sigma(b)}{b}} = \overline{\frac{\sigma(a)}{a}},$$

as required.

Clearly, the pairing is linear on $u(E^\times)/v(F^\times)$. For the linearity on $T(u/v)$ take $\sigma, \tau \in T(u/v)$ and $a \in E^\times$. In particular, $\tau \in Z(u/v)$, so $u(a) = u(\tau(a))$. Taking in the previous argument $a' = \tau(a)$, we obtain that

$$\overline{\frac{\sigma(\tau(a))}{\tau(a)}} = \overline{\frac{\sigma(a)}{a}}.$$

It follows that

$$\overline{\frac{(\sigma\tau)(a)}{a}} = \overline{\frac{\sigma(\tau(a))}{\tau(a)}} \cdot \overline{\frac{\tau(a)}{a}} = \overline{\frac{\sigma(a)}{a}} \cdot \overline{\frac{\tau(a)}{a}},$$

as desired.

Since $u(E^\times)/v(F^\times)$ is a torsion group (Corollary 14.2.3(a)), the pairing is in fact into $(\bar{E}_u^\times)_{\mathrm{tor}} = \bar{\mu}$.

(b) We first recall that $T(u/v)$ is normal in $Z(u/v)$ (Theorem 16.1.1), so the left-hand side is well defined. By assumption $u \circ \sigma^{-1} = u$, so

$$\overline{\frac{\tau(\sigma^{-1}(a))}{\sigma^{-1}(a)}} = \langle \tau, u(\sigma^{-1}(a)) + v(F^\times)\rangle = \langle \tau, u(a) + v(F^\times)\rangle.$$

Hence

$$\begin{aligned} \langle \sigma\tau\sigma^{-1}, u(a) + v(F^\times)\rangle &= \overline{\frac{\sigma\tau\sigma^{-1}(a)}{a}} \\ &= \overline{\sigma\left(\frac{\tau(\sigma^{-1}(a))}{\sigma^{-1}(a)}\right)} \\ &= \bar{\sigma}\left(\overline{\frac{\tau(\sigma^{-1}(a))}{\sigma^{-1}(a)}}\right) \\ &= \bar{\sigma}(\langle \tau, u(a) + v(F^\times)\rangle), \end{aligned}$$

as required. □

We henceforth abbreviate (with notation being as in §1.4)

$$\chi(u/v) = \chi_{\bar{\mu}}(u(E^\times)/v(F^\times)) = \mathrm{Hom}(u(E^\times)/v(F^\times), \bar{\mu}).$$

Proposition 16.2.1 gives rise to a canonical continuous homomorphism $T(u/v) \to \chi(u/v)$, mapping $\tau \in T(u/v)$ to the abelian group homomorphism

$$u(a) + v(F^\times) \mapsto \overline{\tau(a)/a}.$$

Its kernel is

$$V(u/v) = \Big\{ \tau \in T(u/v) \ \Big|\ \forall a \in E^\times : \quad u\Big(1 - \frac{\tau(a)}{a}\Big) > 0 \Big\}.$$

Thus $V(u/v)$ is a normal closed subgroup of $T(u/v)$. We call it the **ramification group** of the extension $(E, u)/(F, v)$. Its fixed field E_V in E is called the **ramification field** of the extension. Notice that the extension E/E_V is Galois. Denote the restriction of u to E_V by u_V. We observe that the condition $u(1 - \tau(a)/a) > 0$ means that $\tau(a)/a$ has residue 1 in $\bar{E}_u^\times$. Just as for the decomposition and inertia groups, we observe (using Remark 16.1.2):

16.2.2 REMARKS.

(a) Let (F_1, v_1) be a valued field with $F \subseteq F_1 \subseteq E$ and where v_1 is the restriction of u to F_1. Then

$$V(u/v_1) = \mathrm{Aut}(E/F_1) \cap V(u/v).$$

(b) For $\sigma \in \mathrm{Aut}(E/F)$ one has

$$V(u \circ \sigma/v) = \sigma^{-1} V(u/v)\sigma.$$

(c) Letting E_1 range over all finite normal extensions of F inside E and denoting the restriction of u to E_1 by u_1, we have $V(u/v) = \varprojlim V(u_1/v)$ as profinite groups.

(d) If u, u' are equivalent valuations on E and v, v' are their restrictions to F, respectively, then $V(u/v) = V(u'/v')$.

Next we identify the ramification group as a characteristic subgroup of the inertia group.

THEOREM 16.2.3. *$V(u/v)$ is the unique p-Sylow subgroup of $T(u/v)$ when $p = \mathrm{char}\,\bar{F}_v > 0$, and is trivial when $\mathrm{char}\,\bar{F}_v = 0$. In particular, $V(u/v)$ is a normal closed subgroup of $Z(u/v)$.*

PROOF. In light of Remarks 15.1.1(c), 16.1.2(c), and 16.2.2(c), we may assume that E/F is finite.

As we have just seen, $T(u/v)/V(u/v)$ embeds in $\chi(u/v)$. When $p = \mathrm{char}\,\bar{F}_v > 0$, the group $\bar{\mu}$ has no elements of order p, and therefore neither does $\chi(u/v)$ (Lemma 1.4.7(a)). Therefore in this case

$$p \nmid (T(u/v) : V(u/v)).$$

Next we show that $V(u/v)$ does not contain automorphisms of prime order $l \neq \mathrm{char}\,\bar{F}_v$. Indeed, suppose that σ is such an automorphism, and let $F_1 = E^{\langle\sigma\rangle}$ be its fixed field in E. The ramification group of the extension $(E, u)/(F_1, \mathrm{Res}_{F_1} u)$ is the cyclic group $\mathrm{Gal}(E/F_1) = \langle\sigma\rangle$ (by Remark 16.2.2(a)). Let $a \in E$ be a primitive element for E/F_1. Set

$$b = a - \frac{1}{l}\,\mathrm{Tr}_{E/F_1}(a),$$

where $\mathrm{Tr}_{E/F_1}: E \to F_1$ is the trace map. Then $b \neq 0$ and $\mathrm{Tr}_{E/F_1}(b) = 0$, whence also

$$\sum_{i=0}^{l-1} \frac{\sigma^i(b)}{b} = 0.$$

By the definition of $V(u/v)$, the residue of $\sigma^i(b)/b$ in $\bar{E}_u$ is 1 for $i = 0, 1, \ldots, l-1$. Therefore $\bar{l} = 0$ in $\bar{E}_u$, contrary to $l \neq \mathrm{char}\, \bar{E}_u$.

We conclude that when the residue characteristic is $p > 0$ (resp., 0), $V(u/v)$ is a p-Sylow subgroup of $T(u/v)$ (resp., is trivial). Since $V(u/v)$ is normal in $T(u/v)$, in the first case it is in fact the *unique* p-Sylow subgroup of $T(u/v)$.

Finally, $V(u/v)$ is normal also in $Z(u/v)$, since it is a characteristic subgroup of the normal subgroup $T(u/v)$ of $Z(u/v)$. □

COROLLARY 16.2.4.
(a) *$u(E^\times)/u_V(E_v^\times)$ is a p-primary group when $p = \mathrm{char}\, \bar{F}_v > 0$, and is trivial when $\mathrm{char}\, \bar{F}_v = 0$.*
(b) *$\bar{E}_u/\overline{(E_V)}_{u_V}$ is a purely inseparable algebraic extension.*

PROOF. If $\mathrm{char}\, \bar{F}_v = 0$, then by Theorem 16.2.3, $E = E_V$ and $u = u_V$, so we are done. Next suppose that $p = \mathrm{char}\, \bar{F}_v > 0$. Let $a \in E$ and let E_1/E_V be a finite normal extension of E_V containing a. By Theorem 16.2.3, $q = [E_1 : E_V]$ is a p-power. For every $\sigma \in V(u/v)$ we have $u(\sigma(a)) = u(a)$ and $\overline{\sigma(a)/a} = 1$. Consequently,

$$qu(a) = u(N_{E_1/E_V}(a)) \in u_V(E_V^\times),$$

and when $u(a) = 0$ also

$$\bar{a}^q = \overline{N_{E_1/E_V}(a)} \in \overline{(E_V)}_{u_V}.$$

This proves both (a) and (b). □

LEMMA 16.2.5. *In the above setup:*
(a) *the functorial map $\chi(u/u_T) \to \chi(u/v)$ is an isomorphism;*
(b) *the functorial map $\chi(u/u_T) \to \chi(u_V/u_T)$ is injective.*

PROOF. By Corollary 15.2.3(a), Proposition 16.1.3(e) and Corollary 16.2.4(a), the groups $u_T(E_T^\times)/v(F^\times)$ and $u(E^\times)/u_V(E_V^\times)$ are p-primary when $p = \mathrm{char}\, \bar{F}_v > 0$, and are trivial otherwise. Now apply Lemma 1.4.8. □

We can now refine Proposition 16.2.1 to get a *perfect* pairing.

THEOREM 16.2.6. *Let $(E, u)/(F, v)$ be a normal extension of valued fields. The pairing of Proposition 16.2.1 induces a canonical pairing*

$$T(u/v)/V(u/v) \ \times\ u_V(E_V^\times)/u_T(E_T^\times) \ \to\ \bar{\mu}$$

with trivial left and right kernels.

PROOF. By Corollary 15.2.3(a) and Proposition 16.1.3(e) again, the group $u_T(E_T^\times)/v(F^\times)$ is p-primary group when $p = \mathrm{char}\, \bar{F}_v > 0$ and is trivial otherwise. Hence it is contained in the right kernel of the pairing of Proposition 16.2.1. Therefore this pairing induces a pairing

$$T(u/v) \ \times\ u_V(E_V^\times)/u_T(E_T^\times) \ \to\ \bar{\mu}. \tag{16.2.1}$$

The original pairing and the induced one give rise to a commutative diagram

$$\begin{array}{ccc} T(u/v) & \longrightarrow & \chi(u/v) \\ & \searrow & \downarrow \\ & & \chi(u_V/u_T) \end{array}$$

where the vertical map is injective by Lemma 16.2.5. By definition, $V(u/v)$ is the kernel of the upper map. Hence it is the kernel of the diagonal map, i.e., it is the left kernel of (16.2.1).

It remains to show that (16.2.1) has trivial right kernel. To this end let $a \in E_V^\times$ and suppose that $(\tau, u(a)+u_T(E_T^\times))$ is mapped trivially to $\bar{\mu}$ under (16.2.1) for every $\tau \in T(u/v)$. Thus $u(\tau(a)-a) > u(a)$ for every $\tau \in T(u/v)$. Take a finite Galois extension E_1 of E_T inside E_V which contains a. By Theorem 16.2.3, $\operatorname{char} \bar{E}_u$ does not divide $m = [E_1 : E_T]$. Hence $u(a) = u(ma)$. It now follows from the ultrametric inequality that

$$u(a) = u\Big(ma + \sum_{\tau \in \mathrm{Gal}(E_1/E_T)} (\tau(a) - a)\Big) = u(\mathrm{Tr}_{E_1/E_T}(a)) \in u_T(E_T^\times),$$

as desired. □

We call the perfect pairing of Theorem 16.2.6 the **ramification pairing** of the extension.

For the following corollary we notice that the group $Z(u/v)$ acts on $\chi(u/v)$ from the left by

$$({}^\sigma h)(\alpha) = \bar{\sigma}(h(\alpha))$$

for $\sigma \in Z(u/v)$, $h \in \chi(u/v)$, and $\alpha \in u(E^\times)/v(F^\times)$. Here $\bar{\sigma} = \Phi(\sigma)$ is the image of σ in $\mathrm{Aut}(\bar{E}_u/\bar{F}_v)$ as in Theorem 16.1.1.

COROLLARY 16.2.7. *Let $(E,u)/(F,v)$ be a normal extension of valued fields. Then:*

(a) $[E_V : E_T] = e(u_V/u_T)$;

(b) $\overline{(E_V)}_{u_V} = \overline{(E_T)}_{u_T}$;

(c) *there is a canonical short exact sequence*

$$1 \to V(u/v) \to T(u/v) \to \chi(u/v) \to 1$$

which is compatible with the actions of $Z(u/v)$ on the first and second terms by conjugation, and on $\chi(u/v)$ as above;

(d) $\mathrm{Gal}(E_V/E_T)$ *is abelian of (supernatural) order not divisible by* $\operatorname{char} \bar{F}_v$.

PROOF. An inverse limit argument reduces all these assertions to the case where the extension E/F is finite.

Now $T(u/v)/V(u/v)$ embeds in $\chi(u/v)$, and by Lemma 16.2.5, $\chi(u/v)$ embeds in $\chi(u_V/u_T)$. Combined with Corollary 1.4.6 and Theorem 14.2.1 this gives

$$\begin{aligned} [E_V : E_T] &= (T(u/v) : V(u/v)) \\ &\le |\chi(u/v)| \\ &\le |\chi(u_V/u_T)| \\ &\le e(u_V/u_T) \\ &\le e(u_V/u_T) f(u_V/u_T) \le [E_V : E_T]. \end{aligned}$$

Therefore these are all equalities. In particular we get (a).

Furthermore, $f(u_V/u_T) = 1$ so we get (b).

As $T(u/v)/V(u/v)$ embeds in $\chi(u/v)$, and both groups have the same finite order, we have the exact sequence of (c). Proposition 16.2.1(b) shows that it is compatible with the actions of $Z(u/v)$ as stated.

Finally, (d) follows from (c). □

REMARK 16.2.8. Since $V(u/v)$ is normal in $Z(u/v)$ (Theorem 16.2.3), the conjugation action of $Z(u/v)$ on $T(u/v)$ induces such an action on $T(u/v)/V(u/v)$. By Corollary 16.2.7(c),

$$T(u/v)/V(u/v) \cong \chi(u/v) \tag{16.2.2}$$

is abelian. Therefore the action of the subgroup $T(u/v)$ of $Z(u/v)$ on the left-hand side of (16.2.2) is trivial. We deduce that $Z(u/v)/T(u/v)$ acts on $T(u/v)/V(u/v)$ by conjugation.

Now $Z(u/v)/T(u/v) \cong \mathrm{Aut}(\bar{E}_u/\bar{F}_v)$ via the epimorphism Φ of Theorem 16.1.1. By the last part of Corollary 16.2.7(c), the conjugation action of $Z(u/v)/T(u/v)$ on the left hand-side of (16.2.2) corresponds under this isomorphism to the natural action of $\mathrm{Aut}(\bar{E}_u/\bar{F}_v)$ on the right-hand side.

From Theorem 16.2.6 we also deduce the following analog of Corollary 16.1.4:

COROLLARY 16.2.9. *The value group map gives an order-preserving bijection between the intermediate fields*

$$E_T \subseteq K \subseteq E_V$$

and the intermediate groups

$$u_T(E_T^\times) \leq \Gamma \leq u_V(E_V^\times).$$

The following diagram describes graphically the tower of fields which was discussed in the last two chapters, and the corresponding towers of the residue fields and value groups. As usual, p is the residue characteristic if this characteristic is positive, and otherwise the relevant extensions are trivial. As before, F_i stands for the maximal inseparable extension of F inside E.

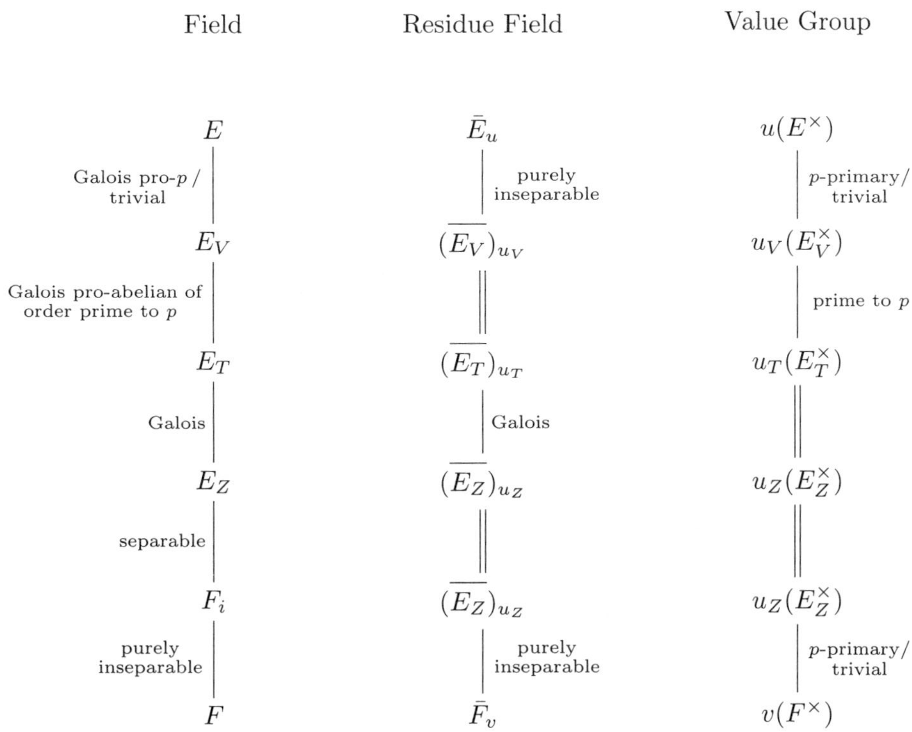
Field
Residue Field
Value Group
E
$\bar{E}_u$
$u(E^\times)$
Galois pro-p / trivial
purely inseparable
p-primary/ trivial
E_V
$\overline{(E_V)}_{u_V}$
$u_V(E_V^\times)$
Galois pro-abelian of order prime to p
prime to p
E_T
$\overline{(E_T)}_{u_T}$
$u_T(E_T^\times)$
Galois
Galois
E_Z
$\overline{(E_Z)}_{u_Z}$
$u_Z(E_Z^\times)$
separable
F_i
$\overline{(E_Z)}_{u_Z}$
$u_Z(E_Z^\times)$
purely inseparable
purely inseparable
p-primary/ trivial
F
$\bar{F}_v$
$v(F^\times)$

CHAPTER 17

The Fundamental Equality

For valued fields which are not relatively Henselian, the valuation extends in more than one way to the extension field. The so-called "fundamental inequality of valuation theory" incorporates the information on the distinct extensions into one quantitative estimate relating the extension degree, the ramification indices, and the residue degrees. This inequality has fundamental importance, e.g., in algebraic number theory. For relatively Henselian valued fields we present a major result due to Ostrowski (following earlier results by Krull [**Kr1**] and Deuring [**D**]), which relates these quantities and the somewhat more mysterious invariant called the "defect".

17.1 The fundamental inequality

Given subgroups H and U of a finite group G, we associate to every $\sigma \in G$ its **double coset** $H\sigma U$. Denote the set of all such double cosets by $H\backslash G/U$. It is a partition of G. We have the following formula:

LEMMA 17.1.1. *For G, H, U as above, and for every system of representatives $\sigma_1, \dots, \sigma_g$ for the double cosets in $H\backslash G/U$, one has*

$$(G:U) = \sum_{i=1}^{g} (H : H \cap \sigma_i U \sigma_i^{-1}).$$

PROOF. For each $1 \le i \le g$ let T_i be a system of representatives for the left cosets of $H \cap \sigma_i U \sigma_i^{-1}$ in H. Then:

$$\begin{aligned} G &= \dot{\bigcup_{i=1}^{g}} H\sigma_i U = \dot{\bigcup_{i=1}^{g}} \dot{\bigcup_{\tau \in T_i}} \tau(H \cap \sigma_i U \sigma_i^{-1})\sigma_i U \\ &= \dot{\bigcup_{i=1}^{g}} \dot{\bigcup_{\tau \in T_i}} \tau\sigma_i(\sigma_i^{-1} H \sigma_i \cap U)U = \dot{\bigcup_{i=1}^{g}} \dot{\bigcup_{\tau \in T_i}} \tau\sigma_i U. \end{aligned}$$

Consequently,

$$(G:U) = \sum_{i=1}^{g} |T_i| = \sum_{i=1}^{g} (H : H \cap \sigma_i U \sigma_i^{-1}). \qquad \square$$

Now fix finite extensions $(E, u)/(F, v)$ and $(\hat{E}, \hat{u})/(E, u)$ of valued fields, with $\hat{E}/F$ normal. Let

$$G = \mathrm{Aut}(\hat{E}/F), \quad H = Z(\hat{u}/v), \quad U = \mathrm{Aut}(\hat{E}/E).$$

LEMMA 17.1.2. *Let $\sigma, \sigma' \in \mathrm{Aut}(\hat{E}/F)$. Then $\hat{u} \circ \sigma$ and $\hat{u} \circ \sigma'$ have the same restriction to E if and only if $H\sigma U = H\sigma' U$.*

PROOF. By Theorem 14.3.2, $\hat{u} \circ \sigma$ and $\hat{u} \circ \sigma'$ have the same restriction to E if and only if there exists $\tau \in U$ such that $\hat{u} \circ \sigma \circ \tau = \hat{u} \circ \sigma'$. Equivalently, there exists $\tau \in U$ such that $\sigma\tau(\sigma')^{-1} \in Z(\hat{u}/v) = H$. But this means that $H\sigma U = H\sigma' U$. □

COROLLARY 17.1.3. *Let $\sigma_1, \ldots, \sigma_g$ be a system of representatives for the double cosets in $H\backslash G/U$. Then the restrictions of $\hat{u} \circ \sigma_i$, $i = 1, \ldots, g$, to E form a system of representatives for the extensions of v to E (up to equivalence).*

PROOF. By Lemma 17.1.2, the restrictions of $\hat{u} \circ \sigma_i$, $i = 1, \ldots, g$, to E are distinct. By Proposition 14.3.1 they are not equivalent.

Conversely, let u' be an arbitrary extension of v to E. In light of Chevalley's theorem (Corollary 14.1.2), it is the restriction of some valuation $\hat{u}'$ on $\hat{E}$. Theorem 14.3.2 yields $\tau \in \mathrm{Aut}(\hat{E}/F)$ such that $\hat{u}'$ and $\hat{u} \circ \tau$ are equivalent. Then u' and the restriction of $\hat{u} \circ \tau$ to E are equivalent. Choose $1 \leq i \leq g$ such that $H\tau U = H\sigma_i U$. By Lemma 17.1.2, $\hat{u} \circ \sigma_i, \hat{u} \circ \tau$ have the same restriction to E. Consequently, u' is equivalent to the restriction of $\hat{u} \circ \sigma_i$ to E. □

LEMMA 17.1.4. *Suppose that $\hat{E}/F$ is Galois. Let $\sigma_1, \ldots, \sigma_g$ be a system of representatives for $H\backslash G/U$. For each $1 \leq i \leq g$ let L_i be the decomposition field of the extension $(\hat{E}, \hat{u} \circ \sigma_i)/(F, v)$, and let u_i be the restriction of $\hat{u} \circ \sigma_i$ to E. Then:*

(a) $[E : F] = \sum_{i=1}^{g} [L_i E : L_i]$.

(b) *$L_i E$ is the decomposition field of the extension $(\hat{E}, \hat{u} \circ \sigma_i)/(E, u_i)$.*

PROOF. We first observe that by Remark 15.1.1(b),

$$\mathrm{Gal}(\hat{E}/L_i) = Z(\hat{u} \circ \sigma_i/v) = \sigma_i^{-1} H \sigma_i.$$

Hence

$$\mathrm{Gal}(\hat{E}/L_i E) = \sigma_i^{-1} H \sigma_i \cap U.$$

Passing to fixed fields we get

$$L_i = \hat{E}^{\sigma_i^{-1} H \sigma_i}, \quad L_i E = \hat{E}^{\sigma_i^{-1} H \sigma_i \cap U}.$$

(a) By Lemma 17.1.1,

$$(G : U) = \sum_{i=1}^{g} (H : H \cap \sigma_i U \sigma_i^{-1}) = \sum_{i=1}^{g} (\sigma_i^{-1} H \sigma_i : \sigma_i^{-1} H \sigma_i \cap U).$$

Passing again to the fixed fields, we deduce that $[E : F] = \sum_{i=1}^{g} [L_i E : L_i]$.

(b) By Remark 15.1.1(a),

$$Z(\hat{u} \circ \sigma_i/u_i) = Z(\hat{u} \circ \sigma_i/v) \cap U = \sigma_i^{-1} H \sigma_i \cap U.$$

Now take the fixed fields in $\hat{E}$. □

We can now deduce the main result of the section, generalizing Theorem 14.2.1.

THEOREM 17.1.5 (The Fundamental Inequality). *Let E/F be a finite extension of fields, let v be a valuation F, and let $u_1, \dots, u_g$ be its extensions to E (up to equivalence). Then*

$$[E:F] \geq \sum_{i=1}^{g} e(u_i/v) f(u_i/v).$$

PROOF. Let F_1 be the maximal purely inseparable extension of F inside E. By Proposition 14.2.5(a), v has a unique extension v_1 to F_1. By Theorem 14.2.1,

$$[F_1 : F] \geq e(v_1/v) f(v_1/v).$$

In light of the tower properties for the degree, the ramification index, and the inertia degree (see Proposition 14.2.4), it is enough to prove the fundamental inequality for the extensions of v_1 to E. We may therefore assume that E/F is separable.

Choose as before an arbitrary finite Galois extension $\hat{E}$ of F containing E, an extension $\hat{u}$ of v to $\hat{E}$, and let

$$G = \mathrm{Gal}(\hat{E}/F), \quad H = Z(\hat{u}/v), \quad U = \mathrm{Gal}(\hat{E}/E).$$

We also choose a system of representatives $\sigma_1, \dots, \sigma_g$ for $H \backslash G/U$. For each $1 \leq i \leq g$ let $\hat{u}_i = \hat{u} \circ \sigma_i$, let L_i be the decomposition field of the extension $\hat{u}_i/v$, and let w_i, w_i' be the restrictions of $\hat{u}_i$ to L_i and L_iE, respectively. In light of Corollary 17.1.3, we may assume that $\hat{u}_i$ extends u_i, $i = 1, \dots, g$. By Lemma 17.1.4(b), (L_iE, w_i') is the decomposition field of $(\hat{E}, \hat{u}_i)/(E, u_i)$. As decomposition fields in a Galois extension, the extensions $(L_i, w_i)/(F, v)$ and $(L_iE, w_i')/(E, u_i)$ are immediate (Theorem 15.2.2). It follows that

$$e(w_i'/w_i) = e(u_i/v), \quad f(w_i'/w_i) = f(u_i/v), \quad i = 1, \dots, g.$$

By Lemma 17.1.4(a) and Theorem 14.2.1,

$$[E:F] = \sum_{i=1}^{g} [L_iE : L_i] \geq \sum_{i=1}^{g} e(w_i'/w_i) f(w_i'/w_i) = \sum_{i=1}^{g} e(u_i/v) f(u_i/v). \qquad \square$$

Now consider the case where E/F is already normal. Examining the proof of Theorem 17.1.5, we observe:

COROLLARY 17.1.6. *Let E/F be a finite normal extension, let v be a valuation F, and let $u_1, \dots, u_g$ be its extensions to E (up to equivalence). For each $1 \leq i \leq g$ let L_i be the decomposition field of u_i and let w_i be the restriction of u_i to L_i. Then the following conditions are equivalent:*

(a) $[E:F] = \sum_{i=1}^{g} e(u_i/v) f(u_i/v)$;
(b) $[E:L_i] = e(u_i/w_i) f(u_i/w_i)$ *for* $i = 1, \dots, g$.

17.2 Ostrowski's theorem

By definition, relatively Henselian valuations extend uniquely to the normal extension. Thus in Theorem 17.1.5 one has $g = 1$. The following fundamental fact gives a more precise information for this situation.

THEOREM 17.2.1 (Ostrowski [**O3**]). *Suppose that $(E,u)/(F,v)$ is a normal extension of valued fields with v Henselian relative to E. Then*

$$[E:F] = d(u/v)\cdot e(u/v)\cdot f(u/v)$$

for some supernatural number $d(u/v)$. Moreover, $d(u/v)$ is a p-power if $p = \mathrm{char}\,\bar{F}_v > 0$ and is 1 if $\mathrm{char}\,\bar{F}_v = 0$.

PROOF. We use the terminology and results of §§14–16. It is enough to consider the case where E/F is a finite extension.

Let

$$q_1 = [E:E_V],\quad q_2 = e(u/u_V),\quad q_3 = e(u_Z/v).$$

By Theorem 16.2.3, Corollary 16.2.4(a) and Corollary 15.2.3(a), respectively, q_1, q_2 and q_3 are powers of the residue characteristic if it is positive, and are 1 otherwise.

By Corollary 16.2.7(a) and Proposition 16.1.3(e),

$$[E_V:E_T] = e(u_V/u_T) = e(u_V/u_Z).$$

The tower property for the ramification index (Proposition 14.2.4(a)) therefore implies that

$$[E_V:E_T] = q_2^{-1}q_3^{-1}e(u/v).$$

The Henselity assumption implies that E_Z/F is purely inseparable (see Remark 15.3.4). Since E_T/E_Z is normal, so is E_T/F. Hence

$$[E_T:F] = q_4|\mathrm{Aut}(E_T/F)|,$$

where q_4 is a p-power if $\mathrm{char}\,F = \mathrm{char}\,\bar{F}_v = p$ and is 1 otherwise. Likewise, since $\bar{E}_u/\bar{F}_v$ is normal (Corollary 15.2.5),

$$f(u/v) = [\bar{E}_u:\bar{F}_v] = q_5|\mathrm{Aut}(\bar{E}_u/\bar{F}_v)|,$$

where q_5 is again a power of the residue characteristic if it is positive, and is 1 otherwise. As $\mathrm{Aut}(E/F) = Z(u/v)$, Theorem 16.1.1 gives an isomorphism $\mathrm{Aut}(E_T/F) \cong \mathrm{Aut}(\bar{E}_u/\bar{F}_v)$. Hence

$$[E_T:F] = q_4q_5^{-1}f(u/v).$$

Putting all this together we obtain

$$[E:F] = [E:E_V]\cdot[E_V:E_T]\cdot[E_T:F] = q_1q_2^{-1}q_3^{-1}q_4q_5^{-1}\cdot e(u/v)\cdot f(u/v).$$

Now let

$$d(u/v) = q_1q_2^{-1}q_3^{-1}q_4q_5^{-1}.$$

It is a power (with integral exponent) of the residue characteristic if this characteristic is positive and is 1 otherwise. By the fundamental inequality, $[E:F] \geq e(u/v)\cdot f(u/v)$, so $d(u/v) \geq 1$. Therefore $d(u/v)$ is a positive integer, i.e., the above exponent is actually non-negative. □

In the situation of Theorem 17.2.1, the positive integer $d(u/v)$ is called the **defect** of the extension $(E,u)/(F,v)$. It is convenient to define the defect $d(u/v)$ also for any finite (but not necessary normal) valued field extension (E,u) of (F,v) by the formula

$$d(u/v) = \frac{[E:F]}{e(u/v)f(u/v)}.$$

When so defined, this is a rational number, and by the fundamental inequality, $d(u/v) \geq 1$. Moreover, if

$$(F, v) \subseteq (E, u) \subseteq (E_1, u_1)$$

is a tower of finite extensions of valued fields, then the tower properties for the extension degree, the ramification index, and the inertia degree (Proposition 14.2.4) imply the following tower property for defects:

$$d(u_1/v) = d(u_1/u) \cdot d(u/v). \tag{17.2.1}$$

In particular, take E_1 to be the normal hull of the finite extension E/F. By Theorem 17.2.1, $d(u_1/v)$ and $d(u_1/u)$ are p-powers. Therefore $d(u/v)$ is also a p-power, so *a fortiori*, it is in fact an integer.

When $(E, u)/(F, v)$ is an arbitrary algebraic extension we define $d(u/v)$ to be the least common multiple of the defects $d(u_1/v)$, where u_1 is the restriction of u to some finite extension E_1 of F inside E. Thus it is a supernatural divisor of p^∞ if $p = \operatorname{char} \bar{F}_v > 0$, and is 1 otherwise. The tower property (17.2.1) for the defect then extends from the finite case to the algebraic case.

For a normal extension $(E, u)/(F, v)$ of valued fields with v is Henselian relative to E, we say that v is **defectless relative to** E if $d(u/v) = 1$. When F_1/F is a subextension of E/F and v_1 is the unique extension of v to F_1, the tower property for defects (17.2.1) implies that v_1 is also defectless relative to E.

LEMMA 17.2.2. *Let $(E, u)/(F, v)$ be a Galois extension of valued fields. If $d(u/v) > 1$, then there exist intermediate valued fields*

$$(F, v) \subseteq (F', v') \subset (E', u') \subseteq (E, u)$$

with $(E', u')/(F', v')$ an immediate Galois extension of degree $p = \operatorname{char} \bar{F}_v > 0$.

PROOF. We may replace E by a sufficiently large Galois finite extension of F to assume that E/F is finite. Since $d(u/v)$ is a p-power, we may also replace F by the fixed field of some p-Sylow subgroup of $\mathrm{Gal}(E/F)$ to assume that $\mathrm{Gal}(E/F)$ is a p-group. By the structure theory of finite p-groups and the Galois correspondence, there is a tower of field extensions

$$F = F_0 \subset F_1 \subset \cdots \subset F_n = E$$

with each F_{i+1}/F_i Galois of degree p. In view of the tower property for defects (17.2.1), there exists i for which the defect of the valuations in F_{i+1}/F_i is not 1. By Ostrowski's formula (Theorem 17.2.1), this extension must be immediate. □

EXAMPLE 17.2.3 (Ostrowski): *An extension with non-trivial defect.*

Our presentation is inspired by [**Ri1**]. Choose c_n, $n = 0, 1, 2, \ldots$, in the algebraic closure $\tilde{\mathbb{Q}}_2$ of $\mathbb{Q}_2$ such that $c_0 = 1/2$ and $c_n^2 = c_{n-1}$ for $n \geq 1$. Set $F_n = \mathbb{Q}_2(c_n)$ and $F = \bigcup_{n=1}^\infty F_n$. Note that

$$F_0 \subseteq F_1 \subseteq F_2 \subseteq \cdots \subseteq F.$$

Let v be an extension to F of the canonical dyadic valuation v_0 on $\mathbb{Q}_2$, and let v_n be its restriction to F_n. As $v(c_n) = -1/2^n$ we have $v_n(F_n^\times) \supseteq \frac{1}{2^n}\mathbb{Z}$. Since $[F_n : \mathbb{Q}_2] \leq 2^n = (\frac{1}{2^n}\mathbb{Z} : \mathbb{Z})$, the fundamental inequality implies that $v_n(F_n^\times) = \frac{1}{2^n}\mathbb{Z}$ and $\overline{(F_n)}_{v_n} = \mathbb{F}_2$. We conclude that $v(F^\times) = \frac{1}{2^\infty}\mathbb{Z}$ is 2-divisible and $\bar{F}_v = \mathbb{F}_2$.

Next let $a = (1+\sqrt{3})/2$, so $a^2 - a = 1/2$. Let u be an extension of v to

$$E = F(a) = F(\sqrt{3}),$$

and let u_n be its restriction to

$$E_n = F_n(a) = F_n(\sqrt{3}).$$

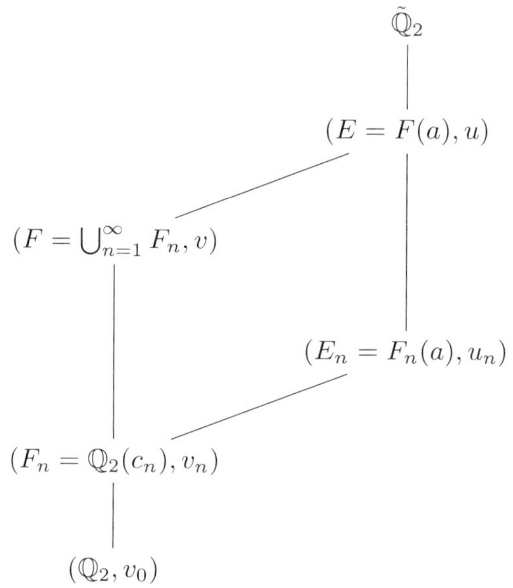

By the fundamental inequality again, $e(u/v) | [E : F] | 2$. But $v(F^\times)$ is 2-divisible, so Lemma 1.1.4(b) shows that $2 \nmid e(u/v)$, whence $e(u/v) = 1$.

Consider $s_n = c_1 + \cdots + c_n$. By Lemma 3.2.1(e), $u(a) = -1/2$. Also, $v(s_n) = v(c_1) = -1/2$, so $u(a - s_n) \geq -1/2$. Now

$$\begin{aligned} a^2 &= \big((a - s_n) + c_1 + \cdots + c_n\big)^2 \\ &= (a - s_n)^2 + c_1^2 + c_2^2 + \cdots + c_n^2 + 2(a - s_n)\sum_{i=1}^{n} c_i + 2\sum_{i<j} c_i c_j \\ &= (a - s_n)^2 + \frac{1}{2} + c_1 + \cdots + c_{n-1} + 2(a - s_n)s_n + 2\sum_{i<j} c_i c_j, \end{aligned}$$

as well as

$$a^2 = a + \frac{1}{2} = a - s_n + \frac{1}{2} + c_1 + \cdots + c_{n-1} + c_n.$$

It follows that

$$(a - s_n)^2 + 2(a - s_n)s_n + 2\sum_{i<j} c_i c_j = a - s_n + c_n.$$

Since

$$u(2(a - s_n)s_n) \geq 0, \quad u(2\sum_{i<j} c_i c_j) > 0, \quad u(c_n) = -\frac{1}{2^n},$$

the ultrametric inequality implies that

$$u((a-s_n)^2-(a-s_n)) = -1/2^n < 0.$$

By Lemma 3.2.1(e) again,

$$u(a-s_n) = -1/2^{n+1} \notin v_n(F_n^\times).$$

Therefore the fundamental inequality for the extension $(E_n, u_n)/(F_n, v_n)$ forces that $e(u_n/v_n) = [E_n : F_n] = 2$, $f(u_n/v_n) = 1$, and furthermore, u_n is the unique extension of v_n to E_n (Theorem 17.1.5). Thus $\overline{(E_n)}_{u_n} = \overline{(F_n)}_{v_n} = \mathbb{F}_2$, so

$$\bar{E}_u = \bigcup_{n=1}^{\infty} \overline{(E_n)}_{u_n} = \mathbb{F}_2 = \bar{F}_v,$$

i.e., $f(u/v) = 1$. Moreover, it follows that u is the unique extension of v to E, i.e., (F, v) is Henselian relative to E. Finally, since a does not belong to any F_n, the extension E/F is proper. Conclude from Theorem 17.2.1 that $d(u/v) = 2$.

In positive characteristics, extensions with non-trivial defect are easier to construct:

EXAMPLE 17.2.4. Let p be a prime number. The Laurent series field $\mathbb{F}_p((t))$ is uncountable. On the other hand, the algebraic closure of $\mathbb{F}_p(t)$ is countable. Therefore we can find $s \in \mathbb{F}_p((t))$ which is transcendental over $\mathbb{F}_p(t)$. Let

$$E = \mathbb{F}_p(t, s), \quad F = \mathbb{F}_p(t, s^p).$$

Finally, let u and v be the restrictions to E and F, respectively, of the canonical discrete valuation on $\mathbb{F}_p((t))$. The extension E/F is purely inseparable of degree p. By Proposition 14.2.5(a), u is therefore the unique extension of v to E. Thus (F, v) is Henselian relative to E. Moreover, the extension $\mathbb{F}_p((t))/\mathbb{F}_p(t)$ is immediate with respect to the canonical valuation, and therefore so is its subextension $(E, u)/(F, v)$. It follows from Ostrowski's formula (Theorem 17.2.1) that $d(u/v) = p$.

A more detailed discussion on the behavior of the defect is given in [**Ku**].

17.3 Defectless fields

Let E/F be a finite extension of fields and let v be a valuation on F. One says that the **fundamental equality** holds for the extensions of v to E if the inequality in Theorem 17.1.5 is in fact an equality, i.e.,

$$[E : F] = \sum_{i=1}^{g} e(u_i/v) f(u_i/v),$$

where $u_1, \dots, u_g$ are the distinct extensions of v to E (up to equivalence).

LEMMA 17.3.1. *Let $(E, u)/(F, v)$ be a finite extension of valued fields and let F_1/F be a subextension of E/F. Let $w_1, \dots, w_k$ be the extensions of v to F_1 (up to equivalence). Suppose that the fundamental equality holds for the extensions of v to E, as well as for the extensions of each w_i to E. Then it holds for the extensions of v to F_1.*

PROOF. For each $1 \le i \le k$ let $u_{i1}, \ldots, u_{ig_i}$ be the extensions of w_i to E (up to equivalence). For any i, j we have by Proposition 14.2.4:

$$e(u_{ij}/v) = e(u_{ij}/w_i)e(w_i/v), \quad f(u_{ij}/v) = f(u_{ij}/w_i)f(w_i/v).$$

Hence the assumptions give:

$$\begin{aligned}[E : F_1][F_1 : F] &= [E : F] \\ &= \sum_{i=1}^{k}\sum_{j=1}^{g_i} e(u_{ij}/v)f(u_{ij}/v) \\ &= \sum_{i=1}^{k}\Big(\sum_{j=1}^{g_i} e(u_{ij}/w_i)f(u_{ij}/w_i)\Big) \cdot e(w_i/v)f(w_i/v) \\ &= \sum_{i=1}^{k}[E : F_1]e(w_i/v)f(w_i/v).\end{aligned}$$

We conclude that $[F_1 : F] = \sum_{i=1}^{k} e(w_i/v)f(w_i/v)$. □

PROPOSITION 17.3.2. *Let E/F be a finite extension and let v be a valuation on F. Assume that:*

(1) *E/F is normal and* $\mathrm{char}\,\bar{F}_v$ *does not divide $[E : F]$; or*

(2) $\mathrm{char}\,\bar{F}_v = 0$.

Then the fundamental equality holds for the extensions of v to E. In particular, if v is Henselian relative to E, then it is defectless relative to E.

PROOF. In light of Lemma 17.3.1, we may take in case (2) $\hat{E}$ to be the normal hull of the extension E/F to reduce the assertion to the case where E/F is normal. It is therefore enough to prove (1).

So let $u_1, \ldots, u_g$ be the extensions of v to E (up to equivalence). For $1 \le i \le g$ let L_i be the decomposition field of u_i, and let w_i be the restriction of u_i to L_i. Since $\mathrm{char}\,\bar{F}_v$ does not divide $[E : L_i]$, Ostrowski's theorem implies that

$$[E : L_i] = e(u_i/w_i)f(u_i/w_i), \quad i = 1, \ldots, g.$$

Now apply Corollary 17.1.6. □

COROLLARY 17.3.3. *Let (F, v) be a Henselian valued field (relative to the algebraic closure) such that $v(F^\times)$ is divisible and $\bar{F}_v$ is algebraically closed and has characteristic 0. Then F is algebraically closed.*

PROOF. Let E be a finite normal extension of F and let u be the unique extension of v to E. Obviously, $f(u/v) = 1$. Since $u(E^\times)/v(F^\times)$ is torsion (Corollary 14.2.3(a)) and $u(E^\times)$ is torsion-free, Lemma 1.1.4(b) implies that $u(E^\times) = v(F^\times)$. Thus $e(u/v) = 1$ as well. By case (2) of Proposition 17.3.2, $(E, u)/(F, v)$ is defectless, so Ostrowski's formula gives $E = F$. □

17.4 Extensions of discrete valuations

Theorem 17.4.3 below shows that Henselian *discrete* valued fields are defectless. To obtain it, we first need the following proposition, which is proved using linear algebra techniques.

PROPOSITION 17.4.1. *Suppose that $(E,u)/(F,v)$ is a finite separable extension of discrete valued fields and that v is Henselian relative to E. Then O_u is a free O_v-module of rank $[E:F]$.*

PROOF. Assume first that E/F is Galois. Let $n=[E:F]$ and let $\sigma_1,\dots,\sigma_n$ be the distinct automorphisms in $\mathrm{Gal}(E/F)$. Let a be a primitive element for the extension E/F. Thus $1,a,a^2,\dots,a^{n-1}$ form an F-linear basis of E. By Corollary 14.2.3(e), $v(F^\times)$ is cofinal in $u(E^\times)$. After multiplying a by a suitable element of $F^\times$ we may therefore assume that $a\in O_u$.

Let d be the discriminant of a, i.e., $d=\prod_{i<j}(\sigma_i(a)-\sigma_j(a))^2$. Then $d\neq 0$. We observe that $d=\det(A)^2$, where A is the Vandermonde matrix

$$A=(\sigma_i(a^j))_{1\le i\le n,0\le j\le n-1}.$$

Now

$$A^tA=\Big(\sum_{k=1}^n\sigma_k(a^i)\sigma_k(a^j)\Big)_{0\le i,j\le n-1}=\Big(\mathrm{Tr}_{E/F}(a^{i+j})\Big)_{0\le i,j\le n-1}.$$

Therefore $d=\det(A^tA)=\det(\mathrm{Tr}_{E/F}(a^{i+j}))$.

Next let M be the O_v-module of rank n spanned by $1,a,a^2,\dots,a^{n-1}$. Then $M\subseteq O_u$. We claim that $dO_u\subseteq M$. Indeed, take $b\in O_u$. Write $b=\sum_{j=0}^{n-1}c_ja^j$ with $c_0,\dots,c_{n-1}\in F$. Multiplying by a^i and taking traces to F, we obtain:

$$\mathrm{Tr}_{E/F}(a^ib)=\sum_{j=0}^{n-1}c_j\mathrm{Tr}_{E/F}(a^{i+j}),\quad i=0,1,\dots,n-1. \tag{17.4.1}$$

By assumption, v is Henselian relative to E, i.e., $\sigma_1,\dots,\sigma_n\in Z(u/v)$. Therefore $\mathrm{Tr}_{E/F}$ maps O_u to $O_u\cap F=O_v$. In particular,

$$\mathrm{Tr}_{E/F}(a^ib),\mathrm{Tr}_{E/F}(a^{i+j})\in O_v,\quad i=0,1,\dots,n-1.$$

Considering (17.4.1) as an $n\times n$ system of non-homogenous linear equations over F in the indeterminates $c_1,\dots,c_n$, we deduce from Cramer's rule that $c_0,\dots,c_{n-1}\in d^{-1}O_v$. Hence $b\in M$, as claimed.

Since v is discrete, O_v is a principal ideal domain (Remark 3.4.2(a)). Moreover, M is a finitely generated torsion-free O_v-module. It follows that M is a free O_v-module [**Lg**, Ch. XV, §2, Th. 2]. Hence so is its submodule dO_u [**Lg**, Ch. XV, §2, Th. 1], and therefore so is O_u. Moreover, we have

$$\mathrm{rank}_{O_v}(O_u)=\mathrm{rank}_{O_v}(dO_u)\le\mathrm{rank}_{O_v}(M)\le\mathrm{rank}_{O_v}(O_u).$$

We conclude that $M=O_u$ and $\mathrm{rank}_{O_v}(O_u)=\mathrm{rank}_{O_v}(M)=n$, as desired.

In the general case of an arbitrary finite separable extension E/F, let $\hat E$ be the Galois hull of E over F, and let $\hat u$ be an extension of u to $\hat E$. By what we have just proved, $O_{\hat u}$ is a free O_v-module of rank $[\hat E:F]$, as well as a free O_u-module of rank $[\hat E:E]$. In light of [**Lg**, Ch. XV, §2, Th. 1] again, the O_v-submodule O_u of $O_{\hat u}$ is also free. By the multiplicativity of the rank,

$$[\hat E:F]=\mathrm{rank}_{O_v}(O_{\hat u})=\mathrm{rank}_{O_u}(O_{\hat u})\cdot\mathrm{rank}_{O_v}(O_u)=[\hat E:E]\cdot\mathrm{rank}_{O_v}(O_u),$$

so $[E:F]=\mathrm{rank}_{O_v}(O_u)$. □

For the following well-known fact we refer, e.g., to [**Mat**, 1.M].

LEMMA 17.4.2 (The Nakayama Lemma). *Let M be a module over a ring R, let N, N' be submodules of M, and let I be an ideal in R. Suppose that N' is finitely generated, that I is contained in all the maximal ideals of R, and that $M = N + IN'$. Then $M = N$.*

We can now prove the desired result.

THEOREM 17.4.3. *Let E/F be a Galois extension and let v be a discrete valuation on F which is Henselian relative to E. Then v is defectless relative to E.*

PROOF. We roughly follow [**N3**, Ch. II, Th. 6.8]. We may assume that the extension E/F is finite.

Let u be the unique extension of v to E. Since v is discrete, so is u (Corollary 14.2.2). Let λ be a uniformizer for u, let $e = e(u/v)$, and let $f = f(u/v)$. Thus $\lambda^e O_u = \mathfrak{m}_v O_u$. Choose $y_1, \ldots, y_f \in O_u^\times$ whose residues $\bar{y}_1, \ldots, \bar{y}_f$ form an $\bar{F}_v$-linear basis of $\bar{E}_u$. Let N_1 be the O_v-module generated by $y_1, \ldots, y_f$, and let

$$N = \sum_{i=0}^{e-1} \lambda^i N_1.$$

The residue map $O_u \to \bar{E}_u$ has a set-theoretic section $\rho\colon \bar{E}_u \to O_u$ with image contained in N_1. By Lemma 12.2.1(b), every $a \in O_u$ has a λ-adic expansion

$$a = \sum_{i=0}^{e-1} \lambda^i b_i + \lambda^e O_u$$

with $b_0, \ldots, b_{e-1} \in \rho(\bar{E}_u) \subseteq N_1$. Thus $O_u = N + \mathfrak{m}_v O_u$. By Proposition 17.4.1, O_u is a finitely generated O_v-module. The Nakayama Lemma 17.4.2 now implies that $O_u = N$.

Since $v(F^\times)$ is cofinal in $u(E^\times)$ (Corollary 14.2.3(e)), for every $x \in E$ we can find $c \in F^\times$ such that $cx \in O_u = N$. It follows that the ef elements

$$\lambda^i y_j, \quad i = 0, 1, \ldots, e-1, \quad j = 1, \ldots, f,$$

span E as an F-linear space. Thus $ef \geq [E : F]$. Since the converse inequality holds for arbitrary finite extensions (Theorem 14.2.1), $ef = [E : F]$, i.e., $d(u/v) = 1$. □

COROLLARY 17.4.4. *Let E/F be a finite separable extension and let v be a discrete valuation on F. Then the fundamental equality holds for the extensions of v to E.*

PROOF. Let $\hat{E}$ be the Galois hull of the extension E/F. In light of Lemma 17.3.1 it suffices to prove the assertion for $\hat{E}/F$ and $\hat{E}/E$. We may therefore assume that E/F is Galois.

Now let $u_1, \ldots, u_g$ be the extensions of v to E (up to equivalence). For $1 \leq i \leq g$ let L_i be the decomposition field of u_i and let w_i be the restriction of u_i to L_i. By Corollary 14.2.2, the w_i are also discrete. By Theorem 17.4.3,

$$[E : L_i] = e(u_i/w_i) f(u_i/w_i), \quad i = 1, \ldots, g.$$

Now apply Corollary 17.1.6. □

CHAPTER 18

Hensel's Lemma

We return to the class of relatively Henselian valued fields defined in §15.3, and characterize it by means of several equivalent Diophantine conditions. Historically, these equivalent formulations were among the main origins of valuation theory. We use these conditions to give several major examples of Henselian field, such as the p-adic fields, quotient fields of rings of Witt vectors, generalized formal power series field, and Puiseux series fields.

18.1 The main variants

We will need the following form of the Taylor expansion formula.

LEMMA 18.1.1. *Let (F, v) be a valued field and let $f(X) \in O_v[X]$. For every $a \in O_v$ and for every non-negative integer k one has:*

$$f(a+X) - \sum_{i=0}^{k} \frac{f^{(i)}(a)}{i!} X^i \in X^{k+1} O_v[X],$$

where $f^{(i)}(X)$ denotes the formal derivative of $f(X)$ of order i.

PROOF. Let $g(X) = f(a+X)$ and expand $g(X) = \sum_{i=0}^{n} c_i X^i$ with $c_i \in O_v$. Then

$$f^{(i)}(a) = g^{(i)}(0) = \begin{cases} i! c_i & \text{if } 0 \leq i \leq n, \\ 0 & \text{if } n < i. \end{cases}$$

For $k' = \min\{k, n\}$ we therefore obtain

$$f(a+X) - \sum_{i=0}^{k} \frac{f^{(i)}(a)}{i!} X^i = g(X) - \sum_{i=0}^{k'} c_i X^i = \sum_{k' < i \leq n} c_i X^i.$$

The assertion follows. □

As before, given a valued field (F, v) and a polynomial $f(X) \in O_v[X]$, we denote the residue polynomial of $f(X)$ in $\bar{F}_v[X]$ by $\bar{f}(X)$.

THEOREM 18.1.2. *Let E/F be a normal extension and let v be a valuation on F. The following conditions are equivalent:*

(a) *v is Henselian relative to E;*
(b) *v is Henselian relative to every finite Galois extension E_1 of F which is contained in E;*
(c) *if $f(X)$ is an irreducible monic polynomial in $O_v[X]$ which splits in E, then $\bar{f}(X)$ is a power of an irreducible monic polynomial in $\bar{F}_v[X]$;*
(d) **Hensel's Lemma**: *if $f(X)$ is a monic polynomial in $O_v[X]$ which splits in E and if $\bar{f}(X) = \tilde{g}(X)\tilde{h}(X)$ for some relatively prime polynomials*

$\tilde{g}(X), \tilde{h}(X)$ in $\bar{F}_v[X]$, then there exist polynomials $g(X), h(X) \in O_v[X]$ such that

$$f(X) = g(X)h(X), \quad \bar{g}(X) = \tilde{g}(X), \quad \bar{h}(X) = \tilde{h}(X);$$

(e) *for every monic polynomial $f(X) \in O_v[X]$ which splits in E and for every simple zero $\bar{a} \in \bar{F}_v$ of $\bar{f}(X)$ there is a zero $a \in O_v$ of $f(X)$ with residue $\bar{a}$ in $\bar{F}_v$;*

(f) **The Hensel–Rychlik condition**: *if $g(X) \in O_v[X]$ splits in E and $a \in O_v$ satisfies $v(g(a)) > 2v(g'(a))$, then there exists a zero c of $g(X)$ such that*

$$v(a - c) = v(g(a)/g'(a)).$$

PROOF. (a)⇔(b): The reduction from the case of normal extensions to the case of Galois extensions is by Proposition 14.2.5(a). The additional reduction from the Galois case to the finite Galois case (as in (b)) is immediate.

(a)⇒(c) By (a), v has a unique extension u to E (up to equivalence). Let $a \in E$ be a zero of $f(X)$. By assumption, $f(X)$ splits in E as

$$f(X) = \prod_{i=1}^{n}(X - \sigma_i(a))$$

for some $\sigma_1, \ldots, \sigma_n \in \mathrm{Aut}(E/F) = Z(u/v)$. Since O_u is integrally closed (Proposition 3.1.3), $\sigma_i(a) \in O_u$, $i = 1, \ldots, n$. We may therefore take residues in $\bar{E}_u$ to obtain

$$\bar{f}(X) = \prod_{i=1}^{n}(X - \overline{\sigma_i(a)}) = \prod_{i=1}^{n}(X - \bar{\sigma}_i(\bar{a})),$$

where $\bar{\sigma}_i$ is the image of σ_i under the epimorphism $\Phi\colon \mathrm{Aut}(E/F) \to \mathrm{Aut}(\bar{E}_u/\bar{F}_v)$ of Theorem 16.1.1. Thus $\bar{f}(X) \in \bar{F}_v[X]$ splits in $\bar{E}_u[X]$ and its zeros are all conjugate to $\bar{a}$ over $\bar{F}_v$. Since $\bar{F}_v[X]$ is a unique factorization domain, this implies that $\bar{f}(X)$ is a power of $\mathrm{irr}(\bar{a}, \bar{F}_v)$.

(c)⇒(d): We may assume that $\tilde{g}(X), \tilde{h}(X)$ are monic. Let u be an extension of v to E. Write

$$f(X) = f_1(X) \cdots f_k(X),$$

with $f_i(X) \in F[X]$ monic and irreducible. Again, since $f(X)$ splits in E and O_u is integrally closed, the zeros of $f(X)$ are in O_u. Hence, so are the zeros of each $f_i(X)$. It follows that

$$f_i(X) \in (O_u \cap F)[X] = O_v[X], \quad i = 1, \ldots, k.$$

By (c), $\bar{f}_i(X) = \bar{r}_i(X)^{m_i}$ for some irreducible monic polynomial $\bar{r}_i(X) \in \bar{F}_v[X]$ and some positive integer m_i. Then

$$\tilde{g}(X)\tilde{h}(X) = \bar{f}(X) = \bar{r}_1(X)^{m_1} \cdots \bar{r}_k(X)^{m_k}$$

in $\bar{F}_v[X]$. Since $\bar{F}_v[X]$ is a unique factorization domain and $\tilde{g}(X)$ and $\tilde{h}(X)$ are relatively prime, we get after renumbering:

$$\tilde{g}(X) = \prod_{i=1}^{l}\bar{r}_i(X)^{m_i} = \prod_{i=1}^{l}\bar{f}_i(X), \quad \tilde{h}(X) = \prod_{i=l+1}^{k}\bar{r}_i(X)^{m_i} = \prod_{i=l+1}^{k}\bar{f}_i(X).$$

Now let

$$g(X)=\prod_{i=1}^{l} f_i(X), \quad h(X)=\prod_{i=l+1}^{k} f_i(X).$$

Then $g(X)$ and $h(X)$ are polynomials in $O_v[X]$ with residues $\tilde{g}(X), \tilde{h}(X)$, respectively, and $f(X)=g(X)h(X)$.

(d)⇒(e): By assumption, $\bar{f}(X)=(X-\bar{a})\tilde{h}(X)$ for some $\tilde{h}(X)\in \bar{F}_v[X]$ with $\tilde{h}(\bar{a})\neq 0$. From (d) we obtain polynomials $g(X), h(X)\in O_v[X]$ such that

$$f(X)=g(X)h(X), \quad \bar{g}(X)=X-\bar{a}, \quad \bar{h}(X)=\tilde{h}(X).$$

Since $f(X)$ is monic, so are $\bar{f}(X)$ and $\tilde{h}(X)$. We obtain:

$$\deg(g)=\deg(f)-\deg(h)=\deg(\bar{f})-\deg(\bar{h})=\deg(\bar{g})=1.$$

The leading coefficient of $g(X)$ is in G_v, so we may assume without loss of generality that $g(X)$ is in fact monic; thus $g(X)=X-a$ for some $a\in O_v$. Then $f(a)=0$ and the residue of a in $\bar{F}_v$ is $\bar{a}$.

(e)⇒(b): We may assume for simplicity that E/F is already a finite Galois extension.

Let u be an extension of v to E. As before, let E_Z be the decomposition field of $(E,u)/(F,v)$. Lemma 15.2.1 yields $a\in O_{u_Z}$ such that $u(1-a)>0$ and $(u\circ\sigma)(a)>0$ whenever $\sigma\in \mathrm{Gal}(E/F)\setminus Z(u/v)$. In particular, $u(a)=0$.

Let $\sigma_1(a),\dots,\sigma_k(a)$ be the distinct F-conjugates of a, where $\sigma_1,\dots,\sigma_k\in \mathrm{Aut}(E/F)$ and $\sigma_1=\mathrm{Id}$. For $2\le i\le k$ the restriction of σ_i to E_Z is non-trivial (e.g., on a). Hence $\sigma_i\notin Z(u/v)$, so $(u\circ\sigma_i)(a)>0$. For the residues in $\bar{E}_u$ we thus have $\overline{\sigma_1(a)}=\bar{a}=\bar{1}$ and $\overline{\sigma_i(a)}=0$ for $2\le i\le k$.

Now let $f(X)=\mathrm{irr}(a,F)$. Write

$$f(X)=\prod_{i=1}^{k}(X-\sigma_i(a))=X^k+c_1X^{k-1}+\cdots+c_{k-1}X+c_k.$$

Expressing the c_j as symmetric polynomials in $\sigma_1(a),\dots,\sigma_k(a)$, we see that

$$c_1,\dots,c_k\in O_v \quad , \quad \bar{c}_1=-\bar{1} \quad , \quad \bar{c}_2=\cdots=\bar{c}_k=0.$$

Hence

$$\bar{f}(X)=X^k-X^{k-1}=X^{k-1}(X-\bar{1}),$$

so $\bar{1}$ is a simple root of $\bar{f}(X)$. By (e), it lifts to a root of $f(X)$ in O_v. But $f(X)$ is irreducible in $F[X]$, so necessarily $\deg(f)=1$, i.e., $a\in F$. By the choice of a, this means that $\mathrm{Gal}(E/F)=Z(u/v)$, as desired.

(f)⇒(e): Let $f(x)$ and $\bar{a}$ be as in (e). Take in (f) $g(X)=f(X)$, and choose an arbitrary $a\in O_v$ with residue $\bar{a}$. Then $v(g(a))>0$ and $v(g'(a))=0$. We obtain a zero c of $f(X)$ such that $v(a-c)=v(f(a))>0$, so $\bar{a}=\bar{c}$.

(e)⇒(f): If $g(a)=0$, then we take $c=a$. So suppose that $g(a)\neq 0$. The assumption then implies that $g'(a)\neq 0$ as well. By the Taylor expansion formula (Lemma 18.1.1),

$$g(a+X)=g(a)+g'(a)X+X^2h(X),$$

with $h(X) \in O_v[X]$. Substitute $X = -(g(a)/g'(a))Y$ to obtain that

$$g\Big(a - \frac{g(a)}{g'(a)}Y\Big) = g(a)\left[1 - Y + \frac{g(a)}{g'(a)^2}h_1(Y)Y^2\right],$$

where

$$h_1(Y) = h\Big(-\frac{g(a)}{g'(a)}Y\Big) \in O_v[Y].$$

Let

$$f(Y) = 1 - Y + \frac{g(a)}{g'(a)^2}h_1(Y)Y^2.$$

The assumption that $v(g(a)) > 2v(g'(a))$ implies that $v(f(1)) > 0$, while $v(f'(1)) = 0$. Thus 1 is a simple root of $\bar{f}(Y)$ in $\bar{F}_v$. Since $g(X)$ splits in E, so does $f(Y)$. Condition (e) therefore gives rise to a zero $b \in G_v$ of $f(Y)$. Then

$$c = a - \frac{g(a)}{g'(a)}b$$

is zero of $g(X)$. Moreover, $v(a - c) = v(g(a)/g'(a))$. □

Another important equivalent condition, the *Krasner-Ostrowski lemma*, will be discussed in §18.5. For other equivalent conditions (in the more classical case, where E is the algebraic closure of F) see [**Ri2**]. The fact that the equivalence of the main classical forms of Henselity can also be formulated in this relative setting was noted by Bröcker [**Br1**].

18.2 nth powers

One important use of Henselity is to show that groups of nth powers in the field are open with respect to the valuation topology.

PROPOSITION 18.2.1. *Let n be a positive integer and let E/F be a normal extension such that E contains all nth roots of elements of F. Let v be a valuation on F which is Henselian relative to E. Then*

$$1 + n^2\mathfrak{m}_v \le (F^\times)^n.$$

In particular, if char $F \nmid n$, *then $(F^\times)^n$ is open with respect to the topology $\mathcal{T}_v$.*

PROOF. Take $d \in 1 + n^2\mathfrak{m}_v$. By assumption, the polynomial $g(X) = X^n - d$ splits in E. Observe that $v(g(1)) > 2v(n) = 2v(g'(1))$. The Hensel-Rychlik condition (condition (f) of Theorem 18.1.2) therefore yields $c \in F$ such that $c^n = d$.

The second assertion follows from the first. □

18.2.2 EXAMPLES. The assumptions of the first part of Proposition 18.2.1 are satisfied in the following situations:

(1) n is arbitrary and E is the algebraic closure of F (so v is Henselian);
(2) n is arbitrary and E is the solvable closure of F (so v is solvably Henselian);
(3) $n = p$ is prime, F contains a root of unity of order p, and $E = F(p)$ is the maximal pro-p Galois extension of F (so v is p-Henselian).

REMARK 18.2.3. Let n be a positive integer. Consider a valuation v on F with residue field of characteristic not dividing n. Then v is $(F^\times)^n$-compatible (in the sense of §11.1) if and only if $1+n^2\mathfrak{m}_v \le (F^\times)^n$. In this sense, the compatibility relation is a weak form of Henselity.

The following result of Wadsworth [**Wad**, Prop. 1.2] shows that when the residue field contains a root of unity of order p, the converse of Example 18.2.2(3) also holds.

PROPOSITION 18.2.4. *Let p be a prime number and let v be a valuation on F such that $\bar F_v$ contains a primitive pth root of unity. Then v is p-Henselian if and only if it is $(F^\times)^p$-compatible.*

PROOF. Note that the assumption implies in particular that $\operatorname{char} \bar F_v \neq p$.

We have just seen the "only if" part.

For the "if" part, assume that v is $(F^\times)^p$-compatible, but that a p-Henselization $(\hat F,\hat v)$ of (F,v) is a proper extension of F. Let E be a minimal proper extension of F inside $\hat F$. Then the extension E/F is finite, and its Galois hull M is contained in $F(p)$. Thus $\mathrm{Gal}(M/E)$ is a maximal proper subgroup of the finite p-group $\mathrm{Gal}(M/F)$. Hence it is normal of index p. Therefore E/F is a Galois extension of degree p. By Kummer's theory, $E=F(\sqrt[p]{a})$ for some $a\in F^\times\setminus(F^\times)^p$.

Let $u_1,\dots,u_m$ be the distinct extensions of v to E (up to equivalence), say with $\hat v$ extending u_1. By Remark 15.1.1(e), $Z(u_1/v)=1$, so v is not Henselian relative to E, i.e., $m\ge 2$. By the fundamental inequality (Theorem 17.1.5),

$$e(u_1/v)f(u_1/v)<[E:F]=p.$$

Now if $v(a)\notin v(F^\times)$, then $p|e(u_1/v)$, a contradiction. Therefore $a\in O_v^\times(F^\times)^p$. Without loss of generality, we may assume that $a\in O_u^\times$. The residue $\bar a$ of a in $\bar F_v$ is then a pth power in $\bar E_{u_1}^\times$. Since

$$[\bar E_{u_1}:\bar F_v]=f(u_1/v)<p$$

and since $\bar F_v$ contains a primitive pth root of unity, Kummer's theory again implies that in fact $\bar a\in(\bar F_v^\times)^p$. Therefore $a\in G_v(F^\times)^p=(F^\times)^p$, by the compatibility assumption. This gives the desired contradiction. □

We conclude with a positive characteristic analog of Proposition 18.2.1.

LEMMA 18.2.5. *Let p be a prime number and let $\wp(x)=x^p-x$. Let E/F be a normal extension of fields of characteristic p such that $F\subseteq\wp(E)$. Let v be a valuation on F which is Henselian relative to E. Then*

$$\mathfrak{m}_v\le\wp(F).$$

In particular, for v nontrivial, $\wp(F)$ is open in $\mathcal{T}_v$.

PROOF. Take $d\in\mathfrak{m}_v$. By assumption, $g(X)=\wp(X)-d$ has a root in E, and therefore splits in E. Its reduction to $\bar F_v[X]$ has p distinct roots in $\bar F_v$, namely the elements of $\mathbb{F}_p$. Condition (e) of Theorem 18.1.2 allows us to lift them to roots of $g(X)$ in F.

The second assertion follows from the first and from the fact that $\wp(F)$ is an additive subgroup of F. □

18.3 Example: complete valued fields

It was the discovery of Kurt Hensel [**He**] that p-adic fields satisfy condition (d) of Theorem 18.1.2 when E is the algebraic closure (whence the names "Hensel's lemma" and "Henselity"). In fact, his argument applies to any field which is complete with respect to a valuation of rank 1. We prove a somewhat sharper version of the equivalent condition (e) of the theorem for such fields.

THEOREM 18.3.1. *Let v be a complete rank 1 valuation on the field F. Let $f(X) \in O_v[X]$ and $a \in O_v$ satisfy $v(f(a)) > 0$ and $v(f'(a)) = 0$. Then there exists $b \in F$ such that $f(b) = 0$ and $v(b-a) \geq v(f(a))$. Moreover, b is the unique zero of $f(X)$ satisfying $v(b-a) > 0$.*

PROOF. For the existence of b set $\alpha = v(f(a))$. Thus $\alpha > 0$. We define inductively a sequence $a_1, a_2, \ldots$ in F such that $a_1 = a$ and for every $n \geq 1$:

(1) $v(a_n - a_{n+1}) \geq n\alpha$;
(2) $v(f(a_n)) \geq n\alpha$;
(3) $v(f'(a_n)) = 0$.

Suppose that a_n has already been defined. By (3), $f'(a_n) \neq 0$. Let $h = f(a_n)/f'(a_n)$ and let $a_{n+1} = a_n - h$. By (2) and (3), $v(a_n - a_{n+1}) = v(h) \geq n\alpha$. The Taylor expansion formula (Lemma 18.1.1) gives:

$$f(a_{n+1}) \in f(a_n) - hf'(a_n) + h^2 O_v = h^2 O_v.$$

Hence

$$v(f(a_{n+1})) \geq 2v(h) \geq 2n\alpha \geq (n+1)\alpha.$$

Furthermore, $v(a_n - a_{n+1}) \geq n\alpha$ implies that

$$v(f'(a_n) - f'(a_{n+1})) \geq n\alpha > 0.$$

Since $v(f'(a_n)) = 0$ (by (3)) also $v(f'(a_{n+1})) = 0$, completing the inductive construction.

It now follows from (1) and the ultrametric inequality that $v(a_n - a_m) \geq n\alpha$ for $n \leq m$. In particular, $\{a_n\}_{n=1}^{\infty}$ is a Cauchy sequence with respect to the metric d_v of §9.1. By the completeness, it converges to a limit $b \in F$. Since $\mathcal{T}_v$ is a ring topology (Proposition 8.1.1), $f(X)$ is $\mathcal{T}_v$-continuous. It therefore follows from (2) that $f(b) = 0$. Also,

$$v(a_n - a) = v(a_n - a_1) \geq \alpha$$

for all n, so $v(b-a) \geq \alpha$. In particular, $v(b) \geq 0$.

For the uniqueness, suppose that c is a zero of $f(X)$ with $v(c-a) > 0$. Then $c \in O_v$. Write $f(X) = (X-b)g(X)$, with $g(X) \in F[X]$. Let $\bar{v}$ be the restricted Gauss valuation on $F(X)$ extending v (see Example 4.3.2). Then

$$\bar{v}(X-b) = \min\{v(1), v(-b)\} = 0.$$

Hence $0 \leq \bar{v}(f) = \bar{v}(g)$, so $g(X) \in O_v[X]$. As

$$f'(X) = g(X) + (X-b)g'(X),$$

we have

$$f'(c) = g(c) + (c-b)g'(c). \tag{18.3.1}$$

Since $v(f'(a)) = 0$ and $v(c-a) > 0$ we also have $v(f'(c)) = 0$. In addition,

$$v(c-b) \geq \min\{v(c-a), v(a-b)\} > 0$$

and $v(g'(c)) \geq 0$. It now follows from (18.3.1) that $g(c) \neq 0$. From $0 = f(c) = (c-b)g(c)$ we deduce that $c = b$. □

COROLLARY 18.3.2. *Let v be a complete rank 1 valuation on the field F and let E be a normal extension of F. Then (F, v) is Henselian relative to E.*

PROOF. Theorem 18.3.1 gives the assertion for the algebraic closure of F. By Corollary 15.3.2, it holds for every normal extension of F. □

COROLLARY 18.3.3. *Let v be a rank 1 valuation on the field F and let $(\hat{F}, \hat{v})$ be the completion of (F, v).*

(a) *Any Henselization of (F, v) embeds over (F, v) in $(\hat{F}, \hat{v})$.*
(b) *If $\operatorname{char} \bar{F}_v = 0$ or v is discrete, then the relative algebraic closure of F in $\hat{F}$ is a Henselization of (F, v).*

PROOF. (a) This follows from Corollary 18.3.2 and the universal property of Henselizations (Theorem 15.3.5).

(b) Let F' be the relative algebraic closure of F in $\hat{F}$, and let v' be the restriction to it of $\hat{v}$. Since $(\hat{F}, \hat{v})$ is Henselian, so is (F', v'), by Proposition 15.3.3. Hence it is an algebraic extension of some Henselization of (F, v). The assumptions imply that this extension is defectless (by Proposition 17.3.2 and Theorem 17.4.3). Moreover, this extension is immediate since $(\hat{F}, \hat{v})/(F, v)$ is immediate. By Ostrowski's formula (Theorem 17.2.1), the extension is trivial, i.e., (F', v') is a Henselization of (F, v). □

When combined with Examples 9.2.1 and 9.2.2, Corollary 18.3.2 above gives some fundamental examples of Henselian valued fields.

18.3.4 EXAMPLES.

(1) (Hensel) The canonical p-adic valuation v_p on the field $\mathbb{Q}_p$ of p-adic numbers is Henselian. Thus it has a unique extension to any algebraic extension F of $\mathbb{Q}_p$, and this valuation is also Henselian, by Corollary 15.3.2.
(2) By (1) and by Proposition 15.3.3, the relative algebraic closure $\mathbb{Q}_{p,\mathrm{alg}}$ of $\mathbb{Q}$ in $\mathbb{Q}_p$ is Henselian.
(3) Let $\bar{F}$ be a perfect field of characteristic $p > 0$ and let F be the quotient field of its ring $W(\bar{F})$ of Witt vectors (see §12.4). By Theorem 12.4.1, F is equipped with a canonical discrete complete valuation with residue field $\bar{F}$. By Corollary 18.3.2, it is Henselian.

From Example 18.3.4(1) one can deduce that on a p-adic field, the p-adic valuation is the only "interesting" valuation, in the following sense.

EXAMPLE 18.3.5. Let $F = \mathbb{Q}_p$ for some prime number p and let v_p be its p-adic valuation. We show that every valuation v on F which is not equivalent to v_p has a divisible value group. This is clear for the trivial valuation. So suppose that v is non-trivial and let n be a positive integer. As we have seen in Example 10.1.2, v_p and v are necessarily independent. Now v_p is Henselian (Example 18.3.4(1)), so by Proposition 18.2.1, $(F^\times)^n$ is $\mathcal{T}_{v_p}$-open. It follows from Proposition 10.3.1(a) that $F^\times = (F^\times)^n O_v^\times$, i.e., $v(F^\times) = nv(F^\times)$.

As an application we obtain:

COROLLARY 18.3.6. *Let p, q be prime numbers and let $\sigma: \mathbb{Q}_p \to \mathbb{Q}_q$ be a field embedding. Then $p = q$ and σ is the identity map.*

PROOF. Denote again the canonical valuations on $\mathbb{Q}_p, \mathbb{Q}_q$ by v_p, v_q, respectively. Then v_p and $v_q \circ \sigma$ are discrete valuations on $\mathbb{Q}_p$. By what we have just seen, this can happen only if they are equivalent. Since σ is the identity on $\mathbb{Q}$, one has $(v_q \circ \sigma)(q) > 0$. Hence also $v_p(q) > 0$, so $p = q$. It follows that σ is an isometry of the metric space $(\mathbb{Q}_p, d_{v_p})$ with itself (see §9.1), and in particular, is continuous in the induced topology $\mathcal{T}_{v_p}$. But $\mathbb{Q}$ is $\mathcal{T}_{v_p}$-dense in $\mathbb{Q}_p$. We conclude that σ is the identity on $\mathbb{Q}_p$. □

Finally, we record the following topological fact which sometimes serves as a partial converse of Proposition 18.2.1.

PROPOSITION 18.3.7. *Let v be a discrete valuation on a field F and let n be a positive integer. Every $\mathcal{T}_v$-open subgroup of $F^\times$ containing $(F^\times)^n$ must contain $(F^\times)^n(1 + n^2\mathfrak{m}_v)$.*

PROOF. Let (E, u) be the completion of (F, v). It is Henselian (Corollary 18.3.2). By Theorem 9.3.2(f), $(E, u)/(F, v)$ is an immediate extension. Hence $\mathfrak{m}_v^i = F \cap \mathfrak{m}_u^i$ for $i \geq 1$.

Now let S be a $\mathcal{T}_v$-open subgroup of $F^\times$ containing $(F^\times)^n$. The subgroups $1 + \mathfrak{m}_v^i$, $i = 1, 2, \ldots$, form a local basis of $\mathcal{T}_v$ at 1. Therefore $1 + \mathfrak{m}_v^i \leq S$ for some $i \geq 1$. By Proposition 3.4.5, $E^\times = F^\times(1 + \mathfrak{m}_u^i)$. Hence $(E^\times)^n \leq (F^\times)^n(1 + \mathfrak{m}_u^i)$. By Proposition 18.2.1, $1 + n^2\mathfrak{m}_u \leq (E^\times)^n$. We conclude that

$$(F^\times)^n(1 + n^2\mathfrak{m}_v) \leq F \cap (E^\times)^n \leq (F^\times)^n(1 + \mathfrak{m}_v^i) \leq S. \qquad \square$$

18.4 Example: power series fields

The Henselity of complete valued fields of rank 1 established in the previous section implies in particular that whenever Γ is an ordered abelian group of rank 1, the field $K((\Gamma))$ of generalized formal power series over K is Henselian with respect to its canonical valuation $v(f) = \min(\mathrm{Supp}(f))$ (see Example 4.2.1). However, this turns out to be valid for ordered abelian groups of arbitrary ranks, as we now show.

THEOREM 18.4.1 (Krull [**Kru2**]). *Let K be a field and Γ an ordered abelian group. Let $F = K((\Gamma))$, and let $v: F^\times \to \Gamma$ be its canonical valuation. Then (F, v) has no proper immediate extensions.*

PROOF. Assume that $(E, u)/(F, v)$ is an immediate extension, and let $a \in E^\times$. We need to show that $a \in F$.

To this end let S be the set of all $f \in F$ such that $u(a - f)$ (as an element of $\Gamma \cup \{\infty\}$) is strictly larger than every element of the support $\mathrm{Supp}(f)$ of f. In particular, $0 \in S$. We partially order S by the relation $\preceq$ defined in §2.7. We recall that $f \preceq f'$ means that $f(\delta) = f'(\delta)$ for every $\delta \in \Gamma$ for which there exists $\gamma \in \mathrm{Supp}(f)$ with $\delta \leq \gamma$.

Assume that C is a non-empty chain in $(S, \preceq)$. By Lemma 2.7.6, it has a least upper bound $f^* \in F$ satisfying $\mathrm{Supp}(f^*) = \bigcup_{f \in C} \mathrm{Supp}(f)$. We claim that $f^* \in S$. Indeed, consider $\gamma \in \mathrm{Supp}(f^*)$. Then $\gamma \in \mathrm{Supp}(f)$ for some $f \in C$. As $f \preceq f^*$ this implies that $\gamma < v(f - f^*)$. Also, $f \in C \subseteq S$ gives $\gamma < u(a - f)$. By the ultrametric inequality, $\gamma < u(a - f^*)$, as claimed.

Consequently Zorn's lemma yields a maximal element f in $(S, \preceq)$. Suppose that $a \neq f$. Then $\gamma = u(a - f) \in \Gamma$. By Lemma 3.2.5, $E^\times = F^\times G_u$. Therefore there exists $g \in F^\times$ with $a - f \in gG_u$. Thus $u(g) = \gamma$. In fact, we may choose $g = bt^\gamma$ for some $b \in K^\times$. Let $f' = f + g$. Since γ is strictly larger than every element of $\mathrm{Supp}(f)$ we have $f \prec f'$. Furthermore,

$$u(a - f') = u(a - f - g) > u(g) = \gamma = \max(\mathrm{Supp}(f')).$$

Hence $f' \in S$. This contradiction to the maximality of f in $(S, \preceq)$ shows that $a = f \in F$, as desired. $\square$

Krull's original proof of Theorem 18.4.1 uses a transfinite construction, which is replaced in the above argument by the application of Zorn's lemma. A proof of an analytic flavor is given in [**Ri1**].

Valued fields having no proper immediate extensions are called **maximally complete**. We refer to [**Kap1**] for an analytic characterization of these fields. Since Henselizations are always immediate extensions (by Proposition 15.3.7), maximally complete fields are necessarily Henselian. In particular, we deduce from Theorem 18.4.1:

COROLLARY 18.4.2. *Let K be a field and let Γ be an ordered abelian group. Then $K((\Gamma))$ is Henselian with respect to its canonical valuation.*

EXAMPLE 18.4.3 (MacLane [**Mac2**]). Let K be an algebraically closed field and Γ a divisible ordered abelian group. By Corollary 14.2.3, every algebraic extension of $K((\Gamma))$ must be immediate with respect to the canonical valuation. In view of Theorem 18.4.1, this extension must be trivial. We conclude that $K((\Gamma))$ is algebraically closed. Of course, when $\mathrm{char}\, K = 0$ this is also a special case of Corollary 17.3.3.

EXAMPLE 18.4.4: *Γ-Puiseux series.*

Let K be an arbitrary field, Γ an ordered abelian group, and

$$K_{\mathrm{Puis}}((\Gamma)) = \varinjlim K((\Gamma'))$$

the field of Γ-Puiseux series over K, where Γ' ranges over all subgroups of Γ_{div} which contain Γ as a subgroup of finite index (§2.8). Let v be the canonical valuation on $K_{\mathrm{Puis}}((\Gamma))$ with respect to the trivial valuation on K, as in Example 4.2.3. Thus the value group of v is Γ_{div} and its residue field is K. It extends the canonical valuation on $F = K((\Gamma))$, which is Henselian by Corollary 18.4.2. As we have observed in §2.8, $K_{\mathrm{Puis}}((\Gamma))$ is algebraic over $K((\Gamma))$. We conclude from Corollary 15.3.2 that v is also Henselian.

PROPOSITION 18.4.5. *Assume that* $\mathrm{char}\, K = 0$ *and let Γ be an ordered abelian group.*

(a) *$K_{\mathrm{Puis}}((\Gamma))$ is the relative algebraic closure of $K((\Gamma))$ in $K((\Gamma_{\mathrm{div}}))$.*
(b) *If K is algebraically closed, then $K_{\mathrm{Puis}}((\Gamma))$ is the algebraic closure of $K((\Gamma))$.*

PROOF. (a) The canonical valuation on $K((\Gamma))$ is Henselian and its residue field K has characteristic 0. Hence it is defectless relative to the algebraic closure (Proposition 17.3.2). By the tower property for the defect (17.2.1), so is its extension v to $K_{\mathrm{Puis}}((\Gamma))$. Furthermore, the extension $K((\Gamma_{\mathrm{div}}))/K_{\mathrm{Puis}}((\Gamma))$ is immediate with respect to the canonical valuations. Therefore the relative algebraic closure of $K_{\mathrm{Puis}}((\Gamma))$ in $K((\Gamma_{\mathrm{div}}))$ is also an immediate extension of $K_{\mathrm{Puis}}((\Gamma))$. Being algebraic, defectless and immediate, this extension is necessarily trivial, i.e., $K_{\mathrm{Puis}}((\Gamma))$ is algebraically closed in $K((\Gamma_{\mathrm{div}}))$, as required.

(b) By Example 18.4.3, $K((\Gamma_{\mathrm{div}}))$ is algebraically closed. By (a), $K_{\mathrm{Puis}}((\Gamma))$ is therefore also algebraically closed. Since it is algebraic over $K((\Gamma))$, it is its algebraic closure. □

The assumption that $\mathrm{char}\, K = 0$ in Proposition 18.4.5(b) (and hence also in Proposition 18.4.5(a)) cannot be omitted even when $\Gamma = \mathbb{Z}$, as the following example demonstrates.

EXAMPLE 18.4.6 (Abhyankar [**Ab**], Chevalley [**Ch**, p. 64]). Let K be a field of characteristic $p > 0$. As usual, let t stand for the map in $K((\mathbb{Z}))$ which is 1 at 1 and 0 elsewhere. We claim that the polynomial equation

$$X^p - X - 1/t = 0$$

has no solutions in $K_{\mathrm{Puis}}((\mathbb{Z})) = \varinjlim K((\frac{1}{m}\mathbb{Z}))$. In particular, the latter field is not algebraically closed.

To prove the claim, suppose that $f \in K_{\mathrm{Puis}}((\mathbb{Z}))$ satisfies $f^p - f = 1/t$. Then $f \in K((\frac{1}{m}\mathbb{Z}))$ for some positive integer m. Let v_m be the canonical valuation on $K((\frac{1}{m}\mathbb{Z}))$ with value group $\frac{1}{m}\mathbb{Z}$. Take i maximal with $p^i | m$, and let

$$g = f - t^{-1/p} - t^{-1/p^2} - \cdots - t^{-1/p^i}$$

(when $i = 0$ we take $g = f$). Observe that $g \in K((\frac{1}{m}\mathbb{Z}))$ and $g^p - g = t^{-1/p^i}$. We obtain that

$$v_m(g) = -1/p^{i+1} \notin \tfrac{1}{m}\mathbb{Z}$$

(Lemma 3.2.1(e)), a contradiction

Still in the case where K is algebraically closed of characteristic $p > 0$, Kedlaya [**Ked**], extending earlier works by Huang, Rayner [**Ra2**], and Ştefănescu [**Ste**], realizes the algebraic closure of $K((\mathbb{Z}))$ explicitly as the *"twist-recurrent"* maps in the algebraically closed field $K((\mathbb{Q}))$. In particular, the support of such maps must be contained in a set of the form

$$\{1/mp^i \mid i = 0, 1, 2, \dots\}$$

for some positive integer m.

18.5 The Krasner–Ostrowski lemma

The following lemma gives another condition which is equivalent to relative Henselity. Its importance lies in the fact that it connects the arithmetic (or better said, analytic) properties of a valued field with its Galois theory. The implication (a)⇒(b) is known in the folklore as "Krasner's Lemma" (see [**Kra**]), however, as pointed out by Roquette in [**Ro**], the argument actually goes back to Ostrowski [**O1**].

LEMMA 18.5.1. *Let $(E,u)/(F,v)$ be a normal extension of valued fields. The following conditions are equivalent:*

(a) *v is Henselian relative to E;*

(b) *if $a,b \in E$ and if $u(b-\sigma(b)) < u(b-a)$ whenever $\sigma \in \mathrm{Aut}(E/F)$ and $\sigma(b) \neq b$, then the extension $F(a,b)/F(a)$ is purely inseparable.*

PROOF. (a)$\Rightarrow$(b): Suppose that $\sigma \in \mathrm{Aut}(E/F)$ fixes a. We need to show that it fixes b as well.

Since v is Henselian relative to E, one has $u = u \circ \sigma$. In particular, $u(b-a) = u(\sigma(b)-\sigma(a))$. From the ultrametric inequality we obtain that

$$u(b-\sigma(b)) = u\big((b-a)-(\sigma(b)-\sigma(a))\big) \geq u(b-a).$$

Our assumption now implies that $\sigma(b) = b$.

(b)$\Rightarrow$(a): In light of condition (b) of Theorem 18.1.2, we may assume that E/F is a finite Galois extension. Lemma 15.2.1 yields $b \in E_Z$ such that $u(1-b) > 0$ and $(u\circ\sigma)(b) > 0$ whenever $\sigma \in \mathrm{Gal}(E/F) \setminus Z(u/v)$. Then $u(b) = 0$.

Now suppose that $\sigma \in \mathrm{Gal}(E/F)$ satisfies $\sigma(b) \neq b$. Then $\sigma \notin Z(u/v)$, so $(u\circ\sigma)(b) > 0$, and therefore $u(b-\sigma(b)) = 0$. Since the extension $(E_Z,u_Z)/(F,v)$ is immediate (Theorem 15.2.2), there exists $a \in O_v$ with the same residue as b, i.e., $u(b-a) > 0$. It follows from (b) that the extension $F(b)/F$ is purely inseparable. But E/F is Galois, so necessarily $b \in F$. By the choice of b, this can happen only if $\mathrm{Gal}(E/F) = Z(u/v)$. We conclude that v is Henselian relative to E. □

The main importance of Lemma 18.5.1 lies in the following consequence of it. It shows that in a relatively Henselian field, minor changes in the coefficients of a separable polynomial do not effect the splitting field.

For a valued field (F,v) we denote again the restricted classical Gauss valuation on $F(X)$ by $\bar{v}$. Thus

$$\bar{v}(f) = \min\{v(c_0),\dots,v(c_n)\}$$

for $f(X) = \sum_{i=0}^n c_i X^i$.

THEOREM 18.5.2. *Let E/F be a normal extension of fields and let v be a Henselian valuation on F relative to E. Let $f(X) \in F[X]$ be a monic polynomial of degree n with n distinct roots $a_1,\dots,a_n$ in E. Then there exists $\alpha \in v(F^\times)$ such that any other monic polynomial $g(X) \in F[X]$ of degree n which splits in E and with $\bar{v}(f-g) > \alpha$ has n distinct roots $b_1,\dots,b_n$ in E such that $F(a_i) = F(b_i)$, $i = 1,\dots,n$. In particular, f and g have the same splitting field over F.*

PROOF. Let u be the unique extension of v to E. We may assume that it is non-trivial and that $n \geq 2$. Choose $\gamma \in u(E^\times)$ with

$$\max\{u(a_i-a_j) \mid i \neq j\} < \gamma.$$

By the continuity of roots in the valuation topology $\mathcal{T}_u$ (Theorem 8.2.3) there exists a positive $\alpha \in u(E^\times)$ such that if $b_1,\dots,b_n \in E$ and

$$g(X) = \prod_{i=1}^n (X-b_i) \in F[X]$$

satisfy $\bar{u}(f - g) > \alpha$ then, after re-numbering, $u(a_i - b_i) > \gamma$, $i = 1, \dots, n$. In light of the cofinality of $v(F^\times)$ in $u(E^\times)$ (Corollary 14.2.3(e)) we may assume that $\alpha \in v(F^\times)$. For $b_1, \dots, b_n$ as above, the ultrametric inequality gives for $i \neq j$:

$$\min\{\gamma, u(b_i - b_j)\} \leq \min\{u(a_i - b_i), u(b_i - b_j), u(b_j - a_j)\} \leq u(a_i - a_j) < \gamma.$$

It follows that

$$\max\{u(b_i - b_j) \mid i \neq j\} < \gamma.$$

In particular, $b_1, \dots, b_n$ are distinct.

Lemma 18.5.1 implies that the extensions $F(a_i, b_i)/F(a_i)$ and $F(a_i, b_i)/F(b_i)$ are purely inseparable. On the other hand, since $a_1, \dots, a_n$ are the distinct zeros of $f(X) \in F[X]$, they are separable over F. Similarly, $b_1, \dots, b_n$ are separable over F. Consequently, $F(a_i) = F(a_i, b_i) = F(b_i)$, $i = 1, \dots, n$. □

COROLLARY 18.5.3. *Let $(\hat{F}, \hat{v})/(F, v)$ be an extension of non-trivially valued fields such that F is $\mathcal{T}_{\hat{v}}$-dense in $\hat{F}$ and $(\hat{F}, \hat{v})$ is Henselian. Let $\hat{F}_{\rm sep}, F_{\rm sep}$ be the separable closures of $\hat{F}, F$, respectively. Then $\hat{F}_{\rm sep}$ is the compositum $F_{\rm sep}\hat{F}$.*

PROOF. Obviously, $F_{\rm sep}\hat{F} \subseteq \hat{F}_{\rm sep}$. Conversely, let $a \in \hat{F}_{\rm sep}$ and let $f(X) = \mathrm{irr}(a, \hat{F})$. Using the density, we may choose a monic polynomial $g(X) \in F[X]$ of the same degree as $f(X)$ and whose coefficients are sufficiently close to those of $f(X)$ with respect to $\mathcal{T}_{\hat{v}}$. It follows from Theorem 18.5.2 that the splitting fields of $f(X)$ and $g(X)$ over $\hat{F}$ coincide. Moreover, $g(X)$ is separable over F, hence it splits in $F_{\rm sep}$. Therefore $f(X)$ splits in $F_{\rm sep}\hat{F}$. We conclude that $a \in F_{\rm sep}\hat{F}$. □

18.5.4 EXAMPLES.

(1) The algebraic closure $\tilde{\mathbb{Q}}_p$ of $\mathbb{Q}_p$ is the compositum $\tilde{\mathbb{Q}}\mathbb{Q}_p$, where $\tilde{\mathbb{Q}}$ is the field of algebraic numbers.

(2) Given a field F, the separable closure of $F((t))$ is $F(t)_{\rm sep}F((t))$. The latter field should not be confused with the Laurent series field $F_{\rm sep}((t))$ over $F_{\rm sep}$, which is usually much larger.

(3) Let F be a separably closed field and let v be a valuation of rank 1 on F. Then the completion $(\hat{F}, \hat{v})$ of (F, v) is also separably closed. Indeed, $(\hat{F}, \hat{v})$ is Henselian by Corollary 18.3.2, and we apply Corollary 18.5.3.

(4) (Kürschák [**Kür**]) Let v_0 be a valuation of rank 1 on a field F_0. Construct inductively an increasing sequence of valued fields of rank 1

$$(F_0, v_0) \subseteq (F_1, v_1) \subseteq (F_2, v_2) \subseteq (F_3, v_3) \subseteq \cdots$$

as follows:

(i) Take (F_{2i+1}, v_{2i+1}) to be the completion of (F_{2i}, v_{2i}). Since this is an immediate extension (Theorem 9.3.2(f)), v_{2i+1} indeed has rank 1.

(ii) Take F_{2i+2} to be the separable closure of F_{2i+1}, and v_{2i+2} to be the extension of v_{2i+1} to it. Notice that this extension is unique by Corollary 18.3.2, and has rank 1 by Corollary 14.2.3(c).

Now since (F_2, v_2) is separably closed, Example (3) above implies that so is its completion (F_3, v_3). Consequently, this sequence of extensions stabilizes from (F_3, v_3) and on.

(5) In particular, start with $F_0 = \mathbb{Q}$ and let v_0 be its p-adic valuation for some prime number p. Then $F_1 = \mathbb{Q}_p$ and $F_2 = \tilde{\mathbb{Q}}_p = \tilde{\mathbb{Q}}\mathbb{Q}_p$. The field F_3 is often denoted by $\mathbb{C}_p$. It is both complete and algebraically closed. Hence it is a convenient "universal domain" for carrying computations of combined algebraic/p-adic analytic nature.

CHAPTER 19

Real Closures

In this chapter we develop the ordered field analog of relative Henselizations, namely, relative real closures. Our presentation in §§19.1–19.2 will follow [**Bou2**, Ch. VI, §2], but in a relative setting. In the absolute case (i.e., relative to the algebraic closure) these results are a part of the classical Artin–Schreier theory ([**AS1**], [**AS2**]). The relative setting was developed by Becker in [**Be1**] and [**Be3**].

19.1 Extensions of orderings

Let K/F be an extensions of fields. We say that an ordering Q on K **extends** an ordering P on F if $P = F \cap Q$. In this case we say that $(K, Q)/(F, P)$ is an **extension of ordered fields**. Given P, we would first wish to know when does it have such an extension to K. In light of Proposition 6.1.5, this is the case if and only if $-1 \notin \sum P(K^\times)^2$. We now apply this fact to obtain more explicit criteria for the existence of extensions.

We say that elements a and b of F have **opposite signs** with respect to an ordering P if $-ab \in P$.

LEMMA 19.1.1. *Let (F, P) be an ordered field and let $f(X) \in F[X]$ be an irreducible polynomial. If there exist $a, b \in F$ such that $f(a), f(b)$ have opposite signs with respect to P, then P extends to an ordering on the field $K = F[X]/(f(X))$.*

PROOF. We argue by induction on $n = \deg(f)$. When $n = 1$ the assertion is trivial.

Next assume that $n \geq 2$ and that P does not extend to K. Then $-1 \in \sum P(K^\times)^2$. Thus

$$-1 = \sum_i c_i f_i(X)^2 \pmod{f(X)}$$

for some $c_i \in P$ and $f_i(X) \in F[X]$. We may further assume that $\deg(f_i) < n$. Then

$$1 + \sum_i c_i f_i(X)^2 = f(X)g(X)$$

for some $g(X) \in F[X]$ with $\deg(g) < n$. Substituting $X = a$ and $X = b$, we obtain that $f(a)g(a), f(b)g(b) \in P$. Since $f(a), f(b)$ have opposite signs with respect to P, this implies that so do $g(a), g(b)$. Therefore, there exists an irreducible factor $h(X)$ of $g(X)$ such that $h(a), h(b)$ have opposite signs with respect to P. The induction hypothesis implies that P extends to an ordering on $F[X]/(h(X))$. However, $-1 = \sum c_i f_i(X)^2 \pmod{h(X)}$ as well, and we got a contradiction. □

LEMMA 19.1.2. *Let (F, P) be an ordered field and let $f(X) = \sum_{i=0}^n c_i X^i$ be a monic polynomial in $F[X]$, where $c_n = 1$. Then there exists $M \in P$ such that*

$f(a) \in a^n P$ on $F \setminus (-M, M)_P$. In fact, one can take

$$M = \max\Big\{1, \sum_{i=0}^{n-1} |c_i|_P\Big\}.$$

PROOF. Take M as above, and suppose that $M <_P |a|_P$. By the triangle inequality,

$$\Big|\sum_{i=0}^{n-1} c_i/a^{n-i}\Big|_P \le_P \sum_{i=0}^{n-1} \Big| c_i/a^{n-i}\Big|_P \le_P \sum_{i=0}^{n-1} \Big| c_i/a\Big|_P <_P 1.$$

Hence

$$f(a) = a^n \Big(1 + \sum_{i=0}^{n-1} c_i/a^{n-i}\Big) \in a^n P. \qquad \square$$

COROLLARY 19.1.3. *Let (F, P) be an ordered field and let K be a finite extension of F. Then P extends to an ordering on K in any of the following situations:*
(1) *$[K : F]$ is odd;*
(2) *$K = F(\sqrt{a})$ for some $a \in P$.*

PROOF. (1) Since $\operatorname{char} F = 0$ (Corollary 6.1.7), there is a primitive element α for the extension K/F. Let $f(X) = \operatorname{irr}(\alpha, F)$. It is monic of odd degree. By Lemma 19.1.2, $f(X)$ changes signs with respect to P. Now identify K with $F[X]/(f(X))$ and use Lemma 19.1.1.

(2) This is trivial when $a \in (F^\times)^2$. When $a \notin (F^\times)^2$, the polynomial $f(X) = X^2 - a$ is irreducible in $F[X]$. By Lemma 19.1.2, some of its values are in P. Also, $f(0) = -a \in -P$. Lemma 19.1.1 therefore implies that P extends to $K = F[X]/(f(X))$. $\square$

We will also need the following computational fact.

LEMMA 19.1.4. *Let (F, P) be an ordered field such that $P = (F^\times)^2$, and let $K = F(\sqrt{-1})$. Then $K^\times = (K^\times)^2$.*

PROOF. Let $a, b \in F$. We need to find $x, y \in F$ such that

$$(x + y\sqrt{-1})^2 = a + b\sqrt{-1}. \tag{19.1.1}$$

Indeed, the assumption yields $c \in F$ with $c^2 = a^2 + b^2$. We may assume that $0 \le_P c$. Then $c \ge_P |a|_P$, i.e., $c \pm a \ge_P 0$. Using again the assumption we get $x, y \in F$ with

$$x^2 = (c + a)/2, \quad y^2 = (c - a)/2$$

(recall that $\operatorname{char} F = 0$). Then

$$x^2 - y^2 = a, \quad (xy)^2 = (c^2 - a^2)/4 = b^2/4.$$

Replace, if necessary, y by $-y$ to get $2xy = b$. This gives (19.1.1). $\square$

19.2 Relative real closures

Now let E/F be a normal extension and let P be an ordering on F. Zorn's lemma yields a maximal ordered field extension $(\hat{F}, \hat{P})$ of (F, P) inside E. We call it a **real closure of** (F, P) **relative to** E. We say that (F, P) is **real closed relative to** E if it is a real closure of itself relative to E, i.e., if P does not extend to any proper algebraic extension of F inside E.

19.2.1 EXAMPLES.

(1) When E is the algebraic closure of F one simply speaks about a **real closure** and a **real closed field**.

(2) When E is the maximal pro-2 Galois extension $F(2)$ of F (i.e., the quadratic closure of F), a real closure of (F, P) is called a **Euclidean closure**. A real closed field relative to $F(2)$ is called a **Euclidean** field [**Be1**] (we shall soon see that this coincides with the previous definition given in Example 11.1.1).

Similarly to Hensel's lemma and its variants in the case of relatively Henselian fields, there are various *"Nullstellensätze"* for relatively real closed fields. For instance, we have the following intermediate value property for polynomials:

PROPOSITION 19.2.2. *Let (F, P) be a real closed field relative to a normal extension E of F. Let $a <_P b$ be elements of F. Suppose that $f(X) \in F[X]$ splits in E, and that $f(a)$ and $f(b)$ have opposite signs with respect to P. Then $f(X)$ has a root in $(a, b)_P$.*

PROOF. There is a monic irreducible factor $g(X) \in F[X]$ of $f(X)$ such that $g(a)$ and $g(b)$ have opposite signs with respect to P. By Lemma 19.1.1, P extends to an ordering on $K = F[X]/(g(X))$. Note that K embeds in E over F. Since (F, P) is real closed relative to E, necessarily $F = K$. Thus $g(X) = X - c$ for some $c \in F$. We obtain $f(c) = 0$ and $a - c, b - c$ have opposite signs with respect to P. As $a <_P b$, this can happen only when $c \in (a, b)_P$. □

As the following proposition shows, relatively real closed fields can even be characterized by means of existence of zeros of polynomials of odd degrees and certain quadratic polynomials.

PROPOSITION 19.2.3. *Let (F, P) be an ordered field and let E be a normal extension of F. The following conditions are equivalent:*

(a) *(F, P) is real closed relative to E;*

(b) (1) *$[E : F]$ is even; and*
(2) *$(F^\times)^2 = P \cap (E^\times)^2$.*

PROOF. (a)⇒(b): (1) follows from Corollary 19.1.3(1).
Next we prove (2). Clearly $(F^\times)^2 \subseteq P \cap (E^\times)^2$. Conversely, suppose that $a \in P \cap (E^\times)^2$. By Corollary 19.1.3(2), P extends to $F(\sqrt{a})$. Since $F(\sqrt{a}) \subseteq E$, the real closedness implies that $\sqrt{a} \in F$, as required.

(b)⇒(a): By (1), every proper extension of F inside E must contain a proper quadratic extension. It thus remains to show that P does not extend to proper quadratic extensions inside E.

To this end, assume that $(F(\sqrt{a}), Q)$ is an ordered field extension of (F, P), with $a \in F \cap (E^\times)^2$. Then

$$a = (\sqrt{a})^2 \in F \cap Q \cap (E^\times)^2 = P \cap (E^\times)^2 = (F^\times)^2,$$

by (2). We conclude that $F(\sqrt{a}) = F$. □

We can now prove that the two definitions of Euclidean fields given earlier coincide:

COROLLARY 19.2.4. *The following conditions on an ordered field (F, P) are equivalent:*
(a) $P = (F^\times)^2$;
(b) *(F, P) is real closed relative to the quadratic closure $F(2)$ of F.*

PROOF. (a)⇒(b): The ordered field (F, P) is clearly real closed relative to $K = F(\sqrt{-1})$. Further, by Lemma 19.1.4, $K^\times = (K^\times)^2$. Hence $K = F(2)$.

(b)⇒(a): One has $(F(2)^\times)^2 = F(2)^\times$. Proposition 19.2.3 therefore gives

$$(F^\times)^2 = P \cap (F(2)^\times)^2 = P \cap F(2)^\times = P.$$ □

Notice that in the situation of Corollary 19.2.4, $(F^\times)^2$ is in fact the unique ordering on F.

A somewhat more general class of ordered fields is described by the following result.

PROPOSITION 19.2.5. *The following conditions on a formally real field F are equivalent:*
(a) $(F^\times)^2 = \sum F^2$;
(b) $F^2 = F^2 + F^2$;
(c) *F is the intersection of all its Euclidean closures.*

PROOF. (a)⇔(b): Immediate.

(a)⇒(c): We need to show that if $a \in F^\times$ and $\sqrt{a}$ belongs to every Euclidean closure of F, then $\sqrt{a} \in F$. To this end, let P be an ordering on F with a Euclidean closure E_P. By Corollary 19.2.4, $(E_P^\times)^2$ is an ordering on E_P. Furthermore, it extends P. Hence $a \in F \cap (E_P^\times)^2 = P$. Since P was arbitrary, $a \in \sum F^2$, by Artin's theorem (Corollary 6.1.9). By (a), $a \in (F^\times)^2$, as desired.

(c)⇒(a): Clearly $(F^\times)^2 \subseteq \sum F^2$. Conversely, take $a \in \sum F^2$. For every Euclidean closure E of F we then have $a \in \sum E^2 = (E^\times)^2$, by Corollary 19.2.4 again. Hence $\pm\sqrt{a} \in \bigcap_E E = F$, so $a \in (F^\times)^2$. □

In view of condition (b) above, formally real fields satisfying the equivalent conditions of Proposition 19.2.5 are called **Pythagorean** fields. A field of characteristic $\neq 2$ which is not formally real is called Pythagorean if and only if it is quadratically closed (this is justified by Proposition 6.1.10).

The next corollary is analogous to Proposition 15.3.3:

COROLLARY 19.2.6. *Let E/F and E'/F' be normal extensions satisfying $F \subseteq F'$ and $E \subseteq E'$. Let P, P' be orderings on F, F', respectively, with P' extending P. Suppose that (F', P') is real closed relative to E'. Then $(E \cap F', E \cap P')$ is real closed relative to E.*

PROOF. Let $F_1 = E \cap F'$ and $P_1 = F_1 \cap P' = E \cap P'$. Clearly, (F_1, P_1) is an ordered field extension of (F, P) inside E. By Proposition 19.2.3, $[E' : F']$ is even and $((F')^\times)^2 = P' \cap ((E')^\times)^2$. Since $[E : F_1]$ divides $[E' : F']$, it is also even. Furthermore,

$$(F_1^\times)^2 = (E^\times)^2 \cap ((F')^\times)^2 = (E^\times)^2 \cap P' = (E^\times)^2 \cap P_1.$$

We use again Proposition 19.2.3 to conclude that (F_1, P_1) is real closed relative to E. □

A stronger version of Proposition 19.2.3 can be obtained when E is quadratically closed. To prove it, we first need the following fact, whose proof is classical:

LEMMA 19.2.7. *Let K be a field of characteristic $\neq 2$ satisfying $K^\times = (K^\times)^2$. Then every polynomial in $K[X]$ of degree 2 splits in K.*

PROOF. Given $a, b, c \in K$ with $a \neq 0$ one has

$$aX^2 + bX + c = a\left[\left(X + \frac{b}{2a}\right)^2 - \frac{b^2 - 4ac}{4a^2}\right].$$

Take $d \in K$ with $d^2 = (b^2 - 4ac)/4a^2$. Then $d - (b/2a)$ is a zero of this quadratic polynomial. □

We now come to the desired characterization.

THEOREM 19.2.8. *Let (F, P) be an ordered field and let E be a normal extension of F such that $E^\times = (E^\times)^2$. The following conditions are equivalent:*
(a) *$E = F(\sqrt{-1})$;*
(b) *(F, P) is real closed relative to E;*
(c) (1) *$[E : F]$ is even; and*
(2) *$P = (F^\times)^2$.*

PROOF. (a)⇒(b): Just observe that there are no orderings on $E = F(\sqrt{-1})$.

(b)⇒(c): This follows from Proposition 19.2.3 and the assumption that $E^\times = (E^\times)^2$.

(c)⇒(a): Let $K = F(\sqrt{-1})$. Then $K \subseteq E$. We need to show that every irreducible polynomial $f(X) \in K[X]$ which splits in E has a zero in K (hence is linear).

We first prove this when $f(X) \in F[X]$. Write $\deg(f) = 2^n k$ with k odd. We argue by induction on n.

When $n = 0$, (1) implies that $\deg(f) = 1$, and we are done.

So suppose that $n \geq 1$. Let $\alpha_1, \ldots, \alpha_m \in E$ be the distinct roots of $f(X)$, where $m = 2^n k$ is even. Given $b \in F$, we consider the polynomial

$$g_b(X) = \prod_{i<j} (X - \alpha_i - \alpha_j - b\alpha_i\alpha_j) \in E[X].$$

Being a symmetric polynomial in the roots of $f(X) \in F[X]$, the polynomial $g_b(X)$ is in fact in $F[X]$ [**Lg**, Ch. V, §9]. Moreover,

$$\deg(g_b) = m(m-1)/2 = 2^{n-1}k(m-1),$$

and $k(m-1)$ is odd. The induction hypothesis implies that $g_b(X)$ has a zero in K, i.e., $\alpha_i+\alpha_j+b\alpha_i\alpha_j \in K$ for some $i<j$ (which may depend on b). But since K has characteristic 0, it is infinite. Therefore we can find $b \neq b'$ in K and indices $i<j$ such that

$$\alpha_i+\alpha_j+b\alpha_i\alpha_j, \alpha_i+\alpha_j+b'\alpha_i\alpha_j \in K.$$

Then $\alpha_i+\alpha_j, \alpha_i\alpha_j \in K$. Thus the quadratic polynomial

$$(X-\alpha_i)(X-\alpha_j) = X^2-(\alpha_i+\alpha_j)X+\alpha_i\alpha_j$$

belongs to $K[X]$. By Lemma 19.1.4 and Lemma 19.2.7, it splits in K, so $\alpha_i, \alpha_j \in K$, and the induction is complete.

Finally consider an arbitrary polynomial $f(X) \in K[X]$ which splits in E. Let σ be the automorphism in $\mathrm{Aut}(K/F)$ given by $\sigma(\sqrt{-1}) = -\sqrt{-1}$. The polynomial $(\sigma f)(X)\cdot f(X)$ is in $F[X]$ and splits in E. By what we have just seen, it has a zero c in K. Then either $f(c)=0$ or $(\sigma f)(c)=0$. In the latter case, $f(\sigma(c))=0$, so either way we get a zero of $f(X)$ in K. □

We can now strengthen Proposition 19.2.3 as follows.

COROLLARY 19.2.9. *Let (F,P) be an ordered field which is real closed relative to a normal extension E of F. Then $[E:F]\leq 2$.*

PROOF. Let $(\hat{F},\hat{P})$ be the real closure of (F,P) relative to the algebraic closure $\tilde{F}$ of F. By Theorem 19.2.8, $[\tilde{F}:\hat{F}]=2$. By the real closedness, $F=E\cap\hat{F}$. Hence $[E:F]=[E\hat{F}:\hat{F}]\leq 2$. □

Continuing the analogy between relative real closures and relative Henselizations further, we get in the setup of Theorem 19.2.8 the group $\mathrm{Gal}(E/F)\cong\mathbb{Z}/2$ as the analog of the decomposition group of a valuation. Thus in the ordered field case the Galois-theoretic structure is not as rich as in the valued field case. As a matter of fact, the condition $\mathrm{Gal}(E/F)\cong\mathbb{Z}/2$ *characterizes* the relatively real closed fields in the pro-2 setting:

THEOREM 19.2.10. *Let E/F be a normal extension with $E^\times=(E^\times)^2$. The following conditions are equivalent:*
(a) *$[E:F]=2$;*
(b) *F is Euclidean.*

This is a relative version, due to Becker ([**Be1**], [**Be3**]), of a classical result of Artin and Schreier, who considered the absolute case in which E is the algebraic closure $\tilde{F}$ of F. Moreover, Artin and Schreier proved that if $1<[\tilde{F}:F]<\infty$, then $[\tilde{F}:F]=2$, so F is real closed. While the implication (b)⇒(a) is contained in Theorem 19.2.8, the converse implication (a)⇒(b) is quite deep. We postpone it to §25.1, where we will indicate a short cohomological proof of it.

One of the main applications of relative Henselity was the openness of the nth powers in a valuation topology, when n is not divisible by the characteristic (Proposition 18.2.1). Here is an ordered field analog of this fact:

PROPOSITION 19.2.11. *Suppose that (F,P) is a real closed field relative to a normal extension E of F. Let n be a positive integer such that E contains all nth* *roots of elements of F. Then*

$$(F^\times)^n = \begin{cases} F^\times & \textit{if } n \textit{ is odd,} \\ P & \textit{if } n \textit{ is even.} \end{cases}$$

In particular, $(F^\times)^n$ is open in the ordering topology $\mathcal{T}_P$.

PROOF. First suppose that n is odd. By assumption, for every $a \in F^\times$ the polynomial $X^n - a$ splits in E. Since $[E : F] \leq 2$ (Corollary 19.2.9), this polynomial splits already in F, so $a \in (F^\times)^n$. We conclude that $(F^\times)^n = F^\times$ in this case.

Next suppose that n is even. Write $n = 2^s k$ with $s \geq 1$ and k odd. The assumption implies that E contains the kth roots, as well as square roots, of elements of F. By what we have just seen $(F^\times)^k = F^\times$. Furthermore, by Proposition 19.2.3, $(F^\times)^2 = P \cap (E^\times)^2 = P$. As $P \cdot P = P$, we inductively obtain that $(F^\times)^{2^s} = P$. Therefore

$$(F^\times)^n = ((F^\times)^k)^{2^s} = P.$$

The second assertion follows from the first one. □

19.3 Sturm's theorem

In the next section we will prove that any two real closures of an ordered field (F, P) relative to a normal extension are isomorphic over F. The crucial fact in the proof is that a given polynomial $f(X) \in F[X]$ has the same number of roots in any real closure of F. In fact, one can express this number in an effective way using the signs with respect to P of various functions of the coefficients of $f(X)$. In this section we present a classical method for achieving this, due to Sturm (who considered only the field $\mathbb{R}$ of real numbers). It should be pointed out that there is an alternative and somewhat more structural method, going back to Borchardt, Hermite, Jacobi, and Sylvester. However, this second method is based on quadratic form techniques, which we do not assume here as prerequisites. We refer to [**BeS**], [**KnS**, Ch. I], or [**KreN**] for expositions on these and other related methods.

DEFINITION 19.3.1. Let (F, P) be an ordered field and A a subset of F. A **Sturm sequence for** (F, P) **in** A is a finite sequence

$$f_0(X), f_1(X), \ldots, f_m(X) \in F[X],$$

where $m \geq 1$, such that:

(i) $f_1(X)$ is the formal derivative $f_0'(X)$ of $f_0(X)$;
(ii) $f_m(X)$ is either always positive or always negative on A with respect to P;
(iii) $f_{i-1}(X), f_i(X)$ have no common roots in A, $i = 1, \ldots, m$;
(iv) if $c \in A$ is a root of $f_i(X)$, where $1 \leq i \leq m-1$, then $f_{i-1}(c)$ and $f_{i+1}(c)$ have opposite signs with respect to P.

Note that a Sturm sequence in A is a Sturm sequence in every subset of A.

EXAMPLE 19.3.2. Suppose that $f(X)$ is a non-constant polynomial over an ordered field (F, P) and has no multiple roots. We construct a Sturm sequence $f_i(X)$, $i = 0, 1, \ldots, m$, starting with $f_0(X) = f(X)$. To this end we use the following variant of the Euclidean algorithm: let

$$f_0(X) = f(X), \quad f_1(X) = f'(X),$$

and for $1 \leq i$ with $f_i(X)$ non-zero put

$$f_{i-1}(X) = q_i(X)f_i(X) - f_{i+1}(X),$$

where $q_i(X), f_{i+1}(X) \in F[X]$ and $\deg(f_{i+1}(X)) < \deg(f_i(X))$. This sequence must terminate with $f_{m+1}(X) = 0$. Observe that $f_m(X)$ is the greatest common divisor of $f(X)$ and $f'(X)$ in $F[X]$ (up to a constant factor), whence is a non-zero constant. Conditions (i), (ii) and (iv) are therefore clear. For (iii), we see by induction that any common root of a consecutive pair of polynomials in the sequence is a root of all the polynomials. But by assumption, $f(X), f'(X)$ have no common root, so this is impossible.

To formulate Sturm's theorem we need the following notation. For an ordered field (F, P) and $a \in F^\times$ set

$$\operatorname{sgn}_P(a) = \begin{cases} 1 & \text{if } a \in P, \\ -1 & \text{if } -a \in P. \end{cases}$$

For $a_0, a_1, \ldots, a_m \in F^\times$ we define

$$V_P(a_0, a_1, \ldots, a_m) = \Big|\{1 \leq i \leq m \mid \operatorname{sgn}_P(a_{i-1}) \neq \operatorname{sgn}_P(a_i)\}\Big|.$$

Finally, given polynomials $f_0(X), f_1(X), \ldots, f_m(X) \in F[X]$ and $c \in F$, let

$$W_P(c) = V_P(f_0(c), f_1(c), \ldots, f_m(c)).$$

THEOREM 19.3.3. *Let (F, P) be a real closed field (relative to the algebraic closure of F), and let $a, b \in F$ satisfy $a <_P b$. Let*

$$f_0(X), f_1(X), \ldots, f_m(X)$$

be a Sturm sequence in $[a, b]_P$ such that $f_i(a), f_i(b) \neq 0$ for $i = 0, 1, \ldots, m$. Then $f_0(X)$ has precisely

$$W_P(a) - W_P(b)$$

roots in $[a, b]_P$.

PROOF. If $d \in (a, b)_P$ is not a root of any $f_i(X)$, and if the assertion holds for the sub-intervals $[a, d]_P$ and $[d, b]_P$, then it holds for the interval $[a, b]_P$. We may therefore divide $[a, b]_P$ into sufficiently small sub-intervals to assume, without loss of generality, that there exists $c \in (a, b)_P$ such that all the f_i are non-zero in $[a, c)_P$ and in $(c, b]_P$. Then Proposition 19.2.2 implies that for each $0 \leq i \leq m$ the sign $\operatorname{sgn}_P(f_i(X))$ is constant on $[a, c)_P$ as well as on $(c, b]_P$.

CLAIM I. *If $f_0(c) = 0$, then*

$$V_P\big(f_0(a), f_1(a)\big) - V_P\big(f_0(b), f_1(b)\big) = 1.$$

To see this, write $f_0(X) = (X - c)g(X)$ with $g(X) \in F[X]$. Then, by (i) of Definition 19.3.1,

$$(X - c)g'(X) + g(X) = f_1(X).$$

By (iii), $g(c) = f_1(c) \neq 0$. Hence a continuity argument (see §8.1) yields a sufficiently small $\epsilon >_P 0$ such that on $(c - \epsilon, c)_P$ we have

$$-\operatorname{sgn}_P(f_0(X)) = \operatorname{sgn}_P(g(c)) = \operatorname{sgn}_P(f_1(X)).$$

Because of the fixed signs, this in fact holds on all of $[a,c)_P$, and in particular for $X=a$. Similarly, for a sufficiently small $\epsilon >_P 0$ we have on $(c,c+\epsilon)_P$:

$$\mathrm{sgn}_P(f_0(X)) = \mathrm{sgn}_P(g(c)) = \mathrm{sgn}_P(f_1(X)).$$

Consequently, this holds on all of $(c,b]_P$, and in particular for $X=b$. We conclude that $V_P(f_0(a),f_1(a))=1$ while $V_P(f_0(b),f_1(b))=0$.

CLAIM II. *If $f_i(c)=0$ for $i\geq 1$, then*

$$V_P\big(f_{i-1}(a),f_i(a),f_{i+1}(a)\big) = V_P\big(f_{i-1}(b),f_i(b),f_{i+1}(b)\big).$$

Indeed, by condition (ii) of Definition 19.3.1, $i\leq m-1$. Further, by condition (vi), $f_{i-1}(c)f_{i+1}(c) <_P 0$. Since $f_{i-1}(X), f_{i+1}(X)$ have fixed signs in $[a,c)_P$ and $(c,b]_P$, we get:

$$\begin{aligned}\mathrm{sgn}_P(f_{i-1}(a)) &= \mathrm{sgn}_P(f_{i-1}(c)) = -\mathrm{sgn}_P(f_{i+1}(c)) = -\mathrm{sgn}_P(f_{i+1}(a)),\\ \mathrm{sgn}_P(f_{i-1}(b)) &= \mathrm{sgn}_P(f_{i-1}(c)) = -\mathrm{sgn}_P(f_{i+1}(c)) = -\mathrm{sgn}_P(f_{i+1}(b)).\end{aligned}$$

We conclude that $V_P(f_{i-1}(a),f_i(a),f_{i+1}(a))$ and $V_P(f_{i-1}(b),f_i(b),f_{i+1}(b))$ are both 1.

CLAIM III. *If $f_i(c), f_{i+1}(c)\neq 0$, then*

$$V\big(f_i(a),f_{i+1}(a)\big) = V\big(f_i(b),f_{i+1}(b)\big).$$

Indeed, in this case $f_i(X), f_{i+1}(X)$ do not have roots in $[a,b]_P$, so by Proposition 19.2.2, they have fixed signs on $[a,b]_P$. The claim follows.

To complete the proof, recall that by condition (iii) of Definition 19.3.1, c is not a root of consecutive polynomials in the sequence $f_0(X), f_1(X),\ldots,f_m(X)$. Therefore we may divide this sequence into pairs and triples as in Claims I–III. Altogether we deduce that $W_P(a)-W_P(b)$ is 1 if $f_0(c)=0$ and is 0 otherwise, as required. □

REMARK 19.3.4. One can relax the assumption in Theorem 19.3.3 that (F,P) is real closed, and require only that it is an ordered field containing a,b as well as the coefficients of the $f_i(X)$, $i=0,1,\ldots,m$. For a real closure $(\hat{F},\hat{P})$ of (F,P), the theorem will then say that $f(X)$ has precisely $W_P(a)-W_P(b)$ roots in the closed interval $[a,b]_{\hat{P}}$ of $\hat{F}$.

One can also use Theorem 19.3.3 to compute the *total* number of roots in a real closed field of a polynomial with no multiple roots. Indeed, let (F,P) be an ordered field, and let $(\hat{F},\hat{P})$ be a real closure of (F,P). Given a non-constant polynomial $f(X)\in F[X]$ with no multiple roots, we construct the Sturm sequence $f_i(X)$, $i=0,1,\ldots,m$, in F as in Example 19.3.2. Write

$$f_0(X)f_1(X)\cdots f_m(X) = dg(X)$$

with $d\in F$ and $g(X)=\sum_{i=0}^n c_iX^i\in F[X]$ monic. Let

$$M=\max\Big\{1,\sum_{i=0}^{n-1}|c_i|_P\Big\},\quad M'=M+1.$$

By Lemma 19.1.2, $g(X)$ has no roots in $\hat{F}$ outside $[-M, M]_{\hat{P}}$. By Theorem 19.3.3, $f(X)$ has precisely $W_P(-M') - W_P(M')$ roots in the larger interval $[-M', M']_{\hat{P}}$, whence in all of $\hat{F}$. Since this number depends only on $f(X)$ and on P, we conclude:

COROLLARY 19.3.5. *Let $(\hat{F}_i, \hat{P}_i)$, $i = 1, 2$, be real closures of an ordered field (F, P), and suppose that $f(X) \in F[X]$ has no multiple roots. Then $f(X)$ has the same number of roots in $\hat{F}_1$ and in $\hat{F}_2$.*

19.4 Uniqueness of real closures

We now apply Sturm's method to prove a major ingredient of the (relative) Artin–Schreier theory, namely, the uniqueness of real closures up to an isomorphism.

LEMMA 19.4.1. *Let $(F_1, P_1)/(F, P)$ be an algebraic extension of ordered fields. Let $\varphi: (F, P) \to (L, Q)$ be an embedding of ordered fields, where (L, Q) is real closed. Consider the set Ψ of all field-embeddings $\psi: F_1 \to L$ which extend φ.*

(a) *If F_1/F is finite, then Ψ is finite and non-empty.*
(b) *There exists $\psi \in \Psi$ with $\psi(P_1) \subseteq Q$.*

PROOF. Let $(\hat{F}, \hat{P})$ be a real closure of (F_1, P_1), hence also of (F, P).

(a) As $\operatorname{char} F = 0$ and F_1/F is finite, we can find $\alpha \in F_1$ with $F_1 = F(\alpha)$. Let $f(X) = \operatorname{irr}(\alpha, F)$ and let $(\varphi f)(X)$ be the image of $f(X)$ in $L[X]$ under φ. Since $f(X)$ has the root α in the real closure $\hat{F}$, Corollary 19.3.5 implies that it has a root in every real closure of F. Passing to the isomorphic ordered field $(\varphi(F), \varphi(P))$, we deduce that $(\varphi f)(X)$ also has roots in every real closure of this ordered field. By Corollary 19.2.6, the relative algebraic closure of $\varphi(F)$ in L is such a real closure. Therefore $(\varphi f)(X)$ has a root in L. But the general theory of field extensions gives a bijective correspondence between the roots of $(\varphi f)(X)$ in L and the set Ψ. Conclude that Ψ is finite and non-empty.

(b) A routine application of Zorn's lemma reduces this to the case where F_1/F is finite.

In light of (a), we may list the distinct elements of Ψ as $\psi_1, \dots, \psi_k$, where $k \geq 1$. Suppose that $\psi_i(P_1) \not\subseteq Q$ for all i. Then there exist $a_1, \dots, a_k \in P_1$ such that $\psi_i(a_i) \notin Q$, $i = 1, \dots, k$. Note that $a_i \in P_1 \subseteq \hat{P} = (\hat{F}^\times)^2$, by Theorem 19.2.8. Therefore the field

$$F_2 = F_1(\sqrt{a_1}, \dots, \sqrt{a_k})$$

is contained in $\hat{F}$. Let $P_2 = \hat{P} \cap F_2$. Apply (a) with respect to the finite extension $(F_2, P_2)/(F, P)$ to obtain a field-embedding $\eta: F_2 \to L$ which extends φ. The restriction of η to F_1 must be some ψ_i, $1 \leq i \leq k$. But then

$$\psi_i(a_i) = \eta(a_i) = (\eta(\sqrt{a_i}))^2 \in (L^\times)^2 = Q,$$

a contradiction. □

We now obtain our main result:

THEOREM 19.4.2. *Let $(\hat{F}_i, \hat{P}_i)$, $i = 1, 2$, be real closures of an ordered field (F, P) relative to a normal extension E of F. There is an ordered-field isomorphism $(\hat{F}_1, \hat{P}_1) \xrightarrow{\sim} (\hat{F}_2, \hat{P}_2)$ which is the identity on F.*

PROOF. Let (L,Q) be a real closure of $(\hat{F}_2,\hat{P}_2)$ (relative to the algebraic closure). By Corollary 19.2.6, $(E\cap L, E\cap Q)$ is a real closure of $(\hat{F}_2,\hat{P}_2)$ relative to E, and therefore it coincides with it. Lemma 19.4.1(b) yields an embedding of $\psi_1\colon(\hat{F}_1,\hat{P}_1)\hookrightarrow(L,Q)$ which is the identity on F. Since E/F is normal, the image of ψ_1 is contained in E. As $E\cap L=\hat{F}_2$, the embedding ψ_1 in fact maps $(\hat{F}_1,\hat{P}_1)$ into $(\hat{F}_2,\hat{P}_2)$.

Similarly we get an embedding $\psi_2\colon(\hat{F}_2,\hat{P}_2)\hookrightarrow(\hat{F}_1,\hat{P}_1)$ which is the identity on F.

Now the composition $\psi_2\circ\psi_1$ is a field-embedding of $\hat{F}_1$ into itself which is the identity on F. Since $\hat{F}_1/F$ is algebraic, this composition must be surjective. It follows that ψ_1 and ψ_2 are isomorphisms. □

The isomorphism in Theorem 19.4.2 is in fact unique, provided that $E^\times=(E^\times)^2$. This is an immediate consequence of the following result:

PROPOSITION 19.4.3. *Let $(\hat{F},\hat{P})$ be a real closure of an ordered field (F,P) relative to a normal extension E of F such that $E^\times=(E^\times)^2$. Then:*

(a) $\mathrm{Aut}(\hat{F}/F)=1$*;*

(b) $\mathrm{Gal}(E/\hat{F})$ *is its own normalizer in* $\mathrm{Gal}(E/F)$.

PROOF. (a) Take $\sigma\in\mathrm{Aut}(\hat{F}/F)$. Our assumption on E implies that $\hat{P}=(\hat{F}^\times)^2$ (Theorem 19.2.8). Therefore σ maps $\hat{P}$ onto itself. Consequently, it preserves $<_{\hat{P}}$.

Now take $a\in\hat{F}$, and order its F-conjugates in $\hat{F}$ as

$$a_1<_{\hat{P}}\cdots<_{\hat{P}}a_r.$$

By what we have just observed,

$$\sigma(a_1)<_{\hat{P}}\cdots<_{\hat{P}}\sigma(a_r).$$

Conclude that $\sigma(a_i)=a_i$ for $i=1,\ldots,r$. In particular, $\sigma(a)=a$. Thus $\sigma=\mathrm{id}_{\hat{F}}$.

(b) Let $\sigma\in\mathrm{Gal}(E/F)$. One has $\mathrm{Gal}(E/\hat{F})=\sigma^{-1}\mathrm{Gal}(E/\hat{F})\sigma$ if and only if $\sigma(\hat{F})=\hat{F}$. By (a) this means that σ is the identity on $\hat{F}$, i.e., $\sigma\in\mathrm{Gal}(E/\hat{F})$. □

CHAPTER 20

Coarsening in Algebraic Extensions

This chapter explores the behavior of the tree of localities (in the sense of §7) under field extensions, as well as its interplay with the ramification filtration of decomposition groups.

20.1 Extensions of localities

Given a field extension E/F, we say that a locality μ on E **extends** a locality λ on F if either:

- $\lambda = P$ and $\mu = Q$ are orderings and $P = F \cap Q$; or
- $\lambda = v$ and $\mu = u$ are valuations and u extends v.

THEOREM 20.1.1. *Let λ_1, λ_2 be localities on the field F with λ_1 finer than λ_2. Let E be an algebraic extension of F. Then:*

(a) *given a locality μ_1 on E extending λ_1 there exists a unique locality (up to equivalence of valuations) μ_2 on E which extends λ_2 and is coarser than μ_1;*

(b) *if $\lambda_1 = v_1$ and $\lambda_2 = v_2$ are valuations and u_2 is a valuation on E extending v_2, then there exists a valuation u_1 on E which extends v_1 and is finer than u_2.*

PROOF. (a) We first show the uniqueness. If μ_2, μ_2' are extensions of λ_2 which are coarser than μ_1, then they must be comparable, by the tree property of localities (Proposition 7.3.5). In the ordering case this can happen only when $\mu_2 = \mu_2'$. In the valuation case Proposition 14.2.6 shows that μ_2 and $\mu_1'2$ are equivalent.

We now turn to the existence of μ_2. If $\lambda_1 = \lambda_2$, then we simply take $\mu_1 = \mu_2$. So assume that λ_1 and λ_2 are distinct. In particular, $\lambda_2 = v_2$ is a valuation.

If $\lambda_1 = P$ and $\mu_1 = Q$ are orderings and v_2 corresponds to the finest coarsening $O_{\mathbb{Q}}(P)$ of P (Proposition 7.3.1(b)), then the valuation u_2 corresponding to $O_{\mathbb{Q}}(Q)$ has the desired properties.

Consequently, if $\lambda_1 = P$ is an ordering, we may replace it by the valuation corresponding to $O_{\mathbb{Q}}(P)$ to reduce the assertion to the case where $\lambda_1 = v_1$ is also a valuation.

Assuming this, write $\lambda_1 = u_1$ and let $\Gamma_{v_1}, \Gamma_{v_2}, \Gamma_{u_1}$ be the value groups of v_1, v_2, u_1, respectively. Then $\Gamma_{v_1} \leq \Gamma_{u_1}$, and Proposition 5.1.1 gives rise to an epimorphism of ordered abelian groups $\eta_F\colon \Gamma_{v_1} \to \Gamma_{v_2}$. From Proposition 2.6.1 we obtain an ordered abelian group Γ_{u_2} and morphisms of ordered abelian groups $\eta_E\colon \Gamma_{u_1} \to \Gamma_{u_2}$, $\varphi_1\colon \Gamma_{v_2} \to \Gamma_{u_2}$ such that η_E is surjective, φ_1 is injective, and the following square commutes:

$$\begin{array}{ccc} \Gamma_{v_1} & \xrightarrow{\eta_F} & \Gamma_{v_2} \\ \downarrow & & \downarrow{\scriptstyle \varphi_1} \\ \Gamma_{u_1} & \xrightarrow{\eta_E} & \Gamma_{u_2}. \end{array}$$

The composition of the maps $u_1: E^\times \to \Gamma_{u_1}$ and η_E is then a valuation u_2 on E which is coarser than u_1 (by Proposition 5.1.1 again). We have $O_{v_2} = F \cap O_{u_2}$, i.e., u_2 extends v_2.

(b) Using Zorn's lemma we can reduce the assertion to the case where E/F is a finite extension. Furthermore, by Chevalley's theorem (Corollary 14.1.2), u_2 extends to the normal hull of E/F. Therefore we may replace E by this hull to assume, without loss of generality, that E is a finite normal extension of F.

Chevalley's theorem also yields a valuation u_1' on E extending v_1. From part (a) we obtain a coarsening u_2' of u_1' which extends v_2. Theorem 14.3.2 gives an automorphism $\sigma \in \mathrm{Aut}(E/F)$ such that u_2 is equivalent to $u_2' \circ \sigma$. We take $u_1 = u_1' \circ \sigma$. It is finer than u_2 and extends v_1. □

The valuation u_1 in Theorem 20.1.1(b) need not be unique up to equivalence.

Our next result shows that the construction of the finest common coarsening (infimum) of localities (see §7.3) is compatible with algebraic extensions, as long as we assume Henselity.

COROLLARY 20.1.2. *Let E/F be an algebraic extension of fields, let μ_i, $i \in I$, be localities on E which are not all equal, and let λ_i, $i \in I$, be their restrictions to F, respectively. Let w be the finest common coarsening of the λ_i, and assume that it has a unique extension w' to E. Then w' is the finest common coarsening of the μ_i.*

PROOF. First note that since the μ_i are not all equal, their finest common coarsening w is indeed a valuation. Therefore so is w'. Let w'' be the finest common coarsening of the μ_i.

Theorem 20.1.1(a) implies that w' is coarser than each μ_i. Hence it is coarser than w''. Therefore w is coarser than the restriction w''_F of w'' to F.

On the other hand, since w'' is coarser than every μ_i, its restriction w''_F is coarser than every λ_i. Therefore it is coarser than w, whence is equal to it. □

REMARK 20.1.3. The assumption in Corollary 20.1.2 that w has a unique extension to E cannot be dropped. For instance, suppose that E/F is a normal extension and v is a rank 1 valuation on F which is not Henselian relative to E. Then v has distinct extensions u_1 and $u_2 = u_1 \circ \sigma$ to E, with $\sigma \in \mathrm{Aut}(E/F)$. By Proposition 14.3.1, u_1 and u_2 are incomparable. Since they also have rank 1 (Corollary 14.2.3(c)), their finest common coarsening is trivial, while v is not.

Theorem 20.1.1(b) does not extend from valuations to arbitrary localities: for instance, take $F = \mathbb{Q}$, $E = \mathbb{Q}(\sqrt{-1})$, v the trivial locality on $\mathbb{Q}$, and P the unique ordering on $\mathbb{Q}$. Then v is coarser than P and extends to E, but obviously, P does not extend to E. Nevertheless one has the following partial result.

PROPOSITION 20.1.4. *Suppose that E/F is a normal extension with $(E^\times = (E^\times)^2$. Let v be a valuation on F and let P be an ordering on F which is finer* *than v. Finally, let (E_Z, u_Z) be the decomposition field of some extension u of v to E. Then P extends to an ordering P_Z on E_Z which is finer than u_Z.*

PROOF. As $G_v \leq P$, Lemma 7.1.3 implies that $\bar{P}_v$ is an ordering on $\bar{F}_v$. In particular, $\operatorname{char} \bar{F}_v = 0$ (Corollary 6.1.7). Likewise $\operatorname{char} F = 0$, so E/F is Galois, and $(E_Z, u_Z)/(F, v)$ is an immediate extension (Theorem 15.2.2). By Lemma 3.2.5, $E_{u_Z}^\times = F^\times G_{u_Z}$.

We claim that $-1 \notin \sum (E_{u_Z}^\times)^2 P$. Otherwise $-1 \in \sum G_{u_Z} P$, so we would get $a_1, \dots, a_n \in G_{u_Z} P$ such that $0 = \sum_{i=1}^n a_i$. After dividing by some a_i with $u_Z(a_i)$ minimal, we may assume that $u_Z(a_i) \geq 0$ for all i and $u_Z(a_i) = 0$ for some i. Hence

$$-1 \in \sum G_{u_Z}(P \cap O_{u_Z}).$$

Further, taking residues, we obtain that $-1 \in \sum \bar{P}_v$, contrary to the fact that $\bar{P}_v$ is an ordering.

From the claim and from Proposition 6.1.5 we obtain an ordering P_Z on E_{u_Z} which contains P. As $(F^\times : P) = 2$ and $-1 \notin P_Z$, the restriction of P_Z to F must coincide with P. Since $\operatorname{char} \bar{F}_v = 0$ and u_Z is Henselian relative to E, Proposition 18.2.1 gives $G_{u_Z} \leq (E_{u_Z}^\times)^2 \leq P_Z$. Consequently, P_Z is finer than u_Z. □

20.2 Coarsening and Galois groups

Next we study the effect of coarsening on the Galois-theoretic filtration discussed in §§15–16.

LEMMA 20.2.1. *Let $(E, u)/(F, v)$ be a normal extension of valued fields. Let u' be a valuation on E which is finer than u, and let v' be its restriction to F. Then:*
(a) *$Z(u'/v') \leq Z(u/v)$;*
(b) *$T(u/v) \leq T(u'/v')$;*
(c) *$V(u/v) \leq V(u'/v')$.*

PROOF. (a) Let $\sigma \in Z(u'/v')$. Then both u and $u \circ \sigma$ are coarser than $u' = u' \circ \sigma$. Therefore they are comparable (Proposition 7.3.5). By Proposition 14.3.1 they are in fact equal, i.e., $\sigma \in Z(u/v)$.

(b) Since $O_{u'} \subseteq O_u$ and $\mathfrak{m}_{u'} \supseteq \mathfrak{m}_u$, this follows immediately from the definitions of the inertia groups.

(c) This is immediate from the definitions and from (b). □

PROPOSITION 20.2.2. *Let v and v' be valuations on the field F with v' finer than v. Let E be a normal extension of F. Let u, u' be extensions of v, v', respectively, to E. Then there exists $\sigma \in \operatorname{Aut}(E/F)$ such that $\sigma^{-1} Z(u'/v')\sigma \leq Z(u/v)$.*

PROOF. Theorem 20.1.1(b) yields an extension u_1' of v' to E which is finer than u. By Lemma 20.2.1(a), $Z(u_1'/v') \leq Z(u/v)$. Since both u' and u_1' extend v', there exists $\sigma \in \operatorname{Aut}(E/F)$ such that $u' \circ \sigma$ and u_1' are equivalent (Theorem 14.3.2). Using Remark 15.1.1(b)(d) we get

$$\sigma^{-1} Z(u'/v')\sigma = Z(u' \circ \sigma / v') = Z(u_1'/v') \leq Z(u/v). \qquad \square$$

To achieve more uniformity, we introduce the following terminology.

DEFINITION. Let E/F be a normal extension and let λ be a locality on F. A **local closure of** (F, λ) **relative to** E is:

(i) an E-Henselization of (F, λ) if $\lambda = v$ is a valuation; and

(ii) a real closure of (F, λ) relative to E if $\lambda = P$ is an ordering.

We say that (F, λ) is **locally closed relative to** E if it is a local closure of itself relative to E.

From Corollary 15.3.6 and Theorem 19.4.2 we immediately obtain:

PROPOSITION 20.2.3. *Let E/F be a normal extension and let λ be a locality on F. Then (F, λ) has a unique local closure relative to E up to an isomorphism which is the identity on F.*

Combining the previous results we now obtain an important connection between coarsening of localities and the inclusion lattice of the relative local closures of F.

COROLLARY 20.2.4. *Let E/F be a normal extension and let λ, λ' be localities on F with λ' finer than λ. When λ' is an ordering, assume further that $E^\times = (E^\times)^2$. Then there exist local closures $\hat{F}, \hat{F}'$ of (F, λ), (F, λ'), respectively, relative to E such that $\hat{F} \subseteq \hat{F}'$.*

PROOF. We may assume that $\lambda \neq \lambda'$, so $\lambda = v$ is a valuation.

If $\lambda' = v'$ is also a valuation, then we apply Proposition 20.2.2 with E replaced by the maximal separable extension of F in E (so the local closures are just decomposition fields).

When $\lambda' = P$ is an ordering, we obtain from Proposition 20.1.4 an extension P_Z of P to some E-Henselization E_Z of (F, v). We can then extend P_Z further to a real closure of E_Z relative to E. □

20.3 Local closedness and quotients

Our next objective is to relate relative local closedness of the field at two comparable localities and the quotient locality in the sense of §7.2. This will be achieved at Theorem 20.3.2 below. First we need a technical lemma (see [**P2**, 1.1]).

LEMMA 20.3.1. *Let E/F be a normal extension of valued fields. Let u, u' be valuations on E extending valuations v, v', respectively, on F, and assume that u' is finer than u. Let u^0 be the quotient of u' by u and let v^0 be the quotient of v' by v. Then:*

(a) *u^0 extends v^0;*

(b) *there is a commutative diagram with exact rows*

$$\begin{array}{ccccccccc} 1 & \longrightarrow & T(u/v) & \hookrightarrow & Z(u'/v') & \longrightarrow & Z(u^0/v^0) & \longrightarrow & 1 \\ & & \| & & \downarrow & & \downarrow & & \\ 1 & \longrightarrow & T(u/v) & \hookrightarrow & Z(u/v) & \xrightarrow{\Phi} & \mathrm{Aut}(\bar{E}_u/\bar{F}_v) & \longrightarrow & 1, \end{array}$$

where Φ is as in Theorem 16.1.1 and the vertical maps are the inclusion maps.

PROOF. (a) Recall that

$$\mathfrak{m}_v \subseteq \mathfrak{m}_{v'} \subseteq O_{v'} \subseteq O_v, \quad \mathfrak{m}_u \subseteq \mathfrak{m}_{u'} \subseteq O_{u'} \subseteq O_u,$$

$$\bar{F}_v = O_v/\mathfrak{m}_v, \quad \bar{E}_u = O_u/\mathfrak{m}_u, \quad O_{v^0} = O_{v'}/\mathfrak{m}_v, \quad O_{u^0} = O_{u'}/\mathfrak{m}_u,$$

and $\bar{F}_v$ embeds in $\bar{E}_u$ via the map $a + \mathfrak{m}_v \mapsto a + \mathfrak{m}_u$ for $a \in O_v$. Hence $a + \mathfrak{m}_v \in \bar{F}_v \cap O_{u^0}$ if and only if $a \in O_v \cap O_{u'} = O_{v'}$, i.e., if and only if $a + \mathfrak{m}_v \in O_{v^0}$.

(b) In light of Lemma 20.2.1,

$$T(u/v) \leq T(u'/v') \leq Z(u'/v') \leq Z(u/v).$$

Also, the lower row of the above diagram is exact by Theorem 16.1.1.

It remains to show that $\Phi(Z(u'/v')) = Z(u^0/v^0)$. To see this we consider $\bar{\sigma} \in \mathrm{Aut}(\bar{E}_u/\bar{F}_v)$.

Assume that $\bar{\sigma} \in Z(u^0/v^0)$, i.e., that $u^0(\bar{\sigma}(\bar{a})) = u^0(\bar{a})$ for all $\bar{a} \in \bar{E}_u$. Take $\sigma \in Z(u/v)$ such that $\bar{\sigma} = \Phi(\sigma)$. Given $a \in O_{u'} (\subseteq O_u)$, its residue $\bar{a}$ in $\bar{E}_u$ lies in O_{u^0}, so $\overline{\sigma(a)} = \bar{\sigma}(\bar{a}) \in O_{u^0}$. Thus $\sigma(a) \in O_{u'} + \mathfrak{m}_u = O_{u'}$. This shows that $\sigma(O_{u'}) \subseteq O_{u'}$, i.e., u' is coarser than $u' \circ \sigma^{-1}$. Since both valuations extend v', Proposition 14.3.1 implies that they coincide, i.e., $\sigma \in Z(u'/v')$ and $\bar{\sigma} \in \Phi(Z(u'/v'))$.

Conversely, suppose that $\bar{\sigma} \in \Phi(Z(u'/v'))$. Take $\sigma \in Z(u'/v')$ such that $\bar{\sigma} = \Phi(\sigma)$. Then for every $a \in O_{u'}$ one has $u'(\sigma(a)) = u'(a)$. Hence

$$u^0(\bar{\sigma}(\bar{a})) = u^0(\overline{\sigma(a)}) = u^0(\bar{a}).$$

Thus $u^0 \circ \bar{\sigma}$ and u^0 coincide O_{u^0}, and therefore on its quotient field $\bar{E}_u$, so $\bar{\sigma} \in Z(u^0/v^0)$. □

We deduce the desired connection between relative closedness of a field with respect to comparable localities.

THEOREM 20.3.2. *Let E/F be a normal extension and let λ, λ' be localities on F with $\lambda = v$ a valuation coarser than λ'. Let u be a valuation on E extending v. When λ' is an ordering, assume further that $E^\times = (E^\times)^2$. If (F, λ') is locally closed relative to E, then (F, λ) is locally closed relative to E and $(\bar{F}_v, \lambda'/v)$ is locally closed relative to $\bar{E}_u$. The converse holds when λ' is also a valuation*

PROOF. Assume first that $\lambda' = v'$ is a valuation. By Lemma 20.3.1 we have $\mathrm{Aut}(E/F) = Z(u'/v')$ if and only if both $\mathrm{Aut}(E/F) = Z(u/v)$ and $\mathrm{Aut}(\bar{E}_u/\bar{F}_v) = Z(u^0/v^0)$. This gives the assertions in this case.

Next consider the case where $\lambda' = P$ is an ordering on F. Assume that (F, P) is real closed relative to E. Since P always extends to an E-Henselization of (F, v) (Proposition 20.1.4), this implies that (F, v) is already Henselian relative to E. It follows from Ostrowski's Theorem 17.2.1 that $[\bar{E}_u : \bar{F}_v] | [E : F]$,

Now by Theorem 19.2.8, $[E : F]$ is even and $(F^\times)^2 = P$. It follows that $[\bar{E}_u : \bar{F}_v]$ is also even and $(\bar{F}_v^\times)^2 = \bar{P}_v$. Furthermore, $\bar{E}_u^\times = (\bar{E}_u^\times)^2$. Use once again Theorem 19.2.8 to conclude that $(\bar{F}_v, \bar{P}_v)$ is real closed relative to $\bar{E}_u$. □

20.4 Ramification pairings under coarsening

In this final section we record the behavior of the ramification pairing of Theorem 16.2.6 under coarsening of valuations.

PROPOSITION 20.4.1. *Let $(E,u)/(F,v)$ be a normal extension of valued fields. Let u' be a valuation on E which is finer than u and let v' be its restriction to F. Let u^0 be the quotient of u' by u. Then the ramification pairings give rise to a diagram*

$$\begin{array}{ccccc} T(u/v)/V(u/v) & \times & u(E^\times)/v(F^\times) & \xrightarrow{\langle\cdot,\cdot\rangle} & (\bar{E}_u^\times)_{\rm tor} \\ \downarrow & & \uparrow & & \downarrow{\scriptstyle \pi_{u^0}} \\ T(u'/v')/V(u'/v') & \times & u'(E^\times)/v'(F^\times) & \xrightarrow{\langle\cdot,\cdot\rangle'} & (\bar{E}_{u'}^\times)_{\rm tor} \end{array}$$

which commutes in the natural way.

PROOF. Let $\varphi: u'(E^\times) \to u(E^\times)$ be an epimorphism of ordered abelian groups such that $\varphi \circ u' = u$ (see Proposition 5.1.1(g)). Let $\sigma \in T(u/v)$ and $a \in E^\times$. Then $\sigma \in T(u'/v')$ (Lemma 20.2.1(b)). By the definitions of the two pairings:

$$\langle \sigma V(u/v), u(a) + v(F^\times)\rangle = \pi_u(\sigma(a)/a)$$
$$\langle \sigma V(u'/v'), u'(a) + v'(F^\times)\rangle' = \pi_{u'}(\sigma(a)/a).$$

Since $\pi_u(\sigma(a)/a)$ is a root of unity in $\bar{E}_u$, it lies in $O_{u^0}^\times$. Therefore we may apply π_{u^0} to it. We compute:

$$\begin{aligned} &\pi_{u^0}(\langle \sigma V(u/v), (\varphi\circ u')(a) + v(F^\times)\rangle) \\ &\quad = \pi_{u^0}(\langle \sigma V(u/v), u(a) + v(F^\times)\rangle) \\ &\quad = \pi_{u^0}(\pi_u(\sigma(a)/a)) \\ &\quad = \pi_{u'}(\sigma(a)/a) \\ &\quad = \langle \sigma V(u'/v'), u'(a) + v'(F^\times)\rangle' \end{aligned}$$

as claimed. □

COROLLARY 20.4.2. *In the setup of Proposition 20.4.1 there is a canonical commutative square*

$$\begin{array}{ccc} T(u/v)/V(u/v) & \xrightarrow{\sim} & \chi(u/v) \\ \downarrow & & \downarrow \\ T(u'/v')/V(u'/v') & \xrightarrow{\sim} & \chi(u'/v'), \end{array}$$

where the right vertical homomorphism is induced by π_{u^0}.

CHAPTER 21

Intersections of Decomposition Groups

As we shall see in this chapter, when valuations are "remote" in the sense of the comparability tree of §7 (e.g., are independent or incomparable), then their decomposition groups relative to a fixed normal extension are also "remote", in the sense that their intersection is small. This phenomena gives rise to several Henselity criteria, which we will prove in §21.3.

21.1 The case of independent valuations

The following result was proved by F.K. Schmidt [**Schm**] for separable closures of valued fields of rank 1.

THEOREM 21.1.1. *Let E/F be a normal extension. If there exist two independent non-trivial valuations v_1, v_2 on F which are Henselian with respect to E, then E/F is a purely inseparable extension.*

PROOF. Since F carries non-trivial valuations, it is infinite (Example 3.2.2(1)). Suppose that $a \in E$ is separable over F. Then $f(X) = \mathrm{irr}(a, F)$ has $d = \deg(f)$ distinct roots in E. Pick arbitrary distinct elements $b_1, \ldots, b_d$ of F and set $g(X) = \prod_{i=1}^{d}(X - b_i)$. Using the weak approximation theorem (Theorem 10.1.7) we find a monic polynomial $h(X) \in F[X]$ of degree d whose coefficients are sufficiently close to those of $f(X)$ with respect to the topology $\mathcal{T}_{v_1}$ and to those of $g(X)$ with respect to the topology $\mathcal{T}_{v_2}$. By Theorem 18.5.2, the splitting fields over F of $f(X)$, $g(X)$ and $h(X)$ coincide. But by construction, $g(X)$ splits in F. Hence so does $f(X)$. Therefore $a \in F$ and the assertion follows. □

By definition, non-equivalent valuations of rank 1 are independent, so we obtain in particular:

COROLLARY 21.1.2. *Let E/F be a normal extension, and let v_1 and v_2 be non-equivalent valuations of rank 1 on F which are Henselian with respect to E. Then the extension E/F is purely inseparable.*

Theorem 21.1.1 and Corollary 21.1.2 can be restated in group-theoretic terms as follows.

COROLLARY 21.1.3. *Let E/F be a normal extension, let v_1, v_2 be valuations on F, and let u_1, u_2 be extensions of v_1, v_2, respectively, to E. Suppose that u_1, u_2 are independent. Then*

$$Z(u_1/v_1) \cap Z(u_2/v_2) = 1.$$

PROOF. Let F_1 be the fixed field of $Z(u_1/v_1) \cap Z(u_2/v_2)$ in E. The extension E/F_1 is Galois. Let v_1', v_2' be the restrictions of u_1, u_2, respectively, to F_1. By Corollary 15.3.2, they are Henselian with respect E. Since u_1, u_2, respectively, are their unique extensions to E (up to equivalence) and are independent, so are v'_1, v'_2, by Corollary 20.1.2. It follows from Theorem 21.1.1 that the extension E/F_1 is purely inseparable, whence $E = F_1$ and we are done. □

COROLLARY 21.1.4. *Let E/F be a normal extension, let v_1 and v_2 be valuations of rank 1 on F, and let u_1, u_2 be extensions of v_1, v_2, respectively, to E. Then either $Z(u_1/v_1) = Z(u_2/v_2)$ or $Z(u_1/v_1) \cap Z(u_2/v_2) = 1$.*

PROOF. In light of Remark 15.1.1(d), if u_1, u_2 are equivalent, then $Z(u_1/v_1) = Z(u_2/v_2)$.

So suppose that u_1, u_2 are not equivalent. Note that they also have rank 1, by Corollary 14.2.3(c). Hence they are independent. Conclude from Corollary 21.1.3, $Z(u_1/v_1) \cap Z(u_2/v_2) = 1$. □

21.2 The case of incomparable valuations

The following refinement of Theorem 21.1.1 is essentially due to Engler [**Eg1**].

THEOREM 21.2.1. *Let E/F be a normal extension, let v_1 and v_2 be incomparable valuations on F which are Henselian relative to E, and let u_1 and u_2, respectively, be their unique extensions to E. Then the residue field extensions $\bar{E}_{u_i}/\bar{F}_{v_i}$, $i = 1, 2$, are purely inseparable.*

PROOF. Being incomparable, the valuations v_1 and v_2 are non-trivial. Let w (resp., w') be the finest common coarsening of v_1, v_2 (resp., u_1, u_2). By Theorem 20.3.2, w is Henselian relative to E. By Corollary 20.1.2, w' is its unique extension to E.

For $i = 1, 2$ let v_i^0 be the quotient of v_i by w, and let u_i^0 be the quotient of u_i by w', in the sense of §5.2. Since v_1, v_2 are incomparable, w is strictly coarser than both of them, so v_1^0, v_2^0 are non-trivial. Furthermore, they are independent (Proposition 7.3.4). By Theorem 20.3.2 again, v_1^0, v_2^0 are Henselian with respect to $\bar{E}_{w'}$. It follows from Theorem 21.1.1 that the extension $\bar{E}_{w'}/\bar{F}_w$ is purely inseparable. For $i = 1, 2$ we consider this extension as a valued field extension $(\bar{E}_{w'}, u_i^0)/(\bar{F}_w, v_i^0)$ (see Lemma 20.3.1(a)). We deduce that the corresponding residue field extension is also purely inseparable (Proposition 14.2.5(c)). But one may canonically identify $\bar{E}_{u_i} = \overline{(\bar{E}_{w'})}_{u_i^0}$ and $\bar{F}_{v_i} = \overline{(\bar{F}_w)}_{v_i^0}$ (Proposition 5.2.1(e)). □

Just as for independent valuations, Theorem 21.2.1 can be rephrased in group-theoretic terms.

LEMMA 21.2.2. *Let $(E,u)/(F,v)$ be a normal extension of valued fields. If $\sigma \in \mathrm{Aut}(E/F) \setminus Z(u/v)$, then*

$$Z(u/v) \cap \sigma^{-1} Z(u/v)\sigma \leq T(u/v).$$

PROOF. By assumption, u and $u \circ \sigma$ are distinct, whence incomparable (Proposition 14.3.1). Let w_1, w_2 be their restrictions to the compositum $K = E_Z\sigma^{-1}(E_Z)$. By Corollary 15.3.2, w_1, w_2 are Henselian relative to E (since so are their restrictions to E_Z and $\sigma^{-1}(E_Z)$, respectively). In light of Theorem 20.1.1, they are also incomparable. Therefore Theorem 21.2.1 implies that $\bar{E}_u/\bar{K}_{w_1}$ is purely inseparable, and hence $\mathrm{Aut}(\bar{E}_u/\bar{K}_{w_1}) = 1$. We conclude from Theorem 16.1.1 that

$$T(u/w_1) = Z(u/w_1) = \mathrm{Gal}(E/K) = Z(u/v) \cap \sigma^{-1} Z(u/v)\sigma.$$

Finally, recall that $T(u/w_1) \leq T(u/v)$ (Remark 16.1.2(a)). □

We deduce that decomposition groups are (almost always) self-normalizing in the following sense. Note that this is a valued field counterpart of Proposition 19.4.3.

COROLLARY 21.2.3. *Let $(E,u)/(F,v)$ be a normal extension of valued fields with decomposition field E_Z. Suppose that the extension $\bar{E}_u/\bar{F}_v$ is not purely inseparable.*

(a) *For $\sigma \in \mathrm{Aut}(u/v)$ one has $Z(u/v) = \sigma^{-1}Z(u/v)\sigma$ if and only if $\sigma \in Z(u/v)$.*

(b) $\mathrm{Aut}(E_Z/F) = 1$.

PROOF. (a) The "if" part is trivial.

For the "only if" part suppose that $Z(u/v) = \sigma^{-1}Z(u/v)\sigma$ and $\sigma \notin Z(u/v)$. Then Lemma 21.2.2 implies that $Z(u/v) = T(u/v)$. It therefore follows from Theorem 16.1.1 that $\mathrm{Aut}(\bar{E}_u/\bar{F}_v) = 1$. Thus $\bar{E}_u/\bar{F}_v$ is purely inseparable, contrary to the assumption.

(b) Take $\tau \in \mathrm{Aut}(E_Z/F)$. It extends to an automorphism $\sigma \in \mathrm{Aut}(E/F)$. Then $\sigma(E_Z) = E_Z$. Equivalently, $Z(u/v) = \sigma^{-1}Z(u/v)\sigma$. By (a), $\sigma \in Z(u/v)$, so τ is trivial. □

Recall that an analog of Theorem 21.2.1 for orderings has already been established in §19.2: a real closed field (F,P) relative to a normal extension E with $E^\times = (E^\times)^2$ is Euclidean, whence $P = (F^\times)^2$ is its only ordering. Finally, in the "mixed" case of a valuation and an ordering we have:

COROLLARY 21.2.4. *Let E/F be a normal extension and let v be a valuation on F which is Henselian relative to E. Suppose that $E^\times = (E^\times)^2$. Then v is coarser than any ordering P on F.*

PROOF. We follow [**Lam2**, Lemma 3.15]. Since F is formally real, $\mathrm{char}\, F = 0$ and the polynomial

$$X^2 + X + 2 = \left(X + \frac{1}{2}\right)^2 + \frac{7}{4}$$

has no roots in F. By the Henselity, this polynomial has no simple roots in $\bar{F}_v$ (see condition (d) of Theorem 18.1.2). But the roots of

$$X^2 + X = X(X+1)$$

are simple, so necessarily $\mathrm{char}\, \bar{F}_v \neq 2$. By this and Proposition 18.2.1,

$$G_v = 1 + 4\mathfrak{m}_v \leq (F^\times)^2 \leq P. \qquad \square$$

21.3 Transition properties for Henselity

As we have observed in §15.3, relative Henselity has the following "going-up" property: if a valuation v on a field F is Henselian relative to a normal extension E of F, then so is the unique extension v_1 of v to any subextension $F \subseteq F_1 \subseteq E$. The analogous "going-down" property is obviously false in general; e.g., if $E = F_1$, then v_1 is Henselian relative to E even if v is not. Still, using the results of the previous section one can prove such "going-down" principles in some special situations. We start with two results of this type which are based on Theorem 21.2.1.

THEOREM 21.3.1. *Let $(F,v) \subseteq (F_1,v_1) \subseteq (E,u)$ be a tower of valued field extensions, with E/F and F_1/F normal. Assume that v_1 is Henselian relative to E and that the extension $\bar{E}_u/\overline{(F_1)}_{v_1}$ is not purely inseparable. Then v is Henselian relative to E.*

PROOF. We need to show that for every $\sigma \in \mathrm{Aut}(E/F)$ one has $u = u \circ \sigma$. To this end, let σ_1 be the restriction of σ to F_1. Then $v_1 \circ \sigma_1$ is the restriction of $u \circ \sigma$ to F_1. Using Remarks 15.1.1(a)(b) and the normality of F_1/F we get:

$$\begin{aligned} Z(u\circ\sigma/v_1\circ\sigma_1) &= \mathrm{Aut}(E/F_1) \cap Z(u\circ\sigma/v) \\ &= \mathrm{Aut}(E/F_1) \cap \sigma^{-1}Z(u/v)\sigma \\ &= \sigma^{-1}\Big(\mathrm{Aut}(E/F_1) \cap Z(u/v)\Big)\sigma \\ &= \sigma^{-1}Z(u/v_1)\sigma \\ &= \sigma^{-1}\mathrm{Aut}(E/F_1)\sigma \\ &= \mathrm{Aut}(E/F_1). \end{aligned}$$

Thus $v_1 \circ \sigma_1$ is also Henselian relative to E. Since the extension $\bar{E}_u/\overline{(F_1)}_{v_1}$ is not purely inseparable, Theorem 21.2.1 implies that v_1 and $v_1 \circ \sigma_1$ are comparable. By Proposition 14.3.1, $v_1 = v_1 \circ \sigma_1$. Since both u and $u \circ \sigma$ extend this valuation on F_1, the Henselity assumption implies that they coincide, as desired. □

THEOREM 21.3.2. *Let*

$$(F,v) \subseteq (F_1,v_1) \subseteq (E,u)$$

be a tower of valued field extensions, with E/F normal and F_1/F of finite separability degree. Assume that v_1 is Henselian relative to E and that the extension $\bar{E}_u/\bar{F}_v$ has infinite separability degree. Then v is Henselian relative to E.

PROOF. Let F_2 be the normal hull of F_1/F. Then $F_2 \subseteq E$. Let v_2 be the restriction of u to F_2. It extends v_2 and therefore is also Henselian relative to E (Corollary 15.3.2). Therefore it suffices to show that v is Henselian relative to F_2.

To this end take $\sigma \in \mathrm{Aut}(F_2/F)$. We need to show that $\sigma \in Z(v_2/v)$. Since F_1/F has finite separability degree, so does F_2/F, and therefore so does $\overline{(F_2)}_{v_2}/\bar{F}_v$. But $\bar{E}_u/\bar{F}_v$ has infinite separability degree, so the subextension $\bar{E}_u/\overline{(F_2)}_{v_2}$ cannot be purely inseparable. Also, both v_2 and $v_2\circ\sigma$ are valuations on F_2 which are Henselian relative to E. Deduce from Theorem 21.2.1 that they must be comparable. By Proposition 14.3.1, they coincide, as required. □

The following consequence of Theorem 21.3.2 is (essentially) proved in [**Ef1**].

PROPOSITION 21.3.3. *Let $(E,u)/(F,v)$ be a normal extension of valued fields and let n be a positive integer. Assume that $\mathrm{char}\,\bar{F}_v \nmid n$ and that the extension $\bar{E}_u/\bar{F}_v$ has an infinite separability degree. Suppose that*

$$\sup_K(K^\times : (K^\times)^n) < \infty$$

with K ranging over all finite extensions of F inside E, and that $\bar{K}^\times/n \neq 1$ for every finite extension $\bar{K}$ of $\bar{F}_v$ inside $\bar{E}_u$. Then v is Henselian relative to E.

PROOF. Assume the contrary. Then there exists a finite normal extension F_1 of F inside E such that v has at least two non-equivalent extensions v_1, v_1' to F_1 – say, with u extending v_1. It follows from Theorem 21.3.2 that v_1 cannot be Henselian relative to E. Applying the same argument with (F, v) replaced by (F_1, v_1) we obtain a finite extension F_2 of F_1 inside E such that v_1' has at least two non-equivalent extensions to F_2. Replacing it by its normal hull over F, we may further assume that it is normal over F. Since v_1' as well extends to F_2, the original valuation v has at least 3 non-equivalent extensions to F_2. Continuing inductively in this manner we obtain a tower of finite normal extensions

$$F \subset F_1 \subset F_2 \subset \cdots \subset E$$

of F such that v has at least $m+1$ non-equivalent extensions to F_m.

Now take a sufficiently large m, and let $w_0, \dots, w_m$ be non-equivalent extensions of v to F_m, with u extending w_0. By Proposition 14.2.6 they are incomparable. The approximation theorem for incomparable valuations (Theorem 10.2.1) implies that the residue homomorphism

$$\bigcap_{j=0}^{m} O_{w_j}^\times \to \prod_{j=0}^{m} \overline{(F_m)}_{w_j}^\times$$

is surjective. Hence so is the induced homomorphism

$$\Big(\bigcap_{j=0}^{m} O_{w_j}^\times\Big)/n \to \prod_{j=0}^{m} \overline{(F_m)}_{w_j}^\times/n.$$

Moreover, $(\bigcap_{j=0}^m O_{w_j}^\times)/n$ embeds in $F_m^\times/n$. Since the residue fields

$$\overline{(F_m)}_{w_j}, \quad j = 0, 1, \dots, m,$$

are all isomorphic (Proposition 14.3.3(b)), the assumption therefore implies that $\overline{(F_m)}_{w_j}^\times/n \neq 1$ for each j. Consequently, $(F_m^\times : (F_m^\times)^n) > m$. Thus the supremum in the statement of the theorem is infinite, a contradiction. □

This Henselity criterion is especially useful in the pro-p context, in the following form. As before, we denote the maximal pro-p Galois extension of a field F by $F(p)$.

COROLLARY 21.3.4. *Let p be a prime number, and let (F, v) be a valued field such that $\bar{F}_v$ contains a root of unity of order p and the extension $\bar{F}_v(p)/\bar{F}_v$ is infinite. Assume that*

$$\sup\nolimits_K (K^\times : (K^\times)^p) < \infty$$

with K ranging over all finite extensions of F inside $F(p)$. Then v is p-Henselian.

PROOF. Let u be an extension of v to $F(p)$. One has $\bar{F}_v(p) = \overline{(F(p))}_u$. For any finite p-extension $\bar{K}$ of $\bar{F}_v$, the pro-p extension $\bar{F}_v(p)/\bar{K}$ is infinite. By Kummer's theory, $\bar{K}^\times/p \neq 1$. Now apply Proposition 21.3.3 with $n = p$. □

The approximation theorem for incomparable valuations lies also behind the next fact.

PROPOSITION 21.3.5. *Let*

$$(F, v) \subseteq (F_1, v_1) \subseteq (E, u)$$

be a tower of valued field extensions, with E/F normal. Let n be a positive integer such that $G_{v_1} \leq (F_1^\times)^n$. Suppose that every finite extension (L, w) of (F, v) inside F_1 satisfies $(L^\times)^n = L^\times \cap (F_1^\times)^n$ and $\bar{L}_w^\times/n \neq 1$. Then v_1 is the unique extension of v to F_1.

PROOF. Assume the contrary. Then there exists a finite extension L of F inside F_1 and two non-equivalent extensions w, w' of v to L. We may assume that v_1 extends w. By the assumptions,

$$G_w \leq L^\times \cap G_{v_1} \leq L^\times \cap (F_1^\times)^n = (L^\times)^n.$$

In the notation of §11.1 we therefore have $w, w' \in \mathrm{Val}((L^\times)^n G_{w'})$. Further, by Proposition 14.2.6, w and w' are incomparable. Lemma 11.1.2(c) therefore implies that they are in

$$\mathrm{Val}((L^\times)^n G_{w'}, (L^\times)^n G_{w'}).$$

(where the notation is as in §11.1). In particular, $O_{w'}^\times \leq (L^\times)^n G_{w'}$, so $\bar{L}_{w'}^\times/n = 1$, contrary to our assumption. □

The following transition property is due to J. Koenigsmann [**Koe2**, Prop. 3.1] in the case where E is the separable closure of F, and the absolute Galois group of F_1 is a pro-p Sylow subgroup of the absolute Galois group of F (see §13).

COROLLARY 21.3.6. *Consider a tower of extensions*

$$(F, v) \subseteq (F_1, v_1) \subseteq (E, u)$$

with E/F normal. Let p be a prime number such that $p \nmid [F_1 : F]$, $p^\infty | [\bar{E}_u : \bar{F}_v]$, $E^\times = (E^\times)^p$, and $\bar{F}_v$ contains a primitive pth root of unity. If (F_1, v_1) is Henselian relative to E, then (F, v) is Henselian relative to E.

PROOF. We need to verify that the assumptions of Proposition 21.3.5 are satisfied with $n = p$. Indeed, $G_{v_1} \leq (F_1^\times)^p$ by Proposition 18.2.1 (note that $\mathrm{char}\,\bar{F}_v \neq p$). The two conditions on the finite extensions L follow from our assumptions on the degrees, using Kummer's theory. □

REMARK 21.3.7. In these Henselity criteria, the various assumptions about the residue fields can sometimes be slightly relaxed by considering variants of the core construction of §11.2; we refer to [**P2**, 1A)] for details.

CHAPTER 22

Sections

As we have seen in §16, the tower of normal subgroups $V(u/v) \leq T(u/v) \leq Z(u/v)$ of ramification theory has arithmetically interesting quotients: the character group $\chi(u/v)$ arising from the value groups, and the automorphism group of the residue fields. It turns out that, under some mild assumptions, the projections onto these canonical quotients have homomorphic sections, i.e., the inertia and ramification groups have complements in the decomposition group. As we shall see, in some instances, these sections give a purely group-theoretic description of the decomposition group.

22.1 Complements of inertia groups

Let $(E, u)/(F, v)$ be as before a normal extension of valued fields. We keep the notation $E_Z, E_T, E_V, u_Z, u_T, u_V$, etc., as in §§15–16. Let

$$\Omega = \{ \, l \mid l \text{ prime} , \quad l \neq \operatorname{char} \bar{F}_v , \quad l | [E : E_Z] \, \}.$$

THEOREM 22.1.1. *Assume that $E = E^l$ for every $l \in \Omega$.*

(a) *The canonical epimorphism*

$$Z(u/v)/V(u/v) \to Z(u/v)/T(u/v)$$

has a continuous homomorphic section.

(b) *There is a subextension $E_Z \subseteq L \subseteq E_V$ such that $E_Z = L \cap E_T$ and $E_V = LE_T$.*

PROOF. (a) is an equivalent group-theoretic formulation of the field-theoretic statement in (b). It is therefore enough to prove (b).

Zorn's lemma gives rise to a maximal subextension $E_Z \subseteq L \subseteq E_V$ such that $\bar{L}_{u_L} = \overline{(E_Z)}_{u_Z}$, where u_L is the restriction of u to L. Then $L \cap E_T$ also has the same residue field as E_Z, so by Corollary 16.1.4, $E_Z = L \cap E_T$.

Next let $K = LE_T$ and let u_K be the restriction of u to K. Since K contains E_Z, the valuation u_K is Henselian relative to E_V. If $p = \operatorname{char} \bar{F}_v > 0$, then Ostrowski's Theorem 17.2.1 implies that

$$[E_V : K] = e(u_V/u_K) f(u_V/u_K).$$

It remains to show that the latter two indices are both 1.

Since $E_T \subseteq K \subseteq E_V$ and $f(u_V/u_T) = 1$ (Corollary 16.2.7(b)), one also has $f(u_V/u_K) = 1$.

To compute $e(u_V/u_K)$ we consider a prime number $l \in \Omega$. We claim that $u_L(L^\times) = l u_L(L^\times)$. Otherwise, there would exist $a \in L^\times$ with $u(a) \notin l u_L(L^\times)$. In particular, $a \notin (L^\times)^l$, so the extension $L(\sqrt[l]{a})/L$ has degree l, where $\sqrt[l]{a}$ is some lth root of a in E. Moreover, the ramification index of $L(\sqrt[l]{a})/L$ is l. The fundamental inequality (Theorem 14.2.1) therefore implies that the inertia degree is 1. By Theorem 16.2.3, $l \nmid [E : E_V]$, whence $L \subset L(\sqrt[l]{a}) \subseteq E_V$. This contradiction to the maximality of L proves our claim.

As

$$e(u_V/u_K) \mid [E_V : K] \mid [E : E_Z],$$

any prime divisor l of $e(u_V/u_K)$ is either in Ω or equals char $\bar{F}_v$. But by the claim and by Lemma 1.1.4(b), the torsion group $u_V(E_V^\times)/u_L(L^\times)$ has no elements of order l for $l \in \Omega$. Hence neither has its quotient $u_V(E_V^\times)/u_K(K^\times)$. Moreover, when $p = \mathrm{char}\, \bar{F}_v > 0$, the torsion group $u_V(E_V^\times)/u_T(E_T^\times)$ has no elements of order p (Corollary 16.2.7(a)(d)), and therefore neither has its quotient $u_V(E_V^\times)/u_K(K^\times)$. Consequently, $e(u_V/u_K) = 1$, as desired. □

COROLLARY 22.1.2. *Assume that $E = E^l$ for every $l \in \Omega$. There is a semi-direct product decomposition*

$$Z(u/v)/V(u/v) \cong \chi(u/v) \rtimes \mathrm{Aut}(\bar{E}_u/\bar{F}_v),$$

with respect to the natural (left) action of $\mathrm{Aut}(\bar{E}_u/\bar{F}_v)$ *on* $\chi(u/v)$.

PROOF. It follows from Theorem 22.1.1(a) that there is a semi-direct product decomposition

$$Z(u/v)/V(u/v) \cong (T(u/v)/V(u/v)) \rtimes (Z(u/v)/T(u/v)).$$

As always, the precise structure of this group is determined by the action of $Z(u/v)/T(u/v)$ on $T(u/v)/V(u/v)$ via conjugation.

Now recall from Remark 16.2.8 that

$$T(u/v)/V(u/v) \cong \chi(u/v), \quad Z(u/v)/T(u/v) \cong \mathrm{Aut}(\bar{E}_u/\bar{F}_v),$$

and the conjugation action of $Z(u/v)/T(u/v)$ on $T(u/v)/V(u/v)$ corresponds under these isomorphisms to the natural action of $\mathrm{Aut}(\bar{E}_u/\bar{F}_v)$ on $\chi(u/v)$. The assertion follows. □

Next we analyze the structure of $\chi(u/v)$ as a profinite group. Let Ω_0 be the set of all prime numbers l such that $l | [E : E_Z]$ and $\bar{E}_u$ contains a root of unity of order l. Note that $\Omega_0 \subseteq \Omega$. For $l \in \Omega$ let

$$m_l = \dim_{\mathbb{F}_l}(v(F^\times)/l),$$

considered as a cardinal number. Since the torsion group $u_Z(E_Z^\times)/v(F^\times)$ has trivial l-primary part (by Theorem 15.2.2 and Corollary 15.2.3(a)), Lemma 1.1.4(c) implies that $m_l = \dim_{\mathbb{F}_l}(u_Z(E_Z^\times)/l)$.

PROPOSITION 22.1.3. *Assume that $E = E^l$ for every $l \in \Omega_0$. Then*

$$\chi(u/v) \cong \prod_{l \in \Omega_0} \mathbb{Z}_l^{m_l}.$$

PROOF. Let $\bar{\mu}$ be the group of all roots of unity in $\bar{E}_u^\times$. Let ${}_r\bar{\mu}$ be its subgroup consisting of all roots of unity of order dividing r. For a prime number l let $\bar{\Gamma}_l$ be the l-primary subgroup of $u(E^\times)/u_Z(E_Z^\times)$. We compute $\chi_{\bar{\mu}}(\bar{\Gamma}_l)$:

CASE I: *$\bar{E}_u$ does not contain a root of unity of order l.*
This means that $\bar{\mu}$ has no elements of order l. By Lemma 1.4.7(b), $\chi_{\bar{\mu}}(\bar{\Gamma}_l) = 0$.

CASE II: *$l \notin \Omega_0$ and $\bar{E}_u$ contains a root of unity of order l.*
Then $l \nmid [E : E_Z]$. As $e(u/u_Z) | [E : E_Z]$ (Theorem 17.2.1), $l \nmid e(u/u_Z)$. Hence $\bar{\Gamma}_l = 0$ and therefore $\chi_{\bar{\mu}}(\bar{\Gamma}_l) = 0$ in this case as well.

CASE III: *$l \in \Omega_0$.*
As $E^\times/l = 1$, the preimage Γ' of $\bar{\Gamma}_l$ in $u(E^\times)$ satisfies $\Gamma' = l\Gamma'$. By Lemma 1.2.8, $\Gamma' \cong \frac{1}{l^\infty} u_Z(E_Z^\times)$ over the subgroup $u_Z(E_Z^\times)$. Therefore,

$$\bar{\Gamma}_l \cong \left(\tfrac{1}{l^\infty} u_Z(E_Z^\times)\right)/u_Z(E_Z^\times).$$

Lemma 1.4.3 now implies that $\chi_{\bar{\mu}}(\bar{\Gamma}_l) \cong (\varprojlim_{l^i} \bar{\mu})^{m_l}$ canonically, where the inverse image is with respect to exponentiations by l. As $E^\times/l = 1$, also $\bar{E}_u^\times/l = 1$. Therefore $\bar{E}_u$ contains the roots of unity of order l^i for all $i \geq 1$. It follows that $\varprojlim_{l^i} \bar{\mu} \cong \varprojlim \mathbb{Z}/l^i \cong \mathbb{Z}_l$ in this case.

Putting all this together, and using Lemma 1.4.2, we obtain that

$$\chi(u/u_Z) \cong \prod_l \chi_{\bar{\mu}}(\bar{\Gamma}_l) \cong \prod_{l \in \Omega_0} \mathbb{Z}_l^{m_l}.$$

It remains to recall that $u_Z(E_Z^\times)/v(F^\times)$ is p-primary if $p = \operatorname{char} \bar{F}_v > 0$ and is trivial otherwise (by Theorem 15.2.2 and Corollary 15.2.3(a)). By Lemma 1.4.8(a) one may therefore identify $\chi(u/v) = \chi(u/u_Z)$. □

EXAMPLE 22.1.4. Let (F, v) be a Henselian valued field with residue field $\mathbb{F}_q$, where $q = p^s$ is a prime power. Let u be the unique extension of v to the algebraic closure $E = \tilde{F}$. The absolute Galois group $G_{\mathbb{F}_q} = \mathrm{Gal}(\tilde{\mathbb{F}}_p/\mathbb{F}_q)$ of $\bar{F}_v = \mathbb{F}_q$ is isomorphic to $\hat{\mathbb{Z}} = \varprojlim \mathbb{Z}/n$, and is topologically generated by the Frobenius automorphism

$$\mathrm{Frob}_q \colon x \mapsto x^q.$$

In particular, every prime number divides the supernatural degree $[\tilde{\mathbb{F}}_p : \mathbb{F}_q]$, and therefore also the supernatural degree $[\tilde{F} : F]$. Thus Ω and Ω_0 are the set of all primes $\neq p$. Corollary 22.1.2 and Proposition 22.1.3 give

$$\begin{aligned} G_F/V(u/v) &= Z(u/v)/V(u/v) \\ &\cong \chi(u/v) \rtimes \mathrm{Aut}(\tilde{\mathbb{F}}_p/\mathbb{F}_q) \\ &\cong \Big(\prod_{l \neq p} \mathbb{Z}_l^{m_l}\Big) \rtimes \hat{\mathbb{Z}}. \end{aligned}$$

Moreover, the generator Frob_q of $\hat{\mathbb{Z}}$ acts on every $\tau \in \mathbb{Z}_l^{m_l}$ by

$$\tau^{\mathrm{Frob}_q} = \tau^q$$

(where we use multiplicative notation for $\mathbb{Z}_l^{m_l}$).

EXAMPLE 22.1.5 (Hasse [**Has**], Iwasawa [**I**]). Let F be a non-archimedean local field, i.e., a finite extension of $\mathbb{Q}_p$ or of $\mathbb{F}_p((t))$, where p is a prime number. Let v be the canonical discrete complete valuation on F. Recall that it is Henselian, and has residue field $\mathbb{F}_q$ for some p-power q. As a special case of Example 22.1.4 we obtain that the "maximal tame Galois group" of F is

$$G_F/V(u/v) \cong (\hat{\mathbb{Z}}/\mathbb{Z}_p) \rtimes \hat{\mathbb{Z}}.$$

Moreover, there are topological generators τ, σ of the semi-direct factors $\hat{\mathbb{Z}}/\mathbb{Z}_p$ and $\hat{\mathbb{Z}}$ such that $\sigma\tau\sigma^{-1} = \tau^q$ (where σ corresponds to the Frobenius automorphism).

EXAMPLE 22.1.6. Let p be a prime number and let (F, v) be a p-Henselian valued field. Thus v is Henselian relative to the maximal pro-p Galois extension $E = F(p)$ of F. Let u be the unique extension of v to $F(p)$. We further assume that $\bar{F}_v$ contains a root of unity of order p (in particular, char $\bar{F}_v \neq p$). Then $\overline{F(p)}_u = \bar{F}_v(p)$. By Theorem 16.2.3, the ramification group of $(F(p), u)/(F, v)$ is trivial. In the previous notation, one then has $\Omega = \Omega_0 = \{p\}$. Let $G_F(p) = \mathrm{Gal}(F(p)/F)$ be the maximal pro-p Galois group of F. It follows from Corollary 22.1.2 and Proposition 22.1.3 that

$$G_F(p) = Z(u/v) \cong \chi(u/v) \rtimes \mathrm{Gal}(\overline{F(p)}_u/\bar{F}_v) \cong \mathbb{Z}_p^m \rtimes G_{\bar{F}_v}(p), \tag{22.1.1}$$

where $m = \dim_{\mathbb{F}_p}(v(F^\times)/p)$.

The action in (22.1.1) is as follows. Let μ_{p^∞} be the group of all roots of unity of p-power order in the algebraic closure of $\bar{F}_v$. The assumptions imply that $\mu_{p^\infty} \subseteq \bar{F}_v(p)$. The group $\mathrm{Aut}(\mu_{p^\infty}/\mu_p)$ of automorphisms of μ_{p^∞} fixing μ_p is isomorphic to the subgroup $1+p\mathbb{Z}_p$ of $\mathbb{Z}_p^\times$. More specifically, $\theta \in \mathrm{Aut}(\mu_{p^\infty}/\mu_p)$ corresponds under this isomorphism to the p-adic unit $\chi \in 1+p\mathbb{Z}_p$ such that $\theta(\zeta) = \zeta^\chi$ for all $\zeta \in \mu_{p^\infty}$ in the natural sense. The composition of the restriction map $G_{\bar{F}_v}(p) \to \mathrm{Aut}(\mu_{p^\infty}/\mu_p)$ and this isomorphism is the **pro-p cyclotomic character**

$$\chi_{\bar{F}_v,p}\colon G_{\bar{F}_v}(p) \to 1 + p\mathbb{Z}_p$$

of $\bar{F}_v$.

Now $\bar{\Gamma} = u(F(p)^\times)/v(F^\times)$ is a p-primary group, by Ostrowski's theorem. Hence

$$\chi(u/v) = \mathrm{Hom}(\bar{\Gamma}, \mu_{p^\infty}).$$

The action of $G_{\bar{F}_v}(p)$ on $\chi(u/v)$ in the semi-direct decomposition (22.1.1) is then given by

$$({}^{\bar{\sigma}}\psi)(\bar{\gamma}) = (\psi(\bar{\gamma}))^{\chi_{\bar{F}_v,p}(\bar{\sigma})}$$

for $\bar{\sigma} \in G_{\bar{F}_v}(p)$, $\psi \in \mathrm{Hom}(\bar{\Gamma}, \mu_{p^\infty})$, and $\bar{\gamma} \in \bar{\Gamma}$. Identifying the multiplicative group $\chi(u/v)$ with the additive group $\mathbb{Z}_p^m$, we obtain that the action in (22.1.1) is given by

$$\tau^{\bar{\sigma}} = \chi_{\bar{F}_v,p}(\bar{\sigma}) \cdot \tau$$

for $\bar{\sigma} \in G_{\bar{F}_v}(p)$ and $\tau \in \mathbb{Z}_p^m$.

22.2 Complements of ramification groups

The following result is due to Kuhlmann, Pank and Roquette [**KuPR**] (in the case of separable closures of fields). While the proof uses techniques of Galois cohomology which are more advanced than the main line of this book, we nevertheless present it here since it complements Theorem 22.1.1 nicely (this result and its consequences will not be used later on). We refer to [**NSW**], [**Ris**] or [**Se3**] for comprehensive treatises on Galois cohomology.

THEOREM 22.2.1. *Let $(E,u)/(F,v)$ be a normal extension of valued fields. When $p = \mathrm{char}\,\bar{F}_v > 0$ let $\wp(x) = x^p - x$ and assume that $\bar{E}_u = \wp(\bar{E}_u)$.*

(a) *The canonical epimorphism*

$$Z(u/v) \to Z(u/v)/V(u/v)$$

has a continuous homomorphic section.

(b) *There is a subextension $E_Z \subseteq L \subseteq E$ such that $E_Z = L \cap E_V$ and $E = LE_V$.*

PROOF. Again, (b) is just a field-theoretic translation of (a), so it suffices to prove (a).

When $\mathrm{char}\,\bar{F}_v = 0$ we have $V(u/v) = 1$ (Theorem 16.2.3), and the assertion is trivial. So assume that $p = \mathrm{char}\,\bar{F}_v > 0$. Consider a pro-$p$ Sylow subgroup $\bar{G}$ of $\mathrm{Aut}(\bar{E}_u/\bar{F}_v)$. It follows from the assumption that there is a short exact sequence of discrete (additive) $\bar{G}$-modules

$$0 \to \mathbb{F}_p \to \bar{E}_u \xrightarrow{\wp} \bar{E}_u \to 0,$$

where $\bar{G}$ acts trivially on $\mathbb{F}_p$. This sequence gives rise to an exact sequence of Galois cohomology groups

$$H^1(\bar{G}, \bar{E}_u) \to H^2(\bar{G}, \mathbb{F}_p) \to H^2(\bar{G}, \bar{E}_u).$$

By the additive version of Hilbert's "Theorem 90", $H^i(\bar{G}, \bar{E}_u) = 0$ for all $i \geq 1$ [**NSW**, Ch. VI, §1]. Hence also $H^2(\bar{G}, \mathbb{F}_p) = 0$, i.e., $\bar{G}$ has p-cohomological dimension ≤ 1 [**Se3**, I, §4.1, Prop. 21].

Now $Z(u/v)/T(u/v) \cong \mathrm{Aut}(\bar{E}_u/\bar{F}_v)$ and $p \nmid (T(u/v) : V(u/v))$, by Theorem 16.1.1 and Theorem 16.2.3, respectively. Hence any pro-p Sylow subgroup of $Z(u/v)/V(u/v)$ must be isomorphic to $\bar{G}$. It follows that the p-cohomological dimension of $Z(u/v)/V(u/v)$ is ≤ 1 [**Se3**, I, §3.3, Cor. 1]. Equivalently [**Se3**, I, §3.4, Prop. 16], every extension of $Z(u/v)/V(u/v)$ by a pro-p group splits. Since $V(u/v)$ is a pro-p group, we obtain the desired section. □

Combining Theorem 22.1.1(a) and Theorem 22.2.1(a) we obtain:

COROLLARY 22.2.2. *Let $(E,u)/(F,v)$ be a normal extension of valued fields. Assume that:*

(i) *$E^\times = (E^\times)^l$ for every prime number $l \neq \mathrm{char}\,\bar{F}_v$ such that $l | [E : E_Z]$;*

(ii) *$\bar{E}_u = \wp(\bar{E}_u)$ if $p = \mathrm{char}\,\bar{F}_v$ (where $\wp$ is as above).*

Then the canonical epimorphism $\Phi\colon Z(u/v) \to \mathrm{Aut}(\bar{E}_u/\bar{F}_v)$ of Theorem 16.1.1 has a continuous homomorphic section. Thus

$$Z(u/v) \cong T(u/v) \rtimes \mathrm{Aut}(\bar{E}_u/\bar{F}_v).$$

The following consequence of Corollary 22.2.2 was proved in [**Ef1**]. While its proof relies on a considerable amount of valuation theory, its statement is purely Galois-theoretic.

COROLLARY 22.2.3. *If a profinite group is realizable as an absolute Galois group of a field, then it is realizable as the absolute Galois group of a field of characteristic* 0.

PROOF. Consider an absolute Galois group $G_{\bar F} = \mathrm{Gal}(\bar F_{\mathrm{sep}}/\bar F)$ of a field $\bar F$ of characteristic $p > 0$. We may replace $\bar F$ by its maximal inseparable extension (inside the algebraic closure) to assume that it is perfect. Let F be the quotient field of the ring $W(\bar F)$ of Witt vectors of $\bar F$, and let v be its canonical valuation. Recall that $\mathrm{char}\, F = 0$ and (F, v) is a complete discrete valued field with residue field $\bar F$ (Theorem 12.4.1). In particular, v is Henselian (Corollary 18.3.2). We conclude from Corollary 22.2.2 that the canonical epimorphism $G_F \to G_{\bar F}$ splits. Thus $G_{\bar F}$ is isomorphic to the absolute Galois group of some algebraic extension of F. □

REMARK 22.2.4. Theorem 22.2.1 gives a semi-direct product decomposition

$$Z(u/v) = V(u/v) \rtimes (Z(u/v)/V(u/v)).$$

However, unlike the decomposition in Corollary 22.1.2, the description of the action here is presently known only in some special cases. For instance:

(1) When (F, v) is a discrete Henselian valued field of characteristic $p > 0$ and E is its separable closure, the group $V(u/v)$ is the free pro-p $Z(u/v)/V(u/v)$-operator group on countably many generators, in the sense of [**NSW**, p. 216]. This was proved by Koch [**Ko1**] when $F = \mathbb{F}_p((t))$, and by Mel'nikov and Sharomet [**MelS**] when F is an arbitrary Laurent series field. As observed in [**EfF**], a similar argument applies for arbitrary discrete Henselian valued fields of characteristic p.

(2) When F is a p-adic field with $p \neq 2$ and containing a pth root of unity, and E is its algebraic closure, this action was explicitly described by Jannsen and Wingberg ([**Jn**], [**JnW**], [**Win**]), following earlier results by Jakovlev ([**Jak1**], [**Jak2**]) and Koch [**Ko2**]. See [**Di**] for results in the case $p = 2$. Using class field theory one can show that here $V(u/v)$ is a *free* pro-p group of countable rank [**N2**].

Part IV

K-RINGS

This final part of the book explores cohomological aspects of valuations and orderings. While it is possible to study these aspects by means of Galois cohomology, it appears to be more natural (and certainly more elementary) to carry the discussion using the language of Milnor's K-theory. This reveals intimate connections between many of the results in this part and the analysis in Part II of valuations and orderings from the viewpoint of the multiplicative group: compatible localities, the canonical valuations $v_{(2)}(S, H)$, and the minimal subgroup H_S of §11.4 – to mention a few.

§§23–24 are a general introduction to Milnor K-theory, and many of the results in these chapters are based on the seminal works by Milnor [**Mi**] and Bass–Tate [**BaT**], where the Milnor K-ring functor of a field was first studied. However in our arithmetical context it is convenient to work with a generalized version of this functor, namely, the Milnor K-ring $K_*^M(F)/S$ of a field F modulo a subgroup S of $F^\times$, introduced in [**Ef9**].

We begin with a formal description of the target category of this functor – the so-called "κ-structures" (§23). In §24 we define the relative Milnor K-ring of a field and study its general properties. Then we go back to several arithmetic situations, and study the reflection in this ring of orderings (§25) and of S-compatible valuations (§26.1). In §§26.3–26.6 we show how to detect S-compatible valuations from the structure of $K_*^M(F)/S$. In particular, we obtain a K-theoretic description of the minimal group H_S, which measures the possible "non-triviality" of an S-compatible valuation. Analogous results in this direction regarding the "wild" (mixed characteristic) case are given in §27.

Finally, in §§28–29 we study semi-local situations, and prove several decomposition theorems for the K-ring at the presence of finitely many localities. This is based on the approximation theorems of §10. We prove a K-theoretic version of one of the main theorems of the theory of ordered fields due to Jacob [**J1**] (see also [**Cr1**]); it gives an explicit inductive description of the ring $K_*^M(F)/S$ for a preordering S on F with $(F^\times : S) < \infty$ (§29.2). This result leads to the formulation of the "elementary type conjecture" on the structure of $K_*^M(F)/S$ when S is an arbitrary subgroup of $F^\times$ containing $(F^\times)^p$ for a prime number p, and for which $(F^\times : S) < \infty$ (§29.3).

CHAPTER 23

κ-Structures

In this chapter we describe the natural target category for the Milnor K-ring functor (to be defined in §24): the κ-structures. This is a slight modification of the notion of "κ-Algebras", as defined by Bass and Tate in [**BaT**], the difference being the additional condition (i) of Definition 23.1.1 below. This condition is needed in order to study direct products of K-rings. We describe the two main constructions in this category: direct products and extensions. The latter construction will turn out to be tightly related to tame valuations via the notion of rigid elements, as we will see later on. Finally, we introduce a special class of κ-structures, the so-called Demuškin structures, which will be related to wild valuations.

23.1 Basic notions

Let Γ be an abelian group, considered as a $\mathbb{Z}$-module. Denote the tensor algebra $\bigoplus_{r=0}^{\infty}\Gamma^{\otimes r}$ of Γ by $\mathrm{Tens}(\Gamma)$ [**Lg**, Ch. XVI, §5]. It has a natural structure of a graded ring. Let

$$\kappa = \bigoplus_{r=0}^{\infty} \kappa_r = \mathrm{Tens}(\{\pm 1\}).$$

The non-trivial element of $\kappa_1 \cong \{\pm 1\}$ will be denoted by ϵ. Thus $\kappa_0 = \mathbb{Z}$ and $\kappa_r = \{0, \epsilon^r\} \cong \mathbb{Z}/2$ for all $r \geq 1$. Alternatively, $\kappa = \mathbb{Z}[X]/(2X)$, where $\mathbb{Z}[X]$ is graded by degree, $(2X)$ is the homogeneous ideal generated by $2X$ in degree 1, and ϵ is the image of X in κ.

DEFINITION 23.1.1. A **κ-structure** consists of a graded ring $A = \bigoplus_{r=0}^{\infty} A_r$ and a graded ring homomorphism $\kappa \to A$ such that:

(i) $A_0 = \mathbb{Z}$ and the homomorphism $\kappa \to A$ is the identity in degree 0;
(ii) A_1 generates A as a ring;
(iii) the image ϵ_A of ϵ in A_1 satisfies $a^2 = \epsilon_A a = a\epsilon_A$ for all $a \in A_1$.

Equivalently, a κ-structure is a graded ring $A = \bigoplus_{r=0}^{\infty} A_r$ with a distinguished element $\epsilon_A \in A_1$ such that $2\epsilon_A = 0$, A_1 generate A, and the equalities in (iii) hold.

Notice that condition (iii) above is equivalent to:

(iii') $a(a + \epsilon_A) = 0$ for all $a \in A_1$.

In fact, since $\epsilon_A = -\epsilon_A$, the latter condition is equivalent to $a^2 = a\epsilon_A$. Substituting $a + \epsilon_A$ for a in (iii') we also get $a^2 = \epsilon_A a$.

A κ-structure A is called **trivial** if $A_1 = 0$ (so $A_r = 0$ for all $r \geq 1$, by (ii)). We denote the trivial κ-structure by **0**.

LEMMA 23.1.2. *A κ-structure A is anti-commutative.*

PROOF. Let $a, b \in A_1$. By (iii),

$$\epsilon_A(a+b) = (a+b)^2 = a^2 + ab + ba + b^2 = \epsilon_A a + ab + ba + \epsilon_A b.$$

Therefore $ab = -ba$. The assertion now follows from (ii). $\square$

A **morphism** $A \to B$ of κ-structures is a graded ring homomorphism which commutes with the structural homomorphisms $\kappa \to A$, $\kappa \to B$. As usual, an **epimorphism** (resp., **monomorphism**) is a surjective (resp., injective) morphism.

23.2 Constructions of κ-structures

The **direct product** $\prod_{i\in I} A_i$ of κ-structures A_i, $i \in I$, is defined by

$$\Big(\prod_{i\in I} A_i\Big)_r = \begin{cases} \mathbb{Z} & \text{if } r = 0, \\ \prod_{i\in I}(A_i)_r & \text{if } r \geq 1, \end{cases}$$

with the natural multiplicative structure. The homomorphism $\kappa \to \prod_{i\in I} A_i$ is given by $\epsilon \mapsto (\epsilon_{A_i})_{i\in I}$.

Recall that the tensor product in the category of *graded rings* is defined by

$$A \otimes_{\mathbb{Z}} B = \bigoplus_{r=0}^{\infty}\Big(\bigoplus_{i+j=r} A_i \otimes_{\mathbb{Z}} B_j\Big),$$

with the multiplication law

$$(a \otimes b)(a' \otimes b') = (-1)^{i'j} aa' \otimes bb'$$

for $a \in A_i$, $a' \in A_{i'}$, $b \in B_j$, $b' \in B_{j'}$. Given κ-structures A, B, we define their **tensor product** in the category of *κ-structures* to be

$$A \otimes_\kappa B = (A \otimes_{\mathbb{Z}} B)/I,$$

where I is the homogeneous ideal generated by $\epsilon_A \otimes 1_B - 1_A \otimes \epsilon_B$ (here $1_A, 1_B$ are the unit elements of A_0, B_0, respectively). Define a graded ring homomorphism $\kappa \to A \otimes_\kappa B$ by

$$\epsilon \;\mapsto\; \epsilon_A \otimes 1_B + I = 1_A \otimes \epsilon_B + I.$$

There are canonical morphisms

$$\iota\colon A \to A \otimes_\kappa B, \quad \iota'\colon B \to A \otimes_\kappa B$$

defined by

$$\iota(a) = a \otimes 1_B + I, \quad \iota'(b) = 1_A \otimes b + I.$$

LEMMA 23.2.1. *For κ-structures A, B as above:*
(a) *$A \otimes_\kappa B$ is a κ-structure;*
(b) *$A \otimes_\kappa B$ is the coproduct of A and B in the category of κ-structures with respect to ι, ι';*
(c) *$A \cong A \otimes_\kappa \kappa$ and $B \cong \kappa \otimes_\kappa B$ via ι and ι', respectively.*

PROOF. (a) Conditions (i) and (ii) of Definition 23.1.1 are clear.

Since A, B are anti-commutative, so is $A \otimes_{\mathbb{Z}} B$. Further, given $a \in A_1$ and $b \in B_1$ we have

$$(a \otimes 1_B)^2 = (\epsilon_A \otimes 1_B)(a \otimes 1_B)$$

and

$$(1_A \otimes b)^2 = (1_A \otimes \epsilon_B)(1_A \otimes b) \equiv (\epsilon_A \otimes 1_B)(1_A \otimes b) \mod I.$$

Hence the anti-commutativity gives

$$(a \otimes 1_B + 1_A \otimes b)^2 \equiv (\epsilon_A \otimes 1_B)(a \otimes 1_B + 1_A \otimes b) \mod I$$

in $(A \otimes_\kappa B)_2$. Since

$$(A \otimes_{\mathbb{Z}} B)_1 = (A_1 \otimes_{\mathbb{Z}} \mathbb{Z}) \oplus (\mathbb{Z} \otimes_{\mathbb{Z}} B_1),$$

this implies the first equality in (iii) for $A \otimes_\kappa B$. The second equality is proved similarly.

(b) Any κ-structure morphisms $\varphi\colon A \to C$ and $\psi\colon B \to C$ give rise to a unique graded ring homomorphism $\varphi \otimes_{\mathbb{Z}} \psi\colon A \otimes_{\mathbb{Z}} B \to C$ which commutes with φ, ψ and the canonical maps $A \to A \otimes_{\mathbb{Z}} B$, $B \to A \otimes_{\mathbb{Z}} B$. Namely,

$$(\varphi \otimes_{\mathbb{Z}} \psi)(a \otimes b) = \varphi(a)\psi(b).$$

Furthermore, $\varphi(\epsilon_A) = \epsilon_C = \psi(\epsilon_B)$, so $(\varphi \otimes \psi)(I) = 0$. Therefore $\varphi \otimes_{\mathbb{Z}} \psi$ induces a κ-structure morphism $\varphi \otimes_\kappa \psi\colon A \otimes_\kappa B \to C$ such that

$$\varphi = (\varphi \otimes_\kappa \psi) \circ \iota, \quad \psi = (\varphi \otimes_\kappa \psi) \circ \iota'.$$

(c) Straightforward. □

Next we construct free objects in the category of κ-structures. Let Γ be an abelian group. We define $\kappa[\Gamma]$ to be the quotient of $\mathrm{Tens}(\kappa_1 \oplus \Gamma)$ by the homogeneous ideal generated by all elements

$$\epsilon \otimes \gamma - \gamma \otimes \gamma,$$

where $\gamma \in \Gamma$. Taking $\gamma' = \epsilon + \gamma$ we find (using $\epsilon = -\epsilon$) that

$$\gamma \otimes \epsilon - \gamma \otimes \gamma = \epsilon \otimes \gamma' - \gamma' \otimes \gamma'$$

is also in this ideal. The obvious embedding $\kappa_1 \hookrightarrow \kappa_1 \oplus \Gamma$ induces a graded ring homomorphism $\kappa \to \kappa[\Gamma]$. Then $\kappa[\Gamma]$ is a κ-structure. It has the following universal property (which follows from the universal property of the tensor algebra as a coproduct in the category of graded rings):

For every κ-structure B and every abelian group homomorphism $\theta\colon \Gamma \to B_1$ there exists a unique morphism $\kappa[\Gamma] \to B$ extending θ.

Given a κ-structure A, we call $A[\Gamma] = A \otimes_\kappa \kappa[\Gamma]$ the **extension** of A by Γ. When $A = \kappa$ this coincides with our previous notation, by Lemma 23.2.1(c). One may identify $(A[\Gamma])_1 = A_1 \oplus \Gamma$. Let $\iota\colon A \to A[\Gamma]$ be the canonical morphism as above. We obtain the following universal property for extensions:

LEMMA 23.2.2. *Let $\varphi: A \to B$ be a morphism of κ-structures and let $\theta: \Gamma \to B_1$ be a homomorphism of abelian groups. There exists a unique morphism $A[\Gamma] \to B$ extending θ which commutes with φ and ι.*

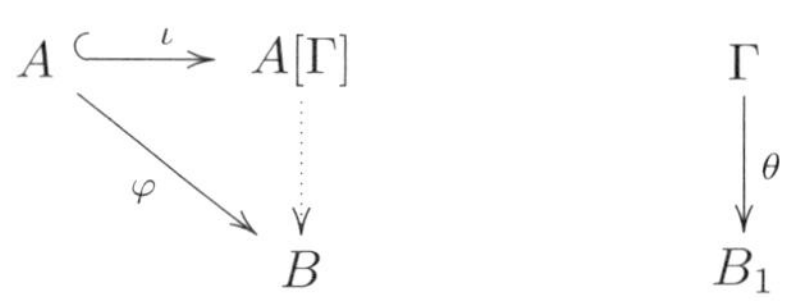

PROOF. The universal property of $\kappa[\Gamma]$ yields a unique morphism $\kappa[\Gamma] \to B$ extending θ. Now use the fact that the tensor product is a coproduct (Lemma 23.2.1(b)). □

LEMMA 23.2.3. *For a κ-structure A and abelian groups Γ_1, Γ_2 one has*

$$(A[\Gamma_1])[\Gamma_2] \cong A[\Gamma_1 \oplus \Gamma_2].$$

PROOF. The degree 1 components of both κ-structures may be identified with $A_1 \oplus \Gamma_1 \oplus \Gamma_2$. By (ii) of Definition 23.1.1, this set generates these κ-structures as rings.

Now the universal property of $A[\Gamma_1 \oplus \Gamma_2]$ gives rise to a unique morphism

$$\varphi: A[\Gamma_1 \oplus \Gamma_2] \to (A[\Gamma_1])[\Gamma_2]$$

which is the identity map on $\Gamma_1 \oplus \Gamma_2$ and which makes the outer square in the diagram below commutative. Thus it is also the identity on $A_1 \oplus \Gamma_1 \oplus \Gamma_2$.

Conversely, the universal property of $A[\Gamma_1]$ yields a unique morphism

$$\psi': A[\Gamma_1] \to A[\Gamma_1 \oplus \Gamma_2]$$

which is the identity on Γ_1 and which makes the upper triangle in the diagram commutative. Finally, the universal property of $(A[\Gamma_1])[\Gamma_2]$ yields a unique morphism

$$\psi: (A[\Gamma_1])[\Gamma_2] \to A[\Gamma_1 \oplus \Gamma_2]$$

which is the identity on Γ_2 and which makes the lower triangle commutative. Hence ψ is the identity map on $A_1 \oplus \Gamma_1 \oplus \Gamma_2$.

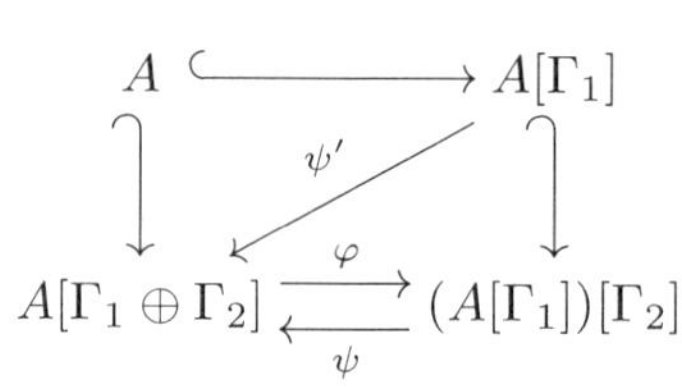

Thus each of the morphisms $\varphi \circ \psi$ and $\psi \circ \varphi$ is the identity map on the generating set $A_1 \oplus \Gamma_1 \oplus \Gamma_2$. Therefore they are the identity morphisms, so φ and ψ are inverse isomorphisms. □

We describe the structure of extensions explicitly in some important situations.

EXAMPLE 23.2.4. For an abelian group Γ, the group $(\mathbf{0}[\Gamma])_r$ is the quotient of $\Gamma^{\otimes r}$ by the subgroup generated by all tensors of the form $\gamma_1 \otimes \cdots \otimes \gamma_r$ such that $\gamma_1, \ldots, \gamma_r \in \Gamma$ and $\gamma_i = \gamma_j$ for some distinct i, j. Thus $\mathbf{0}[\Gamma]$ is the alternating $\mathbb{Z}$-algebra $\bigwedge_*(\Gamma)$ on Γ [**Lg**, Ch. XVI, §6].

EXAMPLE 23.2.5. Let $\bar{A}$ be a κ-structure, let Γ be a cyclic group with generator γ, and let $A = \bar{A}[\Gamma]$. The group A_r is generated by the image of $\bigoplus_{i+j=r}(\bar{A}_i \otimes_{\mathbb{Z}} \Gamma^{\otimes j})$. By (iii) of Definition 23.1.1, $\gamma^j = \epsilon_A^{j-1}\gamma = \epsilon_{\bar{A}}^{j-1}\gamma$ in $\bar{A}[\Gamma]$ for $j \geq 1$. It follows that

$$A_r = \bar{A}_r \oplus (\bar{A}_{r-1} \otimes_{\mathbb{Z}} \Gamma)$$

for $r \geq 1$. In particular, $A_1 = \bar{A}_1 \oplus \Gamma$ (as we have already observed) and $A_2 = \bar{A}_2 \oplus (\bar{A}_1 \otimes_{\mathbb{Z}} \Gamma)$. The product $A_1 \times A_1 \to A_2$ is then given by

$$(\bar{a}, i\gamma) \cdot (\bar{b}, j\gamma) = (\bar{a} \cdot \bar{b}, (j\bar{a} - i\bar{b} + ij\epsilon_{\bar{A}}) \otimes \gamma), \tag{23.2.1}$$

for $\bar{a}, \bar{b} \in \bar{A}_1$ and $i, j \in \mathbb{Z}$.

EXAMPLE 23.2.6. A straightforward computation using Example 23.2.5 gives

$$\kappa[\mathbb{Z}/2] \cong \kappa \times \kappa.$$

23.3 Rigidity

Extensions of κ-structures are intimately related to special elements of degree 1 called "rigid elements". These elements are defined in purely ring-theoretic terms, however, as we shall see later on, they play an arithmetic role similar to that of uniformizers in discrete valued fields. More generally, rigid elements will serve as "cohomological footprints" from which we will detect tame valuations. Variants of this quite subtle notion were defined and studied by Szymiczek [**Szy**], Ware ([**War2**], [**War3**]) and Arason, Elman, and Jacob [**AEJ**].

DEFINITION 23.3.1. Let n be a positive integer, and let A be a κ-structure. We say that $a \in A_1$ is **n-rigid** if:

(i) $\langle a \rangle \cong \mathbb{Z}/n$; and
(ii) for every $b \in A_1$ with $a \cdot b = 0$ in A_2 there exist integers i, j, at least one of which prime to n, such that $j(\epsilon_A + a) = ib$.

We will use this notion mostly when $n = p$ is prime and $pA_1 = 0$. Then (ii) means that for every $b \in A_1$ with $a \cdot b = 0$ the elements $\epsilon_A + a$ and b of A_1 are $\mathbb{F}_p$-linearly dependent. First we show the following basic connection between rigidity and extensions.

LEMMA 23.3.2. *Let n be a positive integer, and let $\bar{A}$ be a κ-structure such that $n\bar{A}_1 = 0$. Let Γ be a free $\mathbb{Z}/n$-module and let $A = \bar{A}[\Gamma]$. Suppose that the projection of $a \in A_1$ in Γ has order n. Then a is n-rigid.*

PROOF. We may decompose $\Gamma = \Gamma_1 \oplus \Gamma_2$, where $\Gamma_2 \cong \mathbb{Z}/n$ and where the projection of a in Γ_2 is of order n. By Lemma 23.2.3, $A \cong (\bar{A}[\Gamma_1])[\Gamma_2]$. Since n-rigidity is clearly invariant under κ-structure isomorphisms, it is therefore enough to consider the case where $\Gamma \cong \mathbb{Z}/n$.

So take a generator γ of $\Gamma \cong \mathbb{Z}/n$. Recall from Example 23.2.5 that we may identify $A_1 = \bar{A}_1 \oplus \Gamma$ and $A_2 = \bar{A}_2 \oplus (\bar{A}_1 \otimes_{\mathbb{Z}} \Gamma)$, with the product given by (23.2.1). We write $a = (\bar{a}, i\gamma)$, with $\bar{a} \in \bar{A}_1$ and $i \in \mathbb{Z}$. By assumption, $\langle a \rangle \cong \mathbb{Z}/n$, and furthermore, n and i are relatively prime. Note that if $\epsilon_{\bar{A}} \neq 0$, then n is even, so i is odd and $i\epsilon_{\bar{A}} = \epsilon_{\bar{A}}$. Obviously, this is also true when $\epsilon_{\bar{A}} = 0$.

Now let $b = (\bar{b}, j\gamma) \in A_1 = \bar{A}_1 \oplus \Gamma$ and suppose that $a \cdot b = 0$ in A_2. By (23.2.1),

$$i\bar{b} = j\bar{a} + ij\epsilon_{\bar{A}} = j(\bar{a} + \epsilon_{\bar{A}}).$$

Therefore

$$ib = (i\bar{b}, ij\gamma) = j(\bar{a} + \epsilon_{\bar{A}}, i\gamma) = j(a + \epsilon_A),$$

as required. □

23.3.3 EXAMPLES.

(1) Let p be a prime number and let Γ be an abelian group of exponent p (i.e., $p\Gamma = 0$). Let $A = \mathbf{0}[\Gamma]$. By Lemma 23.3.2, every non-zero element of $A_1 (= \Gamma)$ is p-rigid.

(2) Let Γ be an abelian group of exponent 2 and let $A = \kappa[\Gamma]$. Then every non-zero $a \in A_1 (= \kappa_1 \oplus \Gamma)$ is 2-rigid. Indeed, for $a = \epsilon_A$ this follows from the definition of rigidity, and for $a \neq 0, \epsilon_A$ it follows from Lemma 23.3.2.

Given a κ-structure A we define a map

$$\mathrm{Bock}_A \colon A_1 \to A_2$$

by

$$\mathrm{Bock}_A(a) = a^2 = \epsilon_A \cdot a.$$

It is clearly a group homomorphism. Following a standard terminology in Galois cohomology, we call Bock_A the **Bockstein operator** of A.

LEMMA 23.3.4. *Let $\bar{A}$ be a κ-structure with $\epsilon_{\bar{A}} \neq 0$, let Γ be an abelian group of exponent* 2, *and let $A = \bar{A}[\Gamma]$. If* $\mathrm{Bock}_{\bar{A}}$ *is injective, then* Bock_A *is also injective.*

PROOF. Take $a \in A_1$ $(= \bar{A}_1 \oplus \Gamma)$ with $a^2 = 0$. We show that $a = 0$.

If $a \notin \bar{A}_1$, then by Lemma 23.3.2, a is 2-rigid. Hence $\epsilon_{\bar{A}} + a$ and a are $\mathbb{F}_2$-linearly dependent. As $\epsilon_{\bar{A}} \neq 0$, this can happen only if $a \in \{0, \epsilon_{\bar{A}}\} \subseteq \bar{A}_1$, a contradiction.

Consequently, $a \in \bar{A}_1$, so the injectivity of $\mathrm{Bock}_{\bar{A}}$ gives $a = 0$, as desired. □

In particular, the Bockstein map of κ is obviously injective. We therefore deduce:

COROLLARY 23.3.5. *Let Γ be an abelian group of exponent* 2. *Then* $\mathrm{Bock}_{\kappa[\Gamma]}$ *is injective.*

23.4 Demuškin κ-structures

Next to the building blocks $\mathbf{0}$ and κ of the category of κ-structures, one also has a more complicated class of "atomic" κ-structures, that arise in valuation-theoretic situations: the Demuškin κ-structures, which we now define. As we shall see in §27, they occur as the generalized K-rings of certain wild valued fields.

DEFINITION 23.4.1. Let n be a positive integer. A κ-structure $A = \bigoplus_{r=0}^{\infty} A_r$ is called n-**Demuškin** if:

(i) A_1 is a finite free $\mathbb{Z}/n$-module;
(ii) $A_2 \cong \mathbb{Z}/n$;
(iii) the product $A_1 \times A_1 \to A_2$ is a **non-degenerate** pairing, i.e., its left and right kernels are trivial.

This terminology is analogous to the one used in Galois cohomology (see [**Se3**, I, §4.5] or [**NSW**, Ch. III, §9]).

PROPOSITION 23.4.2. *Let A be an n-Demuškin κ-structure with $\dim_{\mathbb{Z}/n} A_1 \geq 2$. Then $A_r = 0$ for $r \geq 3$.*

PROOF. One can find $a, a' \in A_1$ of exponent n such that $a \cdot a' = 0$ in $A_2 \cong \mathbb{Z}/n$. Since the pairing $A_1 \times A_1 \to A_2$ is non-degenerate, $aA_1 = A_2 = a'A_1$. Since A_1 generates A as a ring,

$$A_3 = A_2 A_1 = aA_1 A_1 = aA_2 = aa'A_1 = 0,$$

and therefore $A_r = 0$ for all $r \geq 3$. □

The following proposition shows in particular that the assumption in Proposition 23.4.2 that A_1 is non-cyclic cannot be removed.

PROPOSITION 23.4.3. *Let A be a κ-structure with $A_1 \cong \mathbb{Z}/n$. Then A is n-Demuškin if and only if $n = 2$ and either:*

(1) *$A \cong \kappa$; or*
(2) *there exists $r_0 \geq 2$ such that $A_r = \langle \epsilon_A^r \rangle \cong \mathbb{Z}/2$ whenever $1 \leq r \leq r_0$, and $A_r = 0$ whenever $r > r_0$.*

PROOF. It is straightforward to verify that the structures in (1) and (2) are 2-Demuškin.

Conversely, suppose that A is n-Demuškin. Let a be a generator of A_1. Then A_2 is generated by $a^2 = \epsilon_A a$. Hence $\epsilon_A \neq 0$. Since $2\epsilon_A = 0$ and $A_2 \cong \mathbb{Z}/n$, we obtain that $n = 2$ and $A_1 = \langle \epsilon_A \rangle$. It follows that $A_r = \langle \epsilon_A^r \rangle$ for all $r \geq 1$. Now if $\epsilon_A^r \neq 0$ for all r, then $A \cong \kappa$, i.e., (1) holds. If not, take r_0 maximal for which $A_{r_0} \neq 0$. Then (2) holds. □

As the following lemma shows, the n-Demuškin structures are end-objects in the subcategory of κ-structures with $A_2 \cong \mathbb{Z}/n$.

LEMMA 23.4.4. *Let $A, \bar{A}$ be κ-structures. Suppose that A is n-Demuškin, and that $|\bar{A}_2| \geq n$. Then any epimorphism $\varphi\colon A \to \bar{A}$ is an isomorphism.*

PROOF. As $A_2 \cong \mathbb{Z}/n$ and $|\bar{A}_2| \geq n$, the epimorphism φ is an isomorphism in degree 2. For any $a, a' \in A_1$ with $\varphi(a) = 0$ we have $\varphi(a \cdot a') = \varphi(a) \cdot \varphi(a') = 0$, whence $a \cdot a' = 0$. Since the pairing $A_1 \times A_1 \to A_2$ is non-degenerate, this implies that $a = 0$. We conclude that φ is an isomorphism in degree 1 as well.

In light of Proposition 23.4.2, A_r and $\bar{A}_r$ are zero for $r \geq 3$. Therefore φ is an isomorphism in all degrees. □

We conclude this section by showing that Demuškin κ-structures are indeed "atomic" in the category of κ-structures, in the sense that they cannot be constructed in a non-trivial way using the previous operations of direct products and extensions.

PROPOSITION 23.4.5. *Let q be a prime power. A q-Demuškin κ-structure cannot be isomorphic to a direct product of non-trivial κ-structures.*

PROOF. Let A be a q-Demuškin κ-structure, and assume that it has such a decomposition. Then there exist subgroups B_1, C_1 of A_1 such that $A_1 = B_1 + C_1$, $B_1 \cdot C_1 = 0$ in A_2, and $B_2 \oplus C_2 \cong \mathbb{Z}/q$. Therefore, without loss of generality, $B_2 = 0$ and $C_2 \cong \mathbb{Z}/q$. For every $b \in B_1$ we obtain that

$$b \cdot A_1 = b \cdot B_1 + b \cdot C_1 = 0.$$

By the non-degenerateness, $b = 0$. We conclude that $B_1 = 0$, so B is trivial. $\square$

The assumption in Proposition 23.4.5 that q is prime power cannot be removed. Indeed, when n is not a prime power, any non-trivial κ-structure A with A_1 a free $\mathbb{Z}/n$-module has a direct product decomposition arising from the primary decomposition of A_1.

PROPOSITION 23.4.6. *Let n be a positive integer, let $\bar{A}$ be a κ-structure such that $\bar{A}_1$ is a free $\mathbb{Z}/n$-module, and let Γ be a free $\mathbb{Z}/n$-module. Suppose that $A = \bar{A}[\Gamma]$ is an n-Demuškin structure with* $\dim_{\mathbb{Z}/n} A_1 \geq 3$. *Then $\Gamma = 0$.*

PROOF. Suppose that $\Gamma \neq 0$. In light of Lemma 23.2.3, we may assume that $\Gamma \cong \mathbb{Z}/n$. Let γ be a generator of Γ. Since the pairing $A_1 \times A_1 \to A_2$ is non-degenerate, there exists $a \in A_1$ such that $\langle a \rangle \cong \mathbb{Z}/n$ and $a \cdot \gamma \neq 0$. Since $\gamma^2 = \epsilon_A \cdot \gamma$ and $A_1 = \bar{A}_1 \oplus \Gamma$, we may further assume that $a \in \bar{A}_1$.

Now

$$\dim_{\mathbb{Z}/n} \bar{A}_1 = \dim_{\mathbb{Z}/n} A_1 - 1 \geq 2,$$

so there exists $b \in \bar{A}_1$ with $\langle a \rangle \cap \langle b \rangle = \{0\}$. On the other hand, $a \cdot \gamma$ and $b \cdot \gamma$ must be $\mathbb{Z}/n$-dependent in $A_2 \cong \mathbb{Z}/n$, i.e., $ib \cdot \gamma = ja \cdot \gamma$ for some $i, j \in \mathbb{Z}$, at least one of which prime to n. Thus $(ib - ja) \cdot \gamma = 0$. But by Lemma 23.3.2, γ is n-rigid in A. Hence there exist k, l at least one of which being prime to n such that

$$k(ib - ja) = l(\epsilon_A + \gamma).$$

However, $k(ib - ja), \epsilon_A \in \bar{A}_1$, so projecting into Γ, we obtain that $n|l$. Conclude that k is prime to n and $ib = ja$. But this contradicts the choice of b. $\square$

CHAPTER 24

Milnor K-Rings of Fields

After the rather formal discussion in the previous chapter we now return to the field-theoretic situation, and introduce the generalized Milnor K-ring functor modulo a subgroup of the multiplicative group.

24.1 Definition and basic properties

We fix a field F and consider a subgroup S of its multiplicative group $F^\times$. For $r \geq 0$ let

$$(F^\times/S)^{\otimes r} = (F^\times/S) \otimes_{\mathbb{Z}} \cdots \otimes_{\mathbb{Z}} (F^\times/S) \qquad (r \text{ times}).$$

Let $\mathrm{St}_{F,r}(S)$ be the subgroup of $(F^\times/S)^{\otimes r}$ generated by all elements $a_1S \otimes \cdots \otimes a_rS$ such that

$$1 \in a_iS + a_jS$$

for some $i \neq j$. Generalizing a standard terminology (see, e.g., [**FV**, Ch. IX, §1], we call such generators **Steinberg elements**. We set

$$K_r^M(F)/S = (F^\times/S)^{\otimes r}/\mathrm{St}_{F,r}(S).$$

For $t \geq 0$ the sets $\mathrm{St}_{F,r}(S) \otimes_{\mathbb{Z}} (F^\times/S)^{\otimes t}$ and $(F^\times/S)^{\otimes t} \otimes_{\mathbb{Z}} \mathrm{St}_{F,r}(S)$ are contained in $\mathrm{St}_{F,r+t}(S)$. Therefore we may define a graded ring

$$K_*^M(F)/S = \bigoplus_{r=0}^{\infty} K_r^M(F)/S,$$

with the multiplication induced from the tensor algebra

$$\mathrm{Tens}(F^\times/S) = \bigoplus_{r=0}^{\infty} (F^\times/S)^r$$

in the obvious way. We call it the **Milnor K-ring of F modulo S**. In particular,

$$K_0^M(F)/S = \mathbb{Z}, \quad K_1^M(F)/S = F^\times/S.$$

We define a morphism of graded rings $\kappa \to K_*^M(F)/S$ by mapping ϵ to the coset $-S$ in $F^\times/S$.

PROPOSITION 24.1.1. *$K_*^M(F)/S$ with this morphism is a κ-structure.*

PROOF. We only need to verify condition (iii') following Definition 23.1.1, i.e., to show that for every $a \in F^\times$ one has $aS \otimes (-a)S \in \mathrm{St}_{F,2}(S)$. When $a = 1$ this element is zero. When $a \neq 1$ we use the identity $-a = (1-a)(1-a^{-1})^{-1}$ to obtain

$$\begin{aligned} aS \otimes (-a)S &= aS \otimes (1-a)S \; - \; aS \otimes (1-a^{-1})S \\ &= aS \otimes (1-a)S \; + \; a^{-1}S \otimes (1-a^{-1})S \; \in \; \mathrm{St}_{F,2}(S), \end{aligned}$$

as desired.

Given $a_1, \dots, a_r \in F^\times$ we denote the image of $a_1 S \otimes \cdots \otimes a_r S$ in $K_r^M(F)/S$ by

$$\{a_1, \dots, a_r\}_S.$$

These elements are sometimes referred to as **symbols**. We list some of the main properties of the map $(a_1, \dots, a_r) \mapsto \{a_1, \dots, a_r\}_S$.

PROPOSITION 24.1.2.
(a) $\{a, 1-a\}_S = 0$ *whenever* $0, 1 \neq a \in F$;
(b) $\{a, -a\}_S = 0$ *for all* $a \in F^\times$;
(c) $\{a, a\}_S = \{a, -1\}_S = \{-1, a\}_S$ *for all* $a \in F^\times$;
(d) $\{a, b\}_S = -\{b, a\}_S$ *for all* $a, b \in F^\times$ *(anti-commutativity)*;
(e) $\{a_1 s_1, \dots, a_r s_r\}_S = \{a_1, \dots, a_r\}_S$ *for* $a_i \in F^\times$ *and* $s_i \in S$;
(f) $\{a_1, \dots, a_i \cdot a_i', \dots, a_r\}_S = \{a_1, \dots, a_i, \dots, a_r\}_S + \{a_1, \dots, a_i', \dots, a_r\}_S$ *for every* a_i', $a_1, \dots, a_r \in F^\times$ *(multi-linearity)*;
(g) $\{a_1, \dots, a_r\}_S = 0$ *whenever* $a_1, \dots, a_r \in F^\times$ *and* $a_1 + \cdots + a_r = 0$ *or* $a_1 + \cdots + a_r = 1$;
(h) $\{a, b\}_S = \{-a/b, a+b\}_S$ *for any* $a, b \in F^\times$ *with* $a \neq -b$.

PROOF. (a), (e) and (f) follow from the definition of $K_*^M(F)/S$. For (b) and (d) use the fact that it is a κ-structure, and in particular anti-commutative (Lemma 23.1.2). (c) follows from (b) and (f).

To prove (g) we argue by induction on r. When $r = 1$, the assertion is trivial. Suppose that $r \geq 2$. If $a_1 + a_2 = 0$, then

$$\{a_1, a_2, a_3, \dots, a_r\}_S = \{a_1, -a_1, a_3, \dots, a_r\}_S = 0$$

by (b). If $a_1 + a_2 \neq 0$, then

$$\begin{aligned} \{a_1, a_2, a_3, \dots, a_r\}_S = &\{a_1/(a_1+a_2), a_2/(a_1+a_2), a_3, \dots, a_r\}_S \\ &+ \{a_1 + a_2, a_2, a_3, \dots, a_r\}_S \\ &+ \{a_1, a_1 + a_2, a_3, \dots, a_r\}_S \\ &- 2\{a_1 + a_2, a_1 + a_2, a_3, \dots, a_r\}_S. \end{aligned}$$

Now in the last expression, the first and forth summands are zero by (a) and (c), respectively. Also, by the induction hypothesis, $\{a_1 + a_2, a_3, \dots, a_r\}_S = 0$. Therefore the second and third summands are also zero, and we are done.

Finally we prove (h). Let $c = a + b\,(\neq 0)$. We compute using the previous properties:

$$\begin{aligned} \{a, b\}_S &= \{a, b\}_S - \{ac^{-1}, 1 - ac^{-1}\}_S \\ &= \{a, b\}_S - \{ac^{-1}, bc^{-1}\}_S \\ &= \{a, c\}_S + \{c, b\} - \{c, c\}_S \\ &= \{a, c\}_S - \{b, c\}_S - \{-1, c\}_S = \{-a/b, c\}_S, \end{aligned}$$

as desired. □

When $S = \{1\}$, we abbreviate

$$K_*^M(F) = K_*^M(F)/S, \quad K_r^M(F) = K_r^M(F)/S$$

and

$$\{a_1, \dots, a_r\} = \{a_1, \dots, a_r\}_S \quad .$$

These are the classical definitions due to Milnor [**Mi**].

The class of all pairs (F, S), where F is a field and $S \le F^\times$, forms a category, with morphisms being embeddings $(F, S) \hookrightarrow (E, T)$, i.e., field embeddings of F in E which map S into T. Any such embedding induces in a canonical way a κ-structure morphism

$$\mathrm{Res}\colon K_*^M(F)/S \to K_*^M(E)/T, \quad \{a_1, \dots, a_r\}_S \mapsto \{a_1, \dots, a_r\}_T$$

for $a_1, \dots, a_r \in F^\times$. We call it the **restriction** morphism. This makes the map $(F, S) \mapsto K_*^M(F)/S$ a covariant functor from this category to the category of κ-structures.

As another basic example of the structure $K_*^M(F)/S$ one has:

PROPOSITION 24.1.3. *For positive integers m, r and for $S = (F^\times)^m$ we have $K_r^M(F)/S = K_r^M(F)/m$.*

PROOF. Since $F^\times/S$ is an m-torsion group, so is $(F^\times/S)^{\otimes r}$, and therefore so is $K_r^M(F)/S$ for $r \ge 1$. Hence the morphism $\mathrm{Res}\colon K_*^M(F) \to K_*^M(F)/S$ induces a group homomorphism $\varphi\colon K_r^M(F)/m \to K_r^M(F)/S$ which commutes with the canonical projections from $(F^\times)^{\otimes r}$. Conversely, suppose that $a_1, \dots, a_r \in F^\times$ and that $a_1 S \otimes \cdots \otimes a_r S$ is a standard generator of $\mathrm{St}_{F,r}(S)$, i.e., $1 = a_i\alpha^m + a_j\beta^m$ for some $i < j$ and $\alpha, \beta \in F^\times$. By the multi-linearity:

$$\{a_1, \dots, a_r\} \in \{a_1, \dots, a_i\alpha^m, \dots, a_j\beta^m, \dots, a_r\} + mK_r^M(F) = mK_r^M(F).$$

We obtain a projection $\psi\colon K_r^M(F)/S \to K_r^M(F)/m$ which also commutes with the projections from $(F^\times)^{\otimes r}$. Thus φ and ψ are inverse maps, whence are isomorphisms. □

COROLLARY 24.1.4. *If F is a perfect field of characteristic p, then $K_r^M(F)/p = 0$ for all $r \ge 1$.*

PROOF. Taking $S = (F^\times)^p$, the assumptions implies that $F^\times/S = 1$. Hence the tensor algebra $\mathrm{Tens}(F^\times/S)$ is trivial in positive degrees. Therefore so is its epimorphic image $K_*^M(F)/S$. Now apply Proposition 24.1.3. □

24.2 Comparison theorems

The following fact is useful for computing K-rings in arithmetical situations:

PROPOSITION 24.2.1. *Let $\mathcal{T}$ be a ring topology on a field F_1 and let F be a subfield of F_1 which is $\mathcal{T}$-dense in F_1. Let S_1 be a $\mathcal{T}$-open subgroup of $F_1^\times$ and let $S \le S_1 \cap F^\times$. Then*

$$\mathrm{Res}\colon K_*^M(F)/S \to K_*^M(F_1)/S_1$$

is an epimorphism. When $S = S_1 \cap F$ it is an isomorphism.

PROOF. For every $a \in F_1^\times$ we have $F \cap aS_1 \ne \emptyset$ by the density assumption. Hence the natural homomorphism $F^\times/S \to F_1^\times/S_1$ is surjective. Consequently, so is $\mathrm{Res}\colon K_*^M(F)/S \to K_*^M(F_1)/S_1$.

Next suppose that $S = S_1 \cap F$. For each r the induced map $(F^\times/S)^{\otimes r} \to (F_1^\times/S_1)^{\otimes r}$ is then an isomorphism.

We claim that the induced map $\mathrm{St}_{F,r}(S) \to \mathrm{St}_{F_1,r}(S_1)$ is surjective. To this end we take a standard generator $a_1S_1 \otimes \cdots \otimes a_rS_1$ of $\mathrm{St}_{F_1,r}(S_1)$, i.e., $a_1, \dots, a_r \in F_1^\times$ and $1 \in a_iS_1 + a_jS_1$ for some distinct i, j. By continuity, there exist non-empty $\mathcal{T}$-open subsets V, W of S_1 such that $a_iV + a_jW \subseteq S_1$. Using the density assumption we find $x_1, \dots, x_r \in F$ with $x_i \in a_iV$, $x_j \in a_jW$, and $x_l \in a_lS_1$ for all $l \neq i, j$. Then $x_i + x_j \in S_1 \cap F = S$, so

$$x_1S \otimes \cdots \otimes x_rS \in \mathrm{St}_{F,r}(S).$$

Furthermore, $x_1S \otimes \cdots \otimes x_rS$ maps to $a_1S_1 \otimes \cdots \otimes a_rS_1$ under the homomorphism above, as claimed.

Now consider the commutative diagram with exact rows

$$\begin{array}{ccccccccc}
1 & \longrightarrow & \mathrm{St}_{F,r}(S) & \longrightarrow & (F^\times/S)^{\otimes r} & \longrightarrow & K_r^M(F)/S & \longrightarrow & 0 \\
 & & \downarrow & & \downarrow & & \downarrow & & \\
1 & \longrightarrow & \mathrm{St}_{F_1,r}(S_1) & \longrightarrow & (F_1^\times/S_1)^{\otimes r} & \longrightarrow & K_r^M(F_1)/S_1 & \longrightarrow & 0.
\end{array}$$

We have seen that the left vertical map is surjective, and the middle vertical map is bijective. The snake lemma (Lemma 1.1.3) therefore implies that the right vertical map is bijective, as desired. □

Since the defining relations of $K_*^M(F)/S$ are given using only two coordinates of a tensor product, the K-rings are completely determined by their homogeneous components in degrees 1 and 2, as the following proposition shows.

PROPOSITION 24.2.2. *Let F, E be fields, let $S \le F^\times$ and let $T \le E^\times$. Let*

$$\varphi_1 \colon F^\times/S \to E^\times/T, \quad \varphi_2 \colon K_2^M(F)/S \to K_2^M(E)/T$$

be group homomorphisms such that $\varphi_1(\{-1\}_S) = \{-1\}_T$ and such that the following square commutes:

$$\begin{array}{ccc}
(F^\times/S) \times (F^\times/S) & \longrightarrow & K_2^M(F)/S \\
{\scriptstyle \varphi_1 \times \varphi_1}\downarrow & & \downarrow{\scriptstyle \varphi_2} \\
(E^\times/T) \times (E^\times/T) & \longrightarrow & K_2^M(E)/T,
\end{array}$$

where the horizontal maps are the products in the K-rings. Then there exists a unique κ-structure morphism $\varphi \colon K_*^M(F)/S \to K_*^M(E)/T$ extending φ_1 and φ_2. It is surjective (resp., bijective) if and only if φ_1, φ_2 are both surjective (resp., bijective).

PROOF. We have a commutative diagram with exact rows:

$$\begin{array}{ccccccccc}
1 & \longrightarrow & \mathrm{St}_{F,2}(S) & \longrightarrow & (F^\times/S) \otimes_{\mathbb{Z}} (F^\times/S) & \longrightarrow & K_2^M(F)/S & \longrightarrow & 0 \\
 & & & & {\scriptstyle \varphi_1^{\otimes 2}}\downarrow & & \downarrow{\scriptstyle \varphi_2} & & \\
1 & \longrightarrow & \mathrm{St}_{E,2}(T) & \longrightarrow & (E^\times/T) \otimes_{\mathbb{Z}} (E^\times/T) & \longrightarrow & K_2^M(E)/T & \longrightarrow & 0.
\end{array}$$

Hence $\varphi_1^{\otimes 2}(\mathrm{St}_{F,2}(S)) \subseteq \mathrm{St}_{E,2}(T)$. It follows that

$$\varphi_1^{\otimes r}(\mathrm{St}_{F,r}(S)) \subseteq \mathrm{St}_{E,r}(T)$$

for every $r \geq 2$. Consequently, $\varphi_1^{\otimes r}$ induces a homomorphism $\varphi: K_*^M(F)/S \to K_*^M(E)/T$ of graded rings that extends φ_1 and φ_2. By assumption, it maps $\{-1\}_S$ to $\{-1\}_T$, whence is a morphism of κ-structures.

If φ is surjective (resp., bijective), then clearly φ_1, φ_2 are both surjective (resp., bijective). Conversely, if φ_1 is surjective, then so is

$$\varphi_1^{\otimes r}: (F^\times/S)^{\otimes r} \to (E^\times/T)^{\otimes r}$$

for every r, and therefore so is $\varphi: K_*^M(F)/S \to K_*^M(E)/T$. Finally, if φ_1, φ_2 are both bijective, then the previous diagram shows that $\varphi_1^{\otimes 2}$ maps $\mathrm{St}_{F,2}(S)$ bijectively onto $\mathrm{St}_{E,2}(T)$. Hence $\varphi_1^{\otimes r}$ maps $\mathrm{St}_{F,r}(S)$ bijectively onto $\mathrm{St}_{E,r}(T)$ for every r, and we deduce that φ is bijective as well. □

24.3 Connections with Galois cohomology

In this section we will assume familiarity with the language of profinite groups and Galois cohomology. The discussion will be without proofs, and will not be needed later on (except for Remark 25.1.7).

Let F be a field with absolute Galois group $G_F = \mathrm{Gal}(F_{\mathrm{sep}}/F)$, and let n be a positive integer not divisible by $\mathrm{char}\, F$. Hilbert's "Theorem 90" gives the Kummer isomorphism

$$F^\times/n \xrightarrow{\sim} H^1(G_F, \mu_n),$$

where μ_n is the group of roots of unity of order dividing n. For $a \in F^\times$ we denote the element of $H^1(G_F, \mu_n)$ corresponding to the coset $a(F^\times)^n$ by φ_a. For $r \geq 1$ we obtain a homomorphism

$$(F^\times/n)^{\otimes r} \to H^r(G_F, \mu_n^{\otimes r}),$$

given by

$$a_1(F^\times)^n \otimes \cdots \otimes a_r(F^\times)^n \mapsto \varphi_{a_1} \cup \cdots \cup \varphi_{a_r},$$

where $\cup$ is the cup product. An argument of Tate [**NSW**, Th. 6.4.2] shows that the latter homomorphism is trivial on $\mathrm{St}_{F,r}((F^\times)^n)$, hence breaks through a homomorphism

$$K_r^M(F)/n \to H^r(G_F, \mu_n^{\otimes r}), \quad \{a_1 \dots a_r\} + nK_r^M(F) \mapsto \varphi_{a_1} \cup \cdots \cup \varphi_{a_r}.$$

The latter map is called the **Galois symbol** (or the **norm residue map**) of degree r. The *Bloch–Kato Conjecture* predicts that this homomorphism is in fact an isomorphism. This conjecture has been verified, e.g., in the following cases:

(i) $r = 1$ (the Kummer isomorphism);
(ii) $r = 2$ (Merkurjev–Suslin [**MerS1**]);
(iii) $r = 3$, $n = 2$ (Merkurjev–Suslin [**MerS2**], Rost);
(iv) r arbitrary, $n = 2^s$ (Voevodsky [**V1**]).

Moreover, in the preprint [**V2**], Voevodsky announced a proof of the Bloch–Kato conjecture in the general case. However, the proof is based on several facts which at this time have still not appeared in print. The proofs in cases (ii)–(iv) use very deep tools, and are entirely beyond the scope of the book; see, e.g., [**Kah**], [**Ker**] for more details.

Now let us restrict ourselves to the special case where $n = p$ is prime, and $\mu_p \subseteq F$. Then $\mu_p^{\otimes r} \cong \mathbb{Z}/p$ as G_F-modules. Therefore, for $G = G_F$ the Galois symbols give a commutative square

$$
\begin{array}{ccc}
(F^\times/p) \times (F^\times/p) & \longrightarrow & K_2^M(F)/p \\
\downarrow & & \downarrow \\
H^1(G,\mathbb{Z}/p) \times H^1(G,\mathbb{Z}/p) & \xrightarrow{\ \cup\ } & H^2(G,\mathbb{Z}/p)
\end{array}
\tag{24.3.1}
$$

in which the upper horizontal map is the multiplication in $K_*^M(F)/p$, and the vertical maps are isomorphisms (by (i) and (ii) above). It follows that the lower horizontal map is surjective.

As before let $F(p)$ be the maximal pro-p Galois extension of F and let $G_F(p) = \mathrm{Gal}(F(p)/F)$. Then $H^1(G_{F(p)},\mathbb{Z}/p) = 0$, and by what we have just seen, this implies that $H^2(G_{F(p)},\mathbb{Z}/p) = 0$ as well. A spectral sequence argument now shows that the inflation maps

$$H^i(G_F(p),\mathbb{Z}/p) \to H^i(G_F,\mathbb{Z}/p), \quad i = 1,2,$$

are isomorphisms. Consequently, the square (24.3.1) remains the same if we take $G = G_F(p)$ instead of $G = G_F$. Using Proposition 24.2.2, we conclude that the cohomology groups $H^i(G_F(p),\mathbb{Z}/p)$ for $i = 1,2$ and the cup product determine $K_*^M(F)/p$ uniquely.

Furthermore, if the Galois group $G = G_F(p)$ is explicitly presented by means of finitely many pro-p generators and relations, then the structure of its lower cohomology can be computed using the following recipe (for proofs we refer, e.g., to [**Ko3**, §7.8] or to [**NSW**, §3.9]):

Let G be an arbitrary pro-p group of finite generator rank d. Fix (topological) generators $g_1,\dots,g_d$ of G. Let $\hat{F}$ be a free pro-p group of rank d with free generators $f_1,\dots,f_d$. Then $H^1(G,\mathbb{Z}/p) = \mathrm{Hom}(G,\mathbb{Z}/p)$ has a "dual" basis $\chi_1,\dots,\chi_d$ given by $\chi_j(g_k) = \delta_{jk}$. It remains to determine the products $\chi_j \cup \chi_l$. To this end we define inductively a descending sequence $\hat{F}^{(n,p)}$ by

$$\hat{F}^{(1,p)} = \hat{F}, \quad \hat{F}^{(n+1,p)} = \langle (\hat{F}^{(n,p)})^p, [\hat{F}^{(n,p)}, \hat{F}^{(n,p)}] \rangle,$$

where the square brackets denote the commutator map. Let R be the kernel of the continuous epimorphism $\hat{F} \to G$ mapping f_j to g_j, and let r_i, $i \in I$, be a minimal system of generators for R. The group R is contained in the Frattini subgroup $F^{(2,p)}$ of F. Hence for each $i \in I$ we can write

$$r_i \equiv \prod_{j=1}^{d} f_j^{pa_{ij}} \prod_{j<l} [f_j, f_l]^{a_{ijl}} \pmod{F^{(3,p)}},$$

with $a_{ij}, a_{ijl} \in \{0,1,\dots,p-1\}$. Since the inflation map $H^1(G,\mathbb{Z}/p) \to H^1(\hat{F},\mathbb{Z}/p)$ is bijective and $H^2(\hat{F},\mathbb{Z}/p) = 0$, the transgression map

$$\mathrm{trg}\colon H^1(R,\mathbb{Z}/p)^G \to H^2(G,\mathbb{Z}/p)$$

is an isomorphism. For each $i \in I$ we define a map

$$\lambda_i\colon H^2(G,\mathbb{Z}/p) \to \mathbb{Z}/p, \quad \alpha \mapsto (\mathrm{trg}^{-1}(\alpha))(r_i).$$

Thus elements of $H^2(G, \mathbb{Z}/p)$ are determined by their images under all these maps. Now one has

$$\lambda_i(\chi_j \cup \chi_l) = \begin{cases} -\overline{a_{ijl}} & j < l, \\ \overline{a_{ij}} & j = l, \ p = 2, \\ 0 & j = l, \ p \neq 2. \end{cases}$$

CHAPTER 25

Milnor K-Rings and Orderings

After the general study of Milnor K-rings in the previous chapter, we will from now on focus on arithmetical situations, and study the reflection of the ordered and valued structure of a field in its generalized Milnor K-theory. We begin this by describing the reflection of orderings (as well as preorderings) in the structure of K-rings. This extends the results of Bass and Tate [**BaT**, I, §3] and follows [**Ef9**].

25.1 A K-theoretic characterization of orderings

For a field F and a subgroup S of $F^\times$ let $\mathrm{Bock}_{F,S}\colon F^\times/S \to K_2^M(F)/S$ be the Bockstein operator corresponding to the κ-structure $K_*^M(F)/S$. Thus

$$\mathrm{Bock}_{F,S}(\{x\}_S) = \{x\}_S^2 = \{x,-1\}_S$$

for $x \in F^\times$.

LEMMA 25.1.1. *If $\mathrm{Bock}_{F,S}$ is injective, then S is additively closed.*

PROOF. It suffices to show that $1+S \subseteq S$. To this end take $s \in S$. Then

$$\mathrm{Bock}_{F,S}(\{1+s\}_S) = \{1+s,-1\}_S = \{1+s,-s\}_S = 0.$$

By the injectivity, $\{1+s\}_S = 0$ in $K_1^M(F)/S = F^\times/S$, so $1+s \in S$. □

The next theorem characterizes orderings K-theoretically. Recall that a ring is **reduced** if it has no nilpotent elements $\neq 0$. When the ring is graded, this is the same as having no *homogeneous* nilpotent elements $\neq 0$.

THEOREM 25.1.2. *Let F be a field and let S be a subgroup of $F^\times$. The following conditions are equivalent:*
(a) *$K_*^M(F)/S \cong \kappa$ as κ-structures;*
(b) *$F^\times = S \cup -S$ and $K_*^M(F)/S$ is reduced;*
(c) *$F^\times = S \cup -S$ and $\{-1,-1\}_S \neq 0$;*
(d) *S is an ordering on F.*

PROOF. (a)⇒(b)⇒(c): Straightforward.

(c)⇒(d): By Lemma 25.1.1 and the assumptions, S is additively closed. The other properties of an ordering are clear.

(d)⇒(a): Suppose that S is an ordering on F. We first show that $\mathrm{St}_{F,r}(S)$ is trivial for all $r \geq 2$. Indeed, take $a_1,\dots,a_r \in F^\times$ such that $1 \in a_iS + a_jS$ for some distinct i,j. If a_i,a_j were both in $-S$, then we would get $-1 \in S+S \subseteq S$, a contradiction. Hence at least one of a_i,a_j must be in S. It follows that $a_1S \otimes \cdots \otimes a_rS = 1$ in $(F^\times/S)^{\otimes r}$, as claimed.

Consequently,

$$K_*^M(F)/S = \mathrm{Tens}(F^\times/S) \cong \mathrm{Tens}(\{\pm 1\}) = \kappa$$

as graded rings. Furthermore, this is a κ-structure isomorphism. □

COROLLARY 25.1.3. *Let F be a field and let S be a subgroup of $F^\times$. There is a canonical bijective correspondence between the orderings P on F containing S and the epimorphisms $K_*^M(F)/S \to \kappa$.*

PROOF. To an ordering P containing S we associate the composition of the restriction epimorphism $\mathrm{Res}\colon K_*^M(F)/S \to K_*^M(F)/P$ with the canonical isomorphism $K_*^M(F)/P \cong \kappa$ provided by Theorem 25.1.2.

Conversely, let $\varphi\colon K_*^M(F)/S \to \kappa$ be an epimorphism of κ-structures. Take $S \le P \le F^\times$ such that P/S is the kernel of the induced group epimorphism $\varphi_1\colon F^\times/S \to \kappa_1 = \{0, \epsilon\}$. Then $-1 \notin P$. To show that P is an ordering it remains to prove that it is additively closed. Indeed, if $a \in P$, then $1 + a \neq 0$, and $\{-a, 1+a\}_S = 0$ in $K_2^M(F)/S$. Hence

$$\epsilon \cdot \varphi_1(\{1+a\}_S) = \varphi_1(\{-a\}_S) \cdot \varphi_1(\{1+a\}_S) = \varphi_2(0) = 0$$

in κ, so necessarily $\varphi_1(\{1+a\}_S) = 0$. Thus $1 + a \in P$, as desired. □

Preorderings (in the sense of §6.1) can be characterized K-theoretically as follows.

PROPOSITION 25.1.4. *Let F be a field and let S be a subgroup of $F^\times$. The following conditions are equivalent:*

(a) *S is a preordering on F;*

(b) *$F^\times/S$ is a non-trivial group of exponent 2 and $\mathrm{Bock}_{F,S}$ is injective.*

PROOF. (a)⇒(b): Since $-1 \notin S$, the group $F^\times/S$ is non-trivial. As $(F^\times)^2 \le S$, it has exponent 2.

Next let $x \in F^\times$ and suppose that $\{x\}_S^2 = 0$. We need to show that $x \in S$. For every ordering P on F containing S we get by restriction that $\{x\}_P^2 = 0$. By Theorem 25.1.2, $K_*^M(F)/P \cong \kappa$ is reduced, so $\{x\}_P = 0$, i.e., $x \in P$. But by Proposition 6.1.8, the preordering S is the intersection of all the orderings P containing it. Consequently, $x \in S$, as desired.

(b)⇒(a): Since $F^\times/S$ has exponent 2, the group S contains $(F^\times)^2$. Also, in light of Lemma 25.1.1, S is additively closed. The non-triviality assumption yields $x \in F^\times \setminus S$. By the injectivity assumption, $\{x, -1\}_S \neq 0$, hence $-1 \notin S$. Therefore S is a preordering. □

Our next result complements Proposition 6.1.5.

PROPOSITION 25.1.5. *Let S be a subgroup of $F^\times$ containing $(F^\times)^2$. The following conditions are equivalent:*

(a) *there is an ordering P on F containing S;*

(b) *$\{-1\}_S^r \neq 0$ for all $r \ge 1$;*

(c) *$K_*^M(F)/S$ contains a non-nilpotent element $\neq 0$.*

PROOF. (a)⇒(b): Let P be an ordering on F containing S. Then the restriction morphism $\mathrm{Res}\colon K_*^M(F)/S \to K_*^M(F)/P$ is surjective, and by Theorem 25.1.2, $K_*^M(F)/P \cong \kappa$. But for every $r \geq 1$ one has $\epsilon^r \neq 0$. Therefore $\{-1\}_S^r \neq 0$.

(b)⇒(a): In light of Proposition 6.1.5, we need to show that -1 is not the sum of finitely many elements of S. But otherwise

$$1 = (-s_1) + \cdots + (-s_r),$$

with $s_1, \dots, s_r \in S$. Then Proposition 24.1.2(e)(g) would give

$$\{-1\}_S^r = \{-s_1, \dots, -s_r\}_S = 0,$$

contrary to (b).

(b)⇒(c): Trivial.

(c)⇒(b): Suppose that $\{-1\}_S^r = 0$ for some r. Then for every $a \in F^\times$ we have

$$\{a\}_S^{r+1} = \{a\}_S \cdot \{-1\}_S^r = 0.$$

Hence every generator $\varphi = \{a_1, \dots, a_m\}_S$ of $K_m^M(F)/S$ satisfies $\varphi^{r+1} = 0$. Now a general element φ of $K_*^M(F)/S$ can be written as a sum $\varphi_1 + \cdots + \varphi_k$, where each φ_i is a generator as above. Then $\varphi^{k(r+1)} = 0$, so φ is nilpotent. □

As another example, we now compute $K_*^M(F)/S$ for the simplest preorderings S on F which are not orderings.

PROPOSITION 25.1.6. *Let P_1 and P_2 be distinct orderings on the field F and let $S = P_1 \cap P_2$. Then*

$$K_*^M(F)/S \cong (K_*^M(F)/P_1) \times (K_*^M(F)/P_2) \cong \kappa \times \kappa$$

via the restriction morphism.

PROOF. We can choose $a_1 \in P_2 \setminus P_1$. For $a_2 = -a_1$ we have $a_2 \in P_1 \setminus P_2$. Observe that $\{a_i\}_{P_i}$ generates $F^\times/P_i \cong \mathbb{Z}/2$, $i = 1, 2$, and $\{a_1\}_S$ and $\{a_2\}_S$ generate $F^\times/S$; in fact,

$$F^\times/S = \Big\{0, \{-1\}_S, \{a_1\}_S, \{a_2\}_S\Big\} \cong (\mathbb{Z}/2) \times (\mathbb{Z}/2).$$

Therefore the natural projection $F^\times/S \to (F^\times/P_1) \times (F^\times/P_2)$ is an isomorphism. For every $r \geq 1$ it induces a group-epimorphism

$$K_r^M(F)/S \to (K_r^M(F)/P_1) \times (K_r^M(F)/P_2) \cong (\mathbb{Z}/2) \times (\mathbb{Z}/2).$$

Now by Proposition 24.1.2(b)(c),

$$\{a_1, a_2\}_S = 0, \quad \{a_i, a_i\}_S = \{a_i, -1\}_S, \quad i = 1, 2.$$

In light of the structure of $F^\times/S$, the group $K_r^M(F)/S$ therefore consists only of the four elements

$$0, \quad \{-1, -1, \dots, -1\}_S, \quad \{a_1, -1, \dots, -1\}_S, \quad \{a_2, -1, \dots, -1\}_S.$$

We conclude that the epimorphism above must be an isomorphism. □

REMARK 25.1.7. We conclude this section by indicating how Lemma 25.1.1 can be used to prove the more difficult part of the relative Artin-Schreier theorem (Theorem 19.2.10): if E/F is a normal extension of order 2 and $E^\times = (E^\times)^2$, then F is Euclidean. The proof uses some Galois cohomology; namely, let $S = (F^\times)^2$ and consider the diagram:

$$\begin{array}{ccccc} & & K_1^M(F)/S & \xrightarrow{\mathrm{Bock}_{F,S}} & K_2^M(F)/S \\ & & \downarrow & & \downarrow \\ H^1(G_F,\mathbb{Z}/4) & \xrightarrow{\alpha} & H^1(G_F,\mathbb{Z}/2) & \xrightarrow{\beta} & H^2(G_F,\mathbb{Z}/2). \end{array}$$

Here G_F is the absolute Galois group of F, the vertical maps are the Galois symbols (see §24.3), α is functorially induced by the projection $\mathbb{Z}/4 \to \mathbb{Z}/2$, and β is the cohomological Bockstein map $\varphi \mapsto \varphi \cup \varphi$. By the discussion in §24.3, the square commutes, and the left vertical map is bijective (the right vertical map is also bijective by the Merkurjev–Suslin theorem, but we shall not need this deep fact). It is also known that the lower row in the diagram is exact [**Se4**, Th. 1.4.2]. By assumption, $G_F(2) \cong \mathbb{Z}/2$. Identify $H^1(G_F,\mathbb{Z}/4)$ with $\mathrm{Hom}(G_F,\mathbb{Z}/4)$ and similarly for $H^1(G_F,\mathbb{Z}/2)$, to deduce that α is the zero map. Hence β is injective. It follows that $\mathrm{Bock}_{F,S}$ is also injective. Lemma 25.1.1 now implies that S is additively closed. By Kummer's theory, $(F^\times : S) = 2$, and consequently S is an ordering on F. Since $\sqrt{-1} \in E$ and $[E : F] = 2$, necessarily (F, S) is real closed relative to $E = F(2)$, i.e., F is Euclidean.

25.2 Cyclic quotients

In this section we characterize the rings $K_*^M(F)/S$ when $F^\times/S$ is a finite cyclic group. This turns out to be related to the structure of F as an ordered field, which was investigated in the previous section.

THEOREM 25.2.1. *Let F be a field and let S be a subgroup of $F^\times$ such that $F^\times/S$ is finite and cyclic. Then precisely one of following conditions holds:*
(a) *$K_r^M(F)/S = 0$ for all $r \geq 2$;*
(b) *there exists a unique ordering P on F containing S such that*

$$\mathrm{Res}\colon K_*^M(F)/S \to K_*^M(F)/P \quad (\cong \kappa)$$

is an isomorphism in all degrees $r \geq 2$, and $(F^\times : S) = 2m$ with m odd.

PROOF. Let $p_1^{d_1} \cdots p_n^{d_n}$ be the primary decomposition of $(F^\times : S)$. For each $1 \leq i \leq n$ choose $a_i \in F^\times$ such that the coset $\{a_i\}_S$ generates the p_i-primary part of $F^\times/S$. Let $a = a_1 \cdots a_n$. Then the coset $\{a\}_S$ generates $F^\times/S$, and one has

$$\{a,a\}_S = \{a,-1\}_S = \sum_{i=1}^n \{a_i,-1\}_S$$

(Proposition 24.1.2(c)(f)).

Assume that (a) does not hold, i.e., $K_r^M(F)/S \neq 0$ for some $r \geq 2$. Since the canonical map $(F^\times/S)^r \to K_r^M(F)/S$ is multi-linear, $\{a\}_S^r = \{a,\ldots,a\}_S$ generates $K_r^M(F)/S$. Hence this generator is non-zero, and therefore $\{a,a\}_S \neq 0$. It follows that $\{a_i,-1\}_S \neq 0$ for some $1 \leq i \leq n$. We obtain that the orders of $\{a_i,-1\}_S$ and of $\{-1\}_S$ are precisely 2. Furthermore, $p_i^{d_i}\{a_i,-1\}_S = 0$, so we must have $p_i = 2$. Therefore $2^{d_i-1}\{a_i\}_S = \{-1\}_S$, and we get

$$2^{d_i-1}\{a_i,-1\}_S = 2^{d_i-1}\{a_i,a_i\}_S = \{a_i,-1\}_S \neq 0.$$

This implies that $d_i = 1$. Consequently, $(F^\times : S) = 2m$, with m odd.

Let P be the unique subgroup of $F^\times$ of index 2 which contains S. Thus $(P : S) = m$. Since $\{-1\}_S$ has order 2 in $F^\times/S$, it is not in P/S. Therefore $F^\times = P \cup -P$.

Next we claim that $1 + P \subseteq P$. Indeed, suppose that $x \in P$. In particular, $x \neq -1$. Take integers s, t with $-x \in a^s S$ and $1 + x \in a^t S$. Then

$$0 = \{-x, 1+x\}_S = \{a^s, a^t\}_S = st\{a,a\}_S.$$

Now $-x \notin P$, so s is odd. On the other hand, $\{a,a\}_S = \{a,-1\}_S$ has order 2. It follows that t must be even, so $1 + x \in P$. We conclude that P is additively closed, hence an ordering.

Now for every r the canonical map $K_r^M(F)/S \to K_r^M(F)/P$ is clearly surjective. When $r \geq 2$ the group $K_r^M(F)/S$ is generated by

$$\{a, a, \ldots, a\}_S = \{a, -1, \ldots, -1\}_S,$$

and therefore has order at most 2. By Theorem 25.1.2, $K_r^M(F)/P \cong \kappa_r$ has order 2. Consequently, the map above is an isomorphism. This gives the existence part of (b).

For the uniqueness part of (b), assume that P' is another ordering on F containing S. Then

$$4 | (F^\times : P \cap P') | (F^\times : S) = 2m,$$

contrary to the fact that m is odd. □

We list a few special cases.

Corollary 25.2.2. *Let S be a subgroup of $F^\times$ with $F^\times/S$ cyclic of prime power order. Then either $K_r^M(F)/S = 0$ for all $r \geq 2$, or S is an ordering (so $K_*^M(F)/S \cong \kappa$).*

Corollary 25.2.3. *Let F be a field which has no orderings. Let S be a subgroup of $F^\times$ such that $F^\times/S$ is finite and cyclic. Then $K_r^M(F)/S = 0$ for all $r \geq 2$.*

Corollary 25.2.4 (Steinberg [**Mi**, Example 1.5]). *For a finite field F one has $K_r^M(F) = 0$ for all $r \geq 2$.*

Proof. It is well-known that $F^\times$ is cyclic [**Lg**, Ch. VII, §5, Th. 11]. Being a field of positive characteristic, F carries no orderings (Corollary 6.1.7). Now apply Corollary 25.2.3 with $S = \{1\}$. □

CHAPTER 26

K-Rings and Valuations

The reflection of valuations in K-rings is more intricate than that of orderings, discussed in the previous chapter. In the current chapter we consider the case of S-compatible valuations, which may be viewed as the "tame" case (see Remark 18.2.3). Following Bass–Tate and Wadsworth we show that, subject to some splitting assumption, $K_*^M(F)/S$ is then an extension of the corresponding K-ring of the residue field in a way which reflects the value groups as well. Especially interesting is the converse problem: does every extension occur in this way? In §§26.3–26.6 we prove several positive results in this direction, and relate Milnor K-rings, extensions, rigidity, and the canonical valuations studied in §11. In §26.7 we discuss fans, a notion which plays a major (and rather mysterious) role in the theory of ordered fields, and which shows up quite naturally in our K-theoretic context.

26.1 Valuations and extensions

Let F be a field and let S a subgroup of $F^\times$. Let v be a valuation on F and let

$$v^*: F^\times/S \to v(F^\times)/v(S)$$

be the induced homomorphism. The main result of this section (Theorem 26.1.2) relates $K_*^M(F)/S$ to $K_*^M(\bar{F}_v)/\bar{S}_v$ and $v(F^\times)/v(S)$ via the notion of extension (§23.2), once v^* has a homomorphic section. We first note a few cases where such a section exists:

26.1.1 EXAMPLES.

(1) v is discrete and $S = \{1\}$. Indeed, then a section of v^* corresponds to the choice of a uniformizer for v.

(2) $(F^\times)^p \leq S$ for some prime number p. In fact, then a section as required corresponds to a lifting to $F^\times/S$ of a basis of the $\mathbb{F}_p$-linear space $v(F^\times)/v(S)$.

(3) $(F^\times)^q \leq S \leq (F^\times)^q O_v^\times$, where $q = p^s$ is a prime power. Indeed, then $F^\times/S$ is a $\mathbb{Z}/q$-module and $v(F^\times)/v(S) = v(F^\times)/q$ is a free $\mathbb{Z}/q$-module, by Lemma 1.1.6.

(4) $F = K((\Gamma))$ for some ordered abelian group Γ, with the canonical K-trivial valuation (see Example 4.2.1), and S arbitrary. Here a section is given by $\gamma + v(S) \mapsto t^\gamma S$ for $\gamma \in \Gamma$, where as usual, t^γ is the map which is 1 at γ and 0 elsewhere.

(5) $F = K(\Gamma)$ for an ordered abelian group Γ, with the canonical K-trivial valuation (see Example 4.2.4), and S arbitrary. Here one can take the same section as in (4).

Extending results of Bass and Tate [**BaT**] and Wadsworth [**Wad**] (see Remark 26.1.3 below) we now prove the following fundamental fact.

THEOREM 26.1.2. *Let (F, v) be a valued field and let S be a subgroup of $F^\times$. Every section of v^* induces in a canonical way an epimorphism of κ-structures*

$$K_*^M(F)/S \longrightarrow (K_*^M(\bar{F}_v)/\bar{S}_v)[v(F^\times)/v(S)].$$

Moreover, this morphism is an isomorphism if and only if v is S-compatible.

PROOF. Let

$$\theta\colon v(F^\times)/v(S) \to F^\times/S$$

be a section of v^*. Take $S \le \Delta \le F^\times$ with $\Delta/S = \mathrm{Im}(\theta)$. Then

$$F^\times/S = (SO_v^\times/S) \times (\Delta/S).$$

Every $x \in F^\times$ can be written as $x = ab$ with $a \in O_v^\times$ and $b \in \Delta$. Let $\bar{a} = \pi_v(a)$ and let $[v(b)]_S$ be the coset of $v(b)$ in $v(F^\times)/v(S)$. We obtain a well-defined group epimorphism

$$F^\times/S \to (\bar{F}_v^\times/\bar{S}_v) \oplus (v(F^\times)/v(S)), \quad \{x\}_S \mapsto \{\bar{a}\}_{\bar{S}_v} + [v(b)]_S. \tag{26.1.1}$$

This abelian group epimorphism uniquely extends to a graded ring epimorphism

$$\lambda\colon \mathrm{Tens}(F^\times/S) \to (K_*^M(\bar{F}_v)/\bar{S}_v)[v(F^\times)/v(S)].$$

We claim that λ is trivial on $\mathrm{St}_{F,r}(S)$ for all r. It suffices to show that $\lambda(\{x\}_S \otimes \{y\}_S) = 0$ whenever $x, y \in F^\times$ and $1 \in xS + yS$. We may further assume that $1 = x + y$. Now write $x = ab$ and $y = cd$, with $a, c \in O_v^\times$ and $b, d \in \Delta$. Then

$$\begin{aligned}\lambda(xS \otimes yS) &= \big(\{\bar{a}\}_{\bar{S}_v} + [v(b)]_S\big) \cdot \big(\{\bar{c}\}_{\bar{S}_v} + [v(d)]_S\big)\\ &= \{\bar{a}, \bar{c}\}_{\bar{S}_v} + \big(\{\bar{a}\}_{\bar{S}_v} \cdot [v(d)]_S - \{\bar{c}\}_{\bar{S}_v} \cdot [v(b)]_S\big) + [v(b)]_S \cdot [v(d)]_S.\end{aligned}$$

To show that this expression vanishes, we distinguish between four cases.

CASE I: $x \in G_v$. Here we can take $a = x$ and $b = 1$. Then $\{\bar{a}\}_{\bar{S}_v} = 0$ and $[v(b)]_S = 0$, so the assertion is clear.

CASE II: $x \in \mathfrak{m}_v$. Then $y \in G_v$, so we can take $c = y$ and $d = 1$. Hence $\{\bar{c}\}_{\bar{S}_v} = 0$ and $[v(d)]_S = 0$, and we are done again.

CASE III: $x \in O_v^\times \setminus G_v$. Then $y = 1 - x \in O_v \setminus \mathfrak{m}_v = O_v^\times$, so we can take $a = x$, $b = 1$, $c = y$, and $d = 1$. Hence $\lambda(xS \otimes yS) = \{x, 1 - x\}_{\bar{S}_v} = 0$ once again.

CASE IV: $x^{-1} \in \mathfrak{m}_v$. For any a, b as above, $y = a(x^{-1} - 1) \cdot b$, with $a(x^{-1} - 1) \in O_v^\times$. Thus we may take $c = a(x^{-1} - 1)$ and $d = b$. Then $\{\bar{c}\}_{\bar{S}_v} = \{-\bar{a}\}_{\bar{S}_v}$. Further, $\{\bar{a}\}_{\bar{S}_v} - \{-\bar{a}\}_{\bar{S}_v} = \{-\bar{1}\}_{\bar{S}_v}$ and $\{\bar{a}, -\bar{a}\}_{\bar{S}_v} = 0$. It follows that

$$\lambda(xS \otimes yS) = \{-\bar{1}\}_{\bar{S}_v} \cdot [v(b)]_S + [v(b)]_S \cdot [v(b)]_S = 0,$$

using property (iii) of Definition 23.1.1.

This proves the claim. Consequently, λ induces an epimorphism of κ-structures

$$\bar{\lambda}\colon K_*^M(F)/S \longrightarrow (K_*^M(\bar{F}_v)/\bar{S}_v)[v(F^\times)/v(S)],$$

as desired.

For the second assertion of the theorem, suppose that v is S-compatible. Then we have the canonical exact sequence (3.2.7):

$$1 \to \bar{F}_v^\times/\bar{S}_v \xrightarrow{\eta} F^\times/S \xrightarrow{v^*} v(F^\times)/v(S) \to 0,$$

where $\eta(\{\bar{a}\}_{\bar{S}_v}) = \{a\}_S$ for $a \in O_v^\times$. The group monomorphism η induces a morphism

$$\mathrm{Tens}(\bar{F}_v^\times/\bar{S}_v) \to \mathrm{Tens}(F^\times/S)$$

of graded rings. Since $G_v \leq S$, it maps $\mathrm{St}_{\bar{F}_v,r}(\bar{S}_v)$ into $\mathrm{St}_{F,r}(S)$ for every $r \geq 1$. Hence it induces a κ-structure morphism $K_*^M(\bar{F}_v)/\bar{S}_v \to K_*^M(F)/S$. The universal property of extensions (Lemma 23.2.2) yields a unique κ-structure morphism $\bar{\nu}$ which extends the section θ and for which the following diagram commutes (where ι is the canonical monomorphism as in §23.2):

$$\begin{array}{ccc} K_*^M(\bar{F}_v)/\bar{S}_v & \xhookrightarrow{\iota} & (K_*^M(\bar{F}_v)/\bar{S}_v)[v(F^\times)/v(S)] \\ & \searrow & \downarrow \bar{\nu} \\ & & K_*^M(F)/S. \end{array}$$

In degree 1, $\bar{\nu}$ coincides with the isomorphism

$$\eta \oplus \theta \colon (\bar{F}_v^\times/\bar{S}_v) \oplus (v(F^\times)/v(S)) \to F^\times/S.$$

Hence it is surjective in all degrees. By construction, $\bar{\lambda}$ is given in degree 1 by the map (26.1.1). It follows that $\bar{\lambda} \circ \bar{\nu} = \mathrm{id}$ in degree 1, and therefore in all degrees. This proves that $\bar{\nu}$ is injective. Therefore both $\bar{\nu}$ and $\bar{\lambda}$ are isomorphisms.

Conversely, suppose that $\bar{\lambda}$ is an isomorphism. Its definition in degree 1 shows that it maps G_vS/S trivially. Hence $G_v \leq S$, as required. □

REMARK 26.1.3. When v is a discrete valuation and $S = 1$, the first part of Theorem 26.1.2 is due to Bass and Tate [**BaT**, I, Prop. 4.3]. They also prove its second part when (F, v) is a complete discrete valued field with positive residue characteristic prime to m and when $S = (F^\times)^m$ [**BaT**, I, Cor. 4.7]. Note that in the latter case (F, v) is Henselian (Corollary 18.3.2), so v is S-compatible (see Remark 18.2.3).

Wadsworth [**Wad**, §2] proves Theorem 26.1.2 for any valued field (F, v) when $S = (F^\times)^q G_v$ and q is a prime power. Here the splitting assumption is satisfied in light of Example 26.1.1(3).

REMARK 26.1.4. The epimorphism of Theorem 26.1.2 is functorial in the following sense: suppose that (F_1, v_1) is a valued field extension of (F, v), that $S \leq F^\times$, $S_1 \leq F_1^\times$, $S \leq S_1$, and that there exist homomorphic sections θ, θ_1 of the projections

$$v^*\colon F^\times/S \to v(F^\times)/v(S), \quad v_1^*\colon F_1^\times/S_1 \to v_1(F_1^\times)/v_1(S_1)$$

induced by v and v_1, respectively. Moreover, suppose that the following square commutes:

$$\begin{array}{ccc} v(F^\times)/v(S) & \xrightarrow{\theta} & F^\times/S \\ \downarrow & & \downarrow \\ v_1(F_1^\times)/v_1(S_1) & \xrightarrow{\theta_1} & F_1^\times/S_1\ . \end{array}$$

Then the epimorphisms given in Theorem 26.1.2 and the restriction morphisms induce a square:

$$\begin{array}{ccc} K_*^M(F)/S & \longrightarrow & (K_*^M(\bar{F}_v)/\bar{S}_v)[v(F^\times)/v(S)] \\ \downarrow & & \downarrow \\ K_*^M(F_1)/S_1 & \longrightarrow & (K_*^M(\overline{(F_1)}_{v_1})/\overline{(S_1)}_{v_1})[v_1(F_1^\times)/v_1(S_1)] \ . \end{array}$$

This square commutes in degree 1, hence in all degrees.

26.2 The Baer–Krull correspondence

The following consequence of Theorem 26.1.2 generalizes Example 6.2.5, and goes back to Baer and Krull ([**Ba**], [**Kru2**], [**Lam2**, Ch. 3]).

THEOREM 26.2.1. *Let (F,v) be a valued field and suppose that $(F^\times)^2 \leq S \leq F^\times$. There is a bijective correspondence*

$$P \longleftrightarrow (\bar{P},\chi)$$

between the orderings P on F containing SG_v and the pairs $(\bar{P},\chi)$ consisting of an ordering $\bar{P}$ on $\bar{F}_v$ containing $\bar{S}_v$ and a group homomorphism $\chi\colon v(F^\times)/v(S) \to \{\pm 1\}$. This bijection is given by

$$\bar{P} = \bar{P}_v \quad , \quad \mathrm{Ker}(\chi) = v(P)/v(S).$$

PROOF. We may replace S by SG_v to assume that v is S-compatible. Since $(F^\times)^2 \leq S$, the splitting assumption in Theorem 26.1.2 is satisfied (see Example 26.1.1(2)). Hence

$$K_*^M(F)/S \cong (K_*^M(\bar{F}_v)/\bar{S}_v)[v(F^\times)/v(S)].$$

By the universal property of extensions (Lemma 23.2.2), there is a canonical bijective correspondence between the epimorphisms

$$(K_*^M(\bar{F}_v)/\bar{S}_v)[v(F^\times)/v(S)] \to \kappa$$

and the pairs consisting of an epimorphism $K_*^M(\bar{F}_v)/\bar{S}_v \to \kappa$ and a homomorphism

$$\chi\colon v(F^\times)/v(S) \to \kappa_1 \cong \{\pm 1\}.$$

Now by Corollary 25.1.3, epimorphisms $K_*^M(F)/S \to \kappa$ correspond canonically to orderings P on F containing S, while epimorphisms $K_*^M(\bar{F}_v)/\bar{S}_v \to \kappa$ correspond canonically to orderings $\bar{P}$ on $\bar{F}_v$ containing $\bar{S}_v$. The assertion follows. □

In particular, we have the following important special case, which complements Proposition 7.2.1(d).

COROLLARY 26.2.2. *Let (F,v) be a valued field, let $(F^\times)^2 \leq S \leq F^\times$, and suppose that $G_v \leq S$ and $F^\times = SO_v^\times$. Then the maps*

$$P \mapsto \bar{P}_v, \quad \bar{P} \mapsto S\pi_v^{-1}(\bar{P})$$

give a bijective correspondence between the orderings P on F containing S and the orderings $\bar{P}$ on $\bar{F}_v$ containing $\bar{S}_v$.

26.3 Totally rigid subgroups

Let F be a field and S a subgroup of $F^\times$. When, e.g., $(F^\times)^p \leq S$ for some prime number p, any S-compatible valuation on F satisfies the splitting assumption of Theorem 26.1.2, so $K_*^M(F)/S$ is an extension, in which the "free" part $v(F^\times)/v(S)$ measures the non-triviality of v. We are now heading to a converse result, saying that (under some mild assumptions) if $K_*^M(F)/S$ is an extension, then it carries an S-compatible valuation v with $v(F^\times)/v(S)$ sufficiently large. In this and the next section we make some preparatory work on various rigidity notions, which will then be applied in §26.5 to prove this converse result.

By Proposition 24.1.2(b), $\mathrm{St}_{F,r}(S)$ contains any tensor product $a_1S\otimes\cdots\otimes a_rS$ such that $a_iS = -a_jS$ for some distinct i,j. We say that S is **totally rigid** if for every r the group $\mathrm{St}_{F,r}(S)$ is generated by such tensor products.

PROPOSITION 26.3.1. *Let $S \leq F^\times$.*
(a) *Assume that $-1 \in S$. Then S is totally rigid if and only if $K_*^M(F)/S \cong$ $\mathbf{0}[\Gamma]$ for some abelian group Γ.*
(b) *Assume that $-1 \notin S$ and $(F^\times)^2 \leq S$. Then S is totally rigid if and only if $K_*^M(F)/S \cong \kappa[\Gamma]$ for some group Γ.*

PROOF. (a) This follows from Example 23.2.4 with $\Gamma = F^\times/S$.

(b) This follows from the definition of $\kappa[\Gamma]$ (§23.2), where Γ is taken to be a complement of the subgroup generated by $-S$ in $F^\times/S$. □

As in §11.3–11.4 we set:

$$A(S) = \{x \in F^\times \mid 1 - x \notin S \cup -xS\}$$
$$H_S = \langle -1, S, A(S)\rangle.$$

Recall that the condition $A(S) \subseteq S$ was of crucial importance in finding optimal S-compatible valuations (Theorem 11.4.4). It is related to K-theory as follows (see also Remark 11.4.3).

LEMMA 26.3.2. *If $A(S) \subseteq S$, then S is totally rigid.*

PROOF. Let $a_1S \otimes \cdots \otimes a_rS$ be a standard generator of $\mathrm{St}_{F,r}(S)$, i.e., $1 = a_is + a_js'$ for some $i \neq j$ and some $s, s' \in S$. If $a_i \in S$ or $a_j \in S$, then this generator is trivial. If $a_i, a_j \notin S$, then by assumption $a_is \notin A(S)$ and $a_js' \notin S$, so

$$a_js' = 1 - a_is \in -a_iS.$$

Thus in the latter case $a_jS = -a_iS$. □

We also recall that H_S is contained in any intermediate group $S \leq H \leq F^\times$ for which $\mathrm{Val}(S,H) \neq \emptyset$ (Theorem 11.4.4(a)). We can now give a K-theoretic answer to the main problem discussed in §11.3–11.4, namely, whether $\mathrm{Val}(S, H_S)$ itself is non-empty.

THEOREM 26.3.3. *Suppose that S is not totally rigid. Then* $\mathrm{Val}(S, H_S) \neq \emptyset$.

PROOF. Lemma 26.3.2 implies that $A(S) \not\subseteq S$. We can now apply Theorem 11.4.4(b). □

As we shall see in §26.8, the assumption that S is not totally rigid cannot be dropped.

26.4 Sizes of multiplicative subgroups

While Theorem 26.3.3 is a quite satisfactory answer to the above question about existence of "optimal" S-compatible valuations, the minimal group H_S is difficult to compute in practice. Our next task is therefore to interpret it in computable K-theoretic terms. This will be done in the next section. As a preparation, we now derive some technical estimates on the sizes of several subgroups of $F^\times$.

Let p be a prime number and assume that $(F^\times)^p \leq M \leq F^\times$. Let T_M be the subgroup of $F^\times$ generated by M and by all elements $x \in F^\times \setminus M$ such that

$$1 - x \notin \bigcup_{i=0}^{p-1} x^i M.$$

Remark 26.4.1. Suppose $M = T_M$. Then for every $x \in F^\times$ the cosets $\{x\}_M, \{1-x\}_M$ are $\mathbb{F}_p$-linearly dependent. Hence $\mathrm{St}_{F,r}(M)$ is generated by all tensor products $a_1 M \otimes \cdots \otimes a_r M$ such that $a_i M = a_j M$ for some distinct i, j. If in addition $-1 \in M$, then we deduce that M is totally rigid.

When $-1 \in M$ we trivially have $T_M \leq \langle M, A(M) \rangle$. In fact, this is almost an equality, as Corollary 26.4.4 below shows. To prove this, we will need the following two computational facts, which are based on arguments of Hwang and Jacob [**HwJ**] (see also [**Ef6**]). As before, we denote the congruence class modulo p of an integer l by $\bar{l}$.

Lemma 26.4.2. *Let Δ be an abelian group (written multiplicatively) and suppose that $\Delta^p \leq \Delta_0 \leq \Delta$. Let $\omega\colon \Delta \setminus \Delta_0 \to \mathbb{F}_p$ be a map such that:*

(i) *$\bar{1} \notin \mathrm{Im}(\omega)$;*
(ii) *if $a, b \in \Delta \setminus \Delta_0$ have $\mathbb{F}_p$-linearly independent cosets in Δ/Δ_0 and at least one of $\omega(a)$, $\omega(b)$ is non-zero, then $\omega(ab) = \omega(a)\omega(b)$.*

Then every $a, b \in \Delta \setminus \Delta_0$ with $\omega(a), \omega(b) \neq \bar{0}$ have $\mathbb{F}_p$-linearly dependent cosets in Δ/Δ_0.

Proof. Take $a, b \in \Delta \setminus \Delta_0$ with $\omega(a), \omega(b) \neq \bar{0}$ and assume that the cosets of a, b in Δ/Δ_0 are $\mathbb{F}_p$-linearly independent. Then $a^i b \notin \Delta_0$ for every i. From (ii) we obtain inductively that

$$\omega(a^i b) = \omega(a)^i \omega(b) \neq \bar{0}$$

for every $i \geq 0$. Since $\mathbb{F}_p^\times$ has order $p-1$ this gives in particular $\omega(a^{p-1}b) = \omega(b)$. Moreover, the exponent of a in Δ/Δ_0 is not 1 and divides p, so $a^{p-1} \notin \Delta_0$. By (ii) again,

$$\omega(a^{p-1})\omega(b) = \omega(a^{p-1}b).$$

We conclude that $\omega(a^{p-1}) = \bar{1}$, contrary to (i). □

Proposition 26.4.3. *Let M be a subgroup of $F^\times$ containing $(F^\times)^p$ and -1. Assume that $a, b \in F^\times$ satisfy $1 - a \notin M \cup aM$ and $1 - b \notin M \cup bM$. Then a, b have $\mathbb{F}_p$-linearly dependent cosets in $F^\times/T_M$.*

Proof. Assume that the cosets of a and b in $F^\times/T_M$ are $\mathbb{F}_p$-linearly independent. In particular, $a, b, ab \notin T_M$. Let $\Delta = F^\times$ and $\Delta_0 = T_M$. We define a map

$$\omega\colon \Delta \setminus \Delta_0 = F^\times \setminus T_M \to \mathbb{F}_p$$

as follows: given $x \in F^\times \setminus T_M$, there exists a unique $0 \le i \le p-1$ such that $1-x \in x^i M$. When $i \neq 0$ let $\omega(x) = 1 - \bar{i}^{-1}$. When $i = 0$ we set $\omega(x) = \bar{0}$. Note that one has $1 - x \in M \cup xM$ if and only if $\omega(x) = \bar{0}$. Also, $\bar{1} \notin \mathrm{Im}(\omega)$, so condition (i) of Lemma 26.4.2 is satisfied. It remains to verify condition (ii).

To this end take $a, b \in F^\times \setminus T_M$ with $\mathbb{F}_p$-linearly independent cosets in $F^\times/T_M$ and such that at least one of $\omega(a)$, $\omega(b)$ is non-zero. We need to prove that $\omega(ab) = \omega(a)\omega(b)$. As $a, b, ab \notin T_M$, there exist $0 \le i, j, r \le p-1$ such that

$$1 - a \in a^i M, \quad 1 - b \in b^j M, \quad 1 - ab \in (ab)^r M.$$

The assumptions on a and b imply in particular that the pair (i, j) is non of $(1, 0)$, $(0, 1)$. It follows from the linear independence of a, b that $a^{1-i}b^j \notin T_M$. Therefore $a^{1-i}b^j M \cap T_M = \emptyset$, so

$$M - a^{1-i}b^j M \subseteq \bigcup_{k=0}^{p-1} (a^{1-i}b^j)^k M.$$

As $-1 \in M$ we obtain

$$1 - ab = (1-a) + a(1-b) \in a^i M - ab^j M = a^i(M - a^{1-i}b^j M) \subseteq \bigcup_{k=0}^{p-1} a^i (a^{1-i}b^j)^k M.$$

Therefore $(ab)^r M \cap a^i(a^{1-i}b^j)^k M \neq \emptyset$ for some k with $0 \le k \le p-1$. Since $M \le T_M$, the cosets of a, b in $F^\times/M$ are also $\mathbb{F}_p$-linearly independent. We may therefore equate exponents to deduce that

$$r \equiv i + (1-i)k \equiv jk \pmod{p}.$$

We obtain that

$$i \equiv k(i + j - 1) \pmod{p}$$

and

$$r(i+j-1) \equiv jk(i+j-1) \equiv ij \pmod{p}.$$

If $r \neq 0$, then also $j, k \neq 0$, and as $(i, j) \neq (0, 1)$ also $i \neq 0$. We therefore get in this case

$$\bar{1} - \bar{r}^{-1} = (\bar{1} - \bar{i}^{-1})(\bar{1} - \bar{j}^{-1}),$$

i.e., $\omega(ab) = \omega(a)\omega(b)$, as required.

If $r = 0$, then $i = 0$ or $j = 0$. We obtain that $\omega(ab) = \omega(a) = \bar{0}$ or $\omega(ab) = \omega(b) = \bar{0}$, and we are done in this case as well. □

COROLLARY 26.4.4. *Let M be a subgroup of $F^\times$ containing $(F^\times)^p$ and -1. Then*

$$(T_M \cdot \langle A(M) \rangle : T_M) \,|\, p.$$

Now assume that $(F^\times)^p \le S \le F^\times$.

DEFINITION 26.4.5. Let N_S be the subgroup of $F^\times$ generated by -1, S, and by all $x \in F^\times$ such that $\{x\}_S$ is *not* p-rigid in $K_*^M(F)/S$.

Thus N_S is generated by $-1, S$ and by all $x \notin S \cup -S$ for which there exists $y \in F^\times$ with $\{x, y\}_S = 0$ and $y \notin \bigcup_{i=0}^{p-1} (-x)^i S$. Since $\{x, 1-x\}_S = 0$ for $x \notin S$ this gives:

LEMMA 26.4.6.
(a) *If* $-1 \in S$, *then* $T_S \leq N_S$.
(b) *If* $p = 2$, *then* $A(S) \subseteq N_S$.

COROLLARY 26.4.7. *Suppose that* S *is totally rigid. Then:*
(a) $N_S = \langle -1, S\rangle$.
(b) $(H_S : S)|p$.

PROOF. (a) By Proposition 26.3.1, $K_*^M(F)/S$ is isomorphic to either $\mathbf{0}[\Gamma]$ for some abelian group Γ of exponent p, or $\kappa[\Gamma]$ for some abelian group Γ of exponent 2. It follows from Examples 23.3.3(1)(2) that all non-trivial elements of $F^\times/S$ are p-rigid in $K_*^M(F)/S$. The assertion follows.

(b) Assume first that $-1 \in S$. Then $N_S = S$ by (a). By Lemma 26.4.6(a), $S \leq T_S \leq N_S$, so $S = T_S$. It therefore follows from Corollary 26.4.4 that

$$(H_S : S) = (\langle S, A(S)\rangle : S)|p.$$

Next assume that $-1 \notin S$. Then $p = 2$ and $N_S = \langle -1, S\rangle$, by (a) again. By Lemma 26.4.6(b), $A(S) \subseteq N_S$. Therefore $H_S = \langle -1, S\rangle$, whence $(H_S : S) = 2$. □

26.5 H_S and the K-ring

Let again p be a prime number, F a field, and S a subgroup of $F^\times$ containing $(F^\times)^p$. We define N_S as in the previous section. Following [**E10**], we will now show that under some mild assumptions, $H_S = N_S$. The importance of this equality is that unlike H_S, the subgroup N_S is determined by $K_*^M(F)/S$, and in fact only by the pairing

$$(F^\times/S) \times (F^\times/S) \to K_2^M(F)/S$$

and by $\{-1\}_S$. Therefore this gives a *K-theoretic* description of H_S, and consequently a *K-theoretic* determination of the "best" S-compatible valuations one can find on F. Using the connections between Milnor K-theory modulo p and the Galois cohomology of the maximal pro-p Galois group of F (see §24.3), this allows one to detect valuations from *Galois-theoretic* data. As mentioned in the Introduction, this technique plays an important role in the area of birational anabelian geometry.

While the equality $H_S = N_S$ is of an elementary field-theoretic nature, we prove it indirectly using valuations. First we observe the following K-theoretic bound on the size of $(v(F^\times) : v(S))$ for an S-compatible valuation S.

PROPOSITION 26.5.1. *For every* S*-compatible valuation* v *on* F *one has* $N_S \leq SO_v^\times$ *and* $(v(F^\times) : v(S)) \leq (F^\times : N_S)$.

PROOF. Since $(F^\times)^p \leq S$, the epimorphism $F^\times/S \to v(F^\times)/v(S)$ induced by v has a homomorphic section (Example 26.1.1(2)). It therefore follows from Theorem 26.1.2 that

$$K_*^M(F)/S \cong (K_*^M(\bar{F}_v)/\bar{S}_v)[v(F^\times)/v(S)]$$

as κ-structures. By Lemma 23.3.2, elements of $F^\times/S$ with non-zero projection into $v(F^\times)/v(S)$ are p-rigid in $K_*^M(F)/S$. We conclude from the definition of N_S that $v(N_S) \leq v(S)$. Equivalently, $N_S \leq SO_v^\times$. Consequently,

$$(v(F^\times) : v(S)) = (F^\times : SO_v^\times) \leq (F^\times : N_S).$$ □

COROLLARY 26.5.2. *If $S \le H \le F^\times$ and $\mathrm{Val}(S,H) \ne \emptyset$, then $N_S \le H$.*

PROOF. If $v \in \mathrm{Val}(S,H)$, then Proposition 26.5.1 gives $N_S \le SO_v^\times \le H$. □

LEMMA 26.5.3. *Suppose that $\langle -1, S\rangle \le M < N_S$ and $(N_S : M) = p$. Then:*
(a) *$T_M = M$;*
(b) *there exists a subgroup L of $F^\times$ containing M such that $(L : M)|p$ and $\mathrm{Val}(M,L) \ne \emptyset$.*

PROOF. (a) By definition $M \le T_M$. For the converse, it is enough to show that whenever $x \notin M$, the cosets $\{x\}_M$ and $\{1-x\}_M$ are $\mathbb{F}_p$-linearly dependent in $F^\times/M$.

If $\{x\}_S$ and $\{1-x\}_S$ are both not p-rigid in $K_*^M(F)/S$, then $x, 1-x \in N_S$. Since $\dim_{\mathbb{F}_p}(N_S/M) \le 1$, the cosets $\{x\}_M$ and $\{1-x\}_M$ are then $\mathbb{F}_p$-linearly dependent in N_S/M ($\le F^\times/M$), and we are done.

Next suppose that at least one of $\{x\}_S, \{1-x\}_S$ is p-rigid in $K_*^M(F)/S$. From $\{x, 1-x\}_S = 0$ we then deduce that either $\{-x\}_S, \{1-x\}_S$ or $\{x\}_S, \{-(1-x)\}_S$ are $\mathbb{F}_p$-linearly dependent in $F^\times/S$. But $-1 \in M$, so in both cases $\{x\}_M$ and $\{1-x\}_M$ are $\mathbb{F}_p$-linearly dependent in $F^\times/M$.

(b) By (a), and Corollary 26.4.4,

$$(H_M : M) = (\langle M, A(M)\rangle : M)|p.$$

Consequently, when $\mathrm{Val}(M, H_M) \ne \emptyset$ we can take $L = H_M$.

When $\mathrm{Val}(M, H_M) = \emptyset$ we obtain from Theorem 11.4.4 that $p = 2$ and $A(M) \subseteq M$, whence $H_M = M$. Moreover, this theorem yields an intermediate subgroup $M < L \le F^\times$ such that $(L : M) = 2$ and $\mathrm{Val}(M, L) \ne \emptyset$. □

The main step towards proving the equality $H_S = N_S$ is the following proposition.

PROPOSITION 26.5.4. *Suppose that $(F^\times)^p \le S$, $-1 \in S < N_S$ and $(F^\times : S) < \infty$. Then there exists an S-compatible valuation v on F such that $N_S = SO_v^\times$.*

PROOF. Let $\mathcal{M}$ be the collection of all subgroups M of N_S containing S and such that $(N_S : M) = p$. The assumptions imply that $\mathcal{M}$ is finite and non-empty. For every $M \in \mathcal{M}$ Lemma 26.5.3(b) yields an intermediate group $M \le L_M \le F^\times$ with $(L_M : M)|p$ and a valuation $v_M \in \mathrm{Val}(M, L_M)$. Let v be the finest common coarsening of the valuations v_M, $M \in \mathcal{M}$ (Corollary 7.3.2). Then

$$G_v \le \bigcap_{M \in \mathcal{M}} G_{v_M} \le \bigcap_{M \in \mathcal{M}} M = S,$$

so v is S-compatible. By Proposition 26.5.1, $N_S \le SO_v^\times$. To show that this is an equality we distinguish between two cases.

CASE I: *For each $M \in \mathcal{M}$ there exists $M' \in \mathcal{M}$ such that v_M and $v_{M'}$ are incomparable.*

For each M we choose such M' and denote the finest common coarsening of v_M and $v_{M'}$ by u_M. Then v is also the finest common coarsening of the u_M, where $M \in \mathcal{M}$. By Proposition 10.3.2, $O_v^\times$ is generated by the subgroups $O_{u_M}^\times$. By Proposition 10.3.1(d),

$$O_{u_M}^\times = G_{v_M} G_{v_{M'}} \le MM' = N_S.$$

We conclude that $O_v^\times \le N_S$, whence $N_S = SO_v^\times$ in this case.

CASE II: *There exists $M \in \mathcal{M}$ such that v_M is comparable to $v_{M'}$ for all $M' \in \mathcal{M}$.*

Then v is the finest common coarsening of all the $v_{M'}$ which are coarser than v_M. Now the collection of these $v_{M'}$ is totally ordered with respect to coarsening (Proposition 7.3.5). Since $\mathcal{M}$ is finite, $v = v_{M'}$ for some M'. Hence $O_v^\times = O_{v_{M'}}^\times \le L_{M'}$, and therefore

$$M' < N_S \le SO_v^\times \le L_{M'}.$$

Since $(L_{M'} : M')|p$ this implies that $N_S = SO_v^\times$ in this case as well. □

We can now prove our main result:

THEOREM 26.5.5. *Let S be a subgroup of $F^\times$ containing $(F^\times)^p$ and which is not totally rigid. Moreover, assume that $p = 2$ or $(F^\times : S) < \infty$. Then:*

(a) $\mathrm{Val}(S, N_S) \neq \emptyset$*;*

(b) $N_S = H_S$*;*

(c) *there is an S-compatible valuation v on F such that $N_S = SO_v^\times$ and*

$$(v(F^\times) : v(S)) = (F^\times : N_S).$$

PROOF. (a) We break the proof into three cases.

CASE I: $p = 2$. Then $A(S) \subseteq N_S$, by Lemma 26.4.6(b). Hence $H_S \le N_S$. Since S is not totally rigid, Corollary 26.3.3 gives $\emptyset \neq \mathrm{Val}(S, H_S) \subseteq \mathrm{Val}(S, N_S)$.

CASE II: *$p \neq 2$ and $S < N_S$.* Then $-1 \in (F^\times)^p \le S$ and $(F^\times : S) < \infty$. We may therefore apply Proposition 26.5.4.

CASE III: *$p \neq 2$ and $S = N_S$.* Then again $-1 \in S$, so by Lemma 26.4.6(a), $T_S \le N_S = S$, whence $T_S = S$. But according to Remark 26.4.1(b), this can happen only when S is totally rigid, a contradiction.

(b) Since S is not totally rigid, $\mathrm{Val}(S, H_S) \neq \emptyset$ (Theorem 26.3.3). Therefore Corollary 26.5.2 gives $N_S \le H_S$. Conversely, (a) and Theorem 11.4.4(a) show that $H_S \le N_S$.

(c) By (a) there is an S-compatible valuation v on F with $O_v^\times \le N_S$. In view of Proposition 26.5.1, $N_S = SO_v^\times$. Then

$$(v(F^\times) : v(S)) = (F^\times : SO_v^\times) = (F^\times : N_S).$$ □

It would be interesting to find a more direct proof of Theorem 26.5.5, that is, an elementary field-theoretic argument which does not go through auxiliary valuations as in the proof of Proposition 26.5.4.

26.6 Bounds in the totally rigid case

As in the previous section, we assume that p is a prime number, F is a field, and S is a subgroup of $F^\times$ containing $(F^\times)^p$. We now consider the exceptional case were S is totally rigid. Then $N_S = \langle -1, S\rangle$ (Corollary 26.4.7(a)). One has the following slightly weaker version of Theorem 26.5.5.

THEOREM 26.6.1. *Let S be a totally rigid subgroup of $F^\times$ containing $(F^\times)^p$. There exists an S-compatible valuation v on F such that $N_S = \langle -1, S\rangle$ is a subgroup of $SO_v^\times$ of index dividing p. In particular,*

$$(v(F^\times) : v(S)) \geq (F^\times : \langle -1, S\rangle)/p.$$

PROOF. If there exists a valuation $v \in \mathrm{Val}(S, H_S)$, then $\langle -1, S\rangle \leq SO_v^\times \leq H_S$. We obtain from Corollary 26.4.7(b) that

$$(SO_v^\times : \langle -1, S\rangle)|(H_S : S)|p.$$

If $\mathrm{Val}(S, H_S) = \emptyset$, then by Theorem 11.4.4(b) $A(S) \subseteq S$, $H_S = \langle -1, S\rangle$, $p = 2$, and there exists a subgroup H of $F^\times$ such that

$$H_S < H, \quad (H : H_S) = 2, \quad \mathrm{Val}(S, H) \neq \emptyset.$$

Choose $v \in \mathrm{Val}(S, H)$. As

$$H_S = \langle -1, S\rangle \leq SO_v^\times \leq H,$$

we have

$$(SO_v^\times : \langle -1, S\rangle)|(H : H_S)|2.$$

For the last statement of the theorem we compute:

$$\begin{aligned}(v(F^\times) : v(S)) = (F^\times : SO_v^\times) &= (F^\times : \langle -1, S\rangle)/(SO_v^\times : \langle -1, S\rangle)\\ &\geq (F^\times : \langle -1, S\rangle)/p.\end{aligned}$$

□

We can now prove the following converse of Theorem 26.1.2:

COROLLARY 26.6.2. *Let S be a subgroup of $F^\times$ containing $(F^\times)^p$ and suppose that $p = 2$ or $(F^\times : S) < \infty$. Assume that $K_*^M(F)/S \cong \bar{A}[\Gamma]$ for some κ-structure $\bar{A} \not\cong \mathbf{0}, \kappa$ and an abelian group Γ of exponent p. Then there exists an S-compatible valuation v on F such that*

$$(v(F^\times) : v(S)) \geq |\Gamma|.$$

PROOF. From the assumption we obtain a sequence of group epimorphisms

$$F^\times \to F^\times/S \xrightarrow{\sim} \bar{A}_1 \oplus \Gamma \to \Gamma.$$

By Lemma 23.3.2, N_S is mapped trivially to Γ under the composed epimorphism. Thus $(F^\times : N_S) \geq |\Gamma|$.

Now if S is not totally rigid, then Theorem 26.5.5(c) yields an S-compatible valuation v on F with

$$(v(F^\times) : v(S)) = (F^\times : N_S) \geq |\Gamma|.$$

as desired.

So suppose that S is totally rigid. Then $K_*^M(F)/S \cong \bar{A}'[\Gamma']$, where $\bar{A}'$ is either $\mathbf{0}$ or κ and Γ' is an abelian group of exponent p (Proposition 26.3.1). The previous argument shows that $(F^\times : N_S) \geq |\Gamma'|$. Take an S-compatible valuation v as in Theorem 26.6.1. Therefore

$$(v(F^\times) : v(S)) \geq (F^\times : \langle -1, S\rangle)/p = (F^\times : N_S)/p \geq |\Gamma'|/p.$$

Thus it suffices to show that $|\Gamma'|/p \geq |\Gamma|$.

Indeed, we have $\bar{A}'[\Gamma'] \cong \bar{A}[\Gamma]$. If $\bar{A}' = \mathbf{0}$, then $\Gamma' \cong \bar{A}_1 \oplus \Gamma$ and $\bar{A}_1$ is a nontrivial group of exponent p, so the claim is clear. On the other hand, if $\bar{A}' = \kappa$, then $p = 2$ and $\epsilon^r_{\bar{A}} \neq 0$ for all r. As $\bar{A} \neq \mathbf{0}, \kappa$ necessarily $|\bar{A}_1| \geq 4$. Hence $\kappa_1 \oplus \Gamma' \cong \bar{A}_1 \oplus \Gamma$ gives $|\Gamma'|/2 \geq |\Gamma|$, as desired. □

26.7 Fans

Totally rigid *preorderings* S on F were extensively studied in the context of ordered fields and quadratic forms under the name **fans** (see [**BeK**], [**Lam2**]). Of course, in this case $p = 2$ and $-1 \notin S$. Here are some equivalent definitions.

THEOREM 26.7.1. *Let S be a preordering on F. The following conditions are equivalent:*
- (a) *S is totally rigid;*
- (b) *there is an S-compatible valuation v on F such that $(SO_v^\times : S)|4$;*
- (c) *there is an S-compatible valuation v on F such that $(\bar{F}_v^\times : \bar{S}_v)|4$;*
- (d) *there is an S-compatible valuation v on F such that $\bar{S}_v$ is the intersection of at most two orderings on $\bar{F}_v$;*
- (e) *$K_*^M(F)/S \cong \kappa[\Gamma]$ for some abelian group Γ of exponent 2;*
- (f) *for every $a \in F^\times$ at least one of*

$$a, -a, 1-a, 1-a^{-1}$$

  belongs to S;
- (g) *for every $a \in F^\times \setminus (S \cup -S)$ one has $1 - a \in S \cup -aS$;*
- (h) *for $T = S \cup \{0\}$ and for every $b \in F^\times \setminus (-T)$ one has*

$$T + bT = T \cup bT;$$

- (i) *$A(S) \subseteq S$;*
- (j) *$H_S = \langle -1, S\rangle$;*
- (k) *$N_S = \langle -1, S\rangle$.*

PROOF. (a)⇒(b): Apply Theorem 26.6.1.

(b)⇔(c): For v S-compatible, the exact sequence (3.2.6) gives $(\bar{F}_v^\times : \bar{S}_v) = (SO_v^\times : S)$.

(c)⇒(d): Let v be as in (c). By Lemma 7.1.3, $\bar{S}_v$ is a preordering on $\bar{F}_v$. Hence it is an intersection of a family of orderings on $\bar{F}_v$ (Proposition 6.1.8). Since $(\bar{F}_v^\times : \bar{S}_v)\,|\,4$, this can be done using at most two orderings.

(d)⇒(e): For v as in (d), Theorem 26.1.2 gives

$$K_*^M(F)/S \cong (K_*^M(\bar{F}_v)/\bar{S}_v)[v(F^\times)/v(S)].$$

If $\bar{S}_v$ is an ordering, then $K_*^M(\bar{F}_v)/\bar{S}_v \cong \kappa$ (Theorem 25.1.2). If $\bar{S}_v$ is the intersection of two distinct orderings, then

$$K_*^M(\bar{F}_v)/\bar{S}_v \cong \kappa \times \kappa \cong \kappa[\mathbb{Z}/2],$$

by Proposition 25.1.6 and Example 23.2.6. Use Lemma 23.2.3 to conclude that in both cases $K_*^M(F)/S$ is an extension of κ by a group of exponent 2.

(e)⇔(a): This has already been observed in Proposition 26.3.1(b).

(f)⇔(g), (i)⇒(j): Immediate.

(a)⇒(k): This is Corollary 26.4.7(a).

(k)⇒(g): As $p = 2$, Lemma 26.4.6(b) gives $A(S) \subseteq N_S = \langle -1, S\rangle$, and (g) follows.

(g)⇒(i): Suppose that $a \in A(S) \setminus S$. Then $a \neq 0$ and $1 - a \notin S \cup -aS$. Therefore (g) gives $-a \in S$. But then $1 - a \in S + S \subseteq S$, a contradiction.

(j)⇒(b): As $H_S = \langle -1, S\rangle$, Theorem 11.4.4(b) yields an intermediate group $\langle -1, S\rangle \leq H \leq F^\times$ with $(H : \langle -1, S\rangle)|2$ and $\mathrm{Val}(S, H) \neq \emptyset$. Choose $v \in \mathrm{Val}(S, H)$. Then

$$(SO_v^\times : S)|(H : S) = (H : \langle -1, S\rangle)(\langle -1, S\rangle : S)|4.$$

(h)⇒(i): Suppose that $a \in F^\times \setminus S$. By (h),

$$0 \neq 1 - a \in T - aT = T \cup -aT,$$

and therefore $1 - a \in S \cup -aS$, i.e., $a \notin A(S)$.

(g)⇒(h): Take $b \in F^\times \setminus (-T)$ and $t_1, t_2 \in T$. Let $c = t_1 + bt_2$. We need to show that $c \in T \cup bT$. If $b \in S$, then $c \in T + T = T$, and we are done. Also, if $t_1 = 0$ or $t_2 = 0$ then the claim is clear. So suppose that $b \notin S$ and $t_1, t_2 \in S$. Then

$$-bt_1^{-1}t_2 \in F^\times \setminus (S \cup -S).$$

By (g),

$$c = t_1(1 - (-bt_1^{-1}t_2)) \in S \cup bt_1^{-1}t_2S = S \cup bS \subset T \cup bT. \qquad \square$$

Condition (d) above is known as the *"trivialization of fans"* property. This connection between the more naive conditions for fans (as in (f)) and valuations was first discovered by Bröcker [**Br1**].

The following result is due to Jacob [**J2**]. It characterizes the subfamily of fans for which the conclusion of Theorem 26.5.5(a) is nevertheless valid. Such fans are called **valuation fans**. As in §11.3 we set $O^-(S, H) = (1 - S) \setminus H$.

PROPOSITION 26.7.2. *Let S be a preordering on F and let $H = \langle -1, S\rangle$. The following conditions are equivalent:*

(a) $\mathrm{Val}(S, H) \neq \emptyset$;
(b) *there exists an S-compatible valuation v on F such that $(SO_v^\times : S) = 2$;*
(c) *there exists an S-compatible valuation v on F such that $\bar{S}_v$ is an ordering on $\bar{F}_v$;*
(d) *S is a fan and for every $a \in F^\times \setminus H$ we have $1 - a \in S$ if and only if $1 + a \in S$;*
(e) *S is a fan and $O^-(S, H) = -O^-(S, H)$.*

PROOF. (a)⇒(b): For $v \in \mathrm{Val}(S, H)$ one has $(SO_v^\times : S)|(H : S) = 2$. As $-1 \in SO_v^\times \setminus S$, we get (b).

(b)⇒(c): Let v be as in (b). By Lemma 7.1.3, $\bar{S}_v$ is a preordering on $\bar{F}_v$. The exact sequence (3.2.6) again gives $(\bar{F}_v^\times : \bar{S}_v) = (SO_v^\times : S) = 2$, so $\bar{S}_v$ is an ordering on $\bar{F}_v$.

(c)⇒(d): Let v be as in (c). By Theorem 26.7.1, S is totally rigid, whence a fan. Furthermore, $\bar{F}_v^\times = \bar{S}_v \cup -\bar{S}_v$ and $G_v \leq S$ imply that $SO_v^\times = S \cup -S = H$.

Now let $a \in F^\times \setminus H$. Then either $v(a) > 0$ or $v(a) < 0$. In the first case

$$1 - a, 1 + a \in G_v \leq S,$$

and we are done. In the second case

$$-a^{-1}(1 - a) = 1 - a^{-1} \in G_v \leq S$$

as well as

$$a^{-1}(1 + a) = 1 + a^{-1} \in G_v \leq S.$$

Since $a \notin H$ we have $\pm a^{-1} \notin S$. Consequently, $1 - a, 1 + a \notin S$, as desired.

(d)$\Leftrightarrow$(e): Straightforward.

(e)$\Rightarrow$(a): Since S is a fan, $A(S) \subseteq S \leq H$ (condition (i) of Theorem 26.7.1). By definition, $-1 \in H$. In light of Theorem 11.3.6, it remains to show that

$$O^-(S, H)O^-(S, H) \subseteq 1 - S.$$

To this end take $a, b \in O^-(S, H) = -O^-(S, H)$ and assume that $ab \notin 1-S$. Lemma 11.4.1 then implies that $1 - ab \in aS$. Applying this lemma also to $-a$ and $-b$ we obtain that

$$1 - ab = 1 - (-a)(-b) \in -aS.$$

But $aS \cap -aS = \emptyset$, a contradiction. □

26.8 Examples of totally rigid subgroups

The next list of examples shows that for S totally rigid (so $N_S = \langle -1, S\rangle$ by Corollary 26.4.7(a)), the set $\mathrm{Val}(S, \langle -1, S\rangle)$ can be both empty or non-empty. In fact, each of these possibilities occurs when $-1 \in S$ as well as when S is a fan. All our examples are of the following general type: we take a cardinal number m and $\Gamma = \mathbb{Z}^m$ with the lexicographic order, and set $F = \bar{F}((\Gamma))$ for an appropriate field $\bar{F}$ of characteristic $\neq p$. Recall that then F is equipped with a canonical Henselian valuation v with value group Γ and residue field $\bar{F}$ (see Corollary 18.4.2). We will also take an appropriate subgroup $(\bar{F}^\times)^p \leq \bar{S} \leq \bar{F}^\times$ and let $S = (F^\times)^p\bar{S}$. By Proposition 18.2.1, $G_v \leq (F^\times)^p \leq S$. Also $\bar{S}_v = \bar{S}$ and $v(S) = p\Gamma$, so by Theorem 26.1.2

$$K_*^M(F)/S \cong (K_*^M(\bar{F})/\bar{S})[(\mathbb{Z}/p)^m].$$

EXAMPLE 26.8.1: *S is totally rigid, $-1 \in S$, and* $\mathrm{Val}(S, S) \neq \emptyset$.

Take an arbitrary field $\bar{F}$ and let $\bar{S} = \bar{F}^\times$. Then $K_*^M(\bar{F})/\bar{S} = \mathbf{0}$, and therefore $K_*^M(F)/S \cong \mathbf{0}[(\mathbb{Z}/p)^m]$. By Proposition 26.3.1(a), S is totally rigid. For the valuation v as above, the exact sequence (3.2.6) gives $O_v^\times \leq S$. Hence $v \in \mathrm{Val}(S, S)$.

EXAMPLE 26.8.2: *S is a fan and* $\mathrm{Val}(S, \langle -1, S\rangle) \neq \emptyset$.

Take $p = 2$, $\bar{F} = \mathbb{R}$, and $\bar{S} = (\mathbb{R}^\times)^2$. Since $\bar{S}$ is an ordering on $\mathbb{R}$, we have $K_*^M(\bar{F})/\bar{S} = \kappa$ (Theorem 25.1.2). Thus

$$K_*^M(F)/S \cong \kappa[(\mathbb{Z}/2)^m].$$

We conclude from Proposition 26.3.1(b) that S is totally rigid. By Corollary 23.3.5, the Bockstein map here is injective. Therefore S is a preordering (Proposition 25.1.4), whence a fan.

Now by the exact sequence (3.2.6),

$$(SO_v^\times : S) = (\bar{F}_v^\times : \bar{S}_v) = (\bar{F}^\times : \bar{S}) = (\mathbb{R}^\times : (\mathbb{R}^\times)^2) = 2 = (\langle -1, S\rangle : S).$$

As $\langle -1, S\rangle \le SO_v^\times$, necessarily $\langle -1, S\rangle = SO_v^\times$. Conclude that $v \in \mathrm{Val}(S, \langle -1, S\rangle)$.

EXAMPLE 26.8.3: *S totally rigid, $-1 \in S$, and $\mathrm{Val}(S, S) = \emptyset$.*

Dirichlet's theorem on primes in arithmetical progressions gives a prime number l such that $l \equiv 1 \pmod{2p}$. Take $\bar{F} = \mathbb{F}_l$ and $\bar{S} = (\mathbb{F}_l^\times)^p$. Then $S = (F^\times)^p$ and $-1 \in (\mathbb{F}_l^\times)^p \le S$. The group $\mathbb{F}_l^\times/p$ is cyclic and non-trivial, whence of order p. Further, $K_r^M(\mathbb{F}_l)/p = 0$ for $r \ge 2$ (Corollary 25.2.4). Therefore

$$K_*^M(\mathbb{F}_l)/(\mathbb{F}_l^\times)^p \cong \mathbf{0}[\mathbb{Z}/p]$$

(see Example 23.2.5). By Lemma 23.2.3,

$$K_*^M(F)/(F^\times)^p \cong \mathbf{0}[(\mathbb{Z}/p)^{m+1}] \quad .$$

We conclude from Proposition 26.3.1(a) that S is totally rigid.

Now suppose that $u \in \mathrm{Val}(S, S)$. To derive a contradiction we let w be the finest common coarsening of v and u. We claim that $O_w^\times \le S$. Indeed, since there are no non-trivial valuations on $\bar{F}_v = \mathbb{F}_l$ (Example 3.2.2(1)), the valuation u cannot be finer than v (see Proposition 7.2.1(d)). If u is coarser than v, then $w = u$ and obviously $O_w^\times \le S$. Finally, if v and u are incomparable, then Proposition 10.3.1(d) gives $O_w^\times = G_v G_u \le S$. This proves the claim.

Consequently, $\bar{F}_w^\times = \bar{S}_w$. Then, by Proposition 5.2.1(e)(f):

$$\mathbb{F}_l^\times = \bar{F}_v^\times = (\bar{F}_w)_{v/w}^\times = (\bar{S}_w)_{v/w} = \bar{S}_v = (\mathbb{F}_l^\times)^p,$$

a contradiction.

EXAMPLE 26.8.4: *S is a fan and $\mathrm{Val}(S, \langle -1, S\rangle) = \emptyset$.*

Let $p = 2$, let $\bar{F} = \mathbb{Q}(\sqrt{2})$, and let $\bar{S} = \bar{P}_1 \cap \bar{P}_2$, where $\bar{P}_1, \bar{P}_2$ are the distinct orderings on $\bar{F}$ (Example 6.2.3). By Proposition 25.1.6 and Example 23.2.6,

$$K_*^M(\bar{F})/\bar{S} \cong \kappa \times \kappa \cong \kappa[\mathbb{Z}/2].$$

Therefore

$$K_*^M(F)/S \cong \kappa[(\mathbb{Z}/2)^{m+1}].$$

By Proposition 26.3.1(b), S is totally rigid. Moreover, Corollary 23.3.5 again shows that the Bockstein operator here is injective. It follows that S is a preordering (Proposition 25.1.4), whence a fan.

Now suppose that $u \in \mathrm{Val}(S, \langle -1, S\rangle)$. To get a contradiction, we break again the discussion into three cases.

If v and u are incomparable, then we define w to be their finest common coarsening. By Proposition 10.3.1(d), $-1 \in O_w^\times = G_v G_u \le S$, which is not the case.

If v is finer than u, then $O_v^\times \le O_u^\times \le \langle -1, S\rangle$, so the exact sequence (3.2.6) gives the obvious contradiction

$$4 = (\bar{F}^\times : \bar{P}_1 \cap \bar{P}_2) = (\bar{F}_v^\times : \bar{S}_v) = (SO_v^\times : S) \mid (\langle -1, S\rangle : S) = 2.$$

Finally suppose that u is strictly finer than v. Then u/v is a non-trivial valuation on $\bar{F} = \bar{F}_v$ and

$$G_{u/v} \le \bar{S}_v = \bar{S} = \bar{P}_1 \cap \bar{P}_2.$$

Since $O_{u/v}$ is integrally closed (Proposition 3.1.3), $\mathbb{Z}[\sqrt{2}] \subseteq O_{u/v}$. Lemma 4.1.1 shows that u/v is a π-adic valuation for some prime element π of $\mathbb{Z}[\sqrt{2}]$. Hence $l \in \mathfrak{m}_{u/v}$ for the prime number l lying under π, so $1 - l \in G_{u/v} \leq \bar{P}_1$, and we got again a contradiction.

CHAPTER 27

K-Rings of Wild Valued Fields

We continue our investigation from the previous chapter, this time focusing on the reflection of wild valuations in the generalized Milnor K-theory. As usual, this case is more subtle than the tame one, and complete analogs of the main results we had in the tame case are still unknown. Therefore we will mostly restrict ourselves to the better understood case of subgroups S of $F^\times$ containing $(F^\times)^p$ (with p prime) with $(F^\times : S) < \infty$. Then the corresponding K-rings are either Demuškin κ-structures, or some quite degenerate structures. Our analysis will follow [**Ef3**], [**Ef8**], and [**Ef9**].

27.1 The discrete case

Throughout this chapter p will be a fixed prime number and $q = p^s$ a p-power, with $s \geq 1$. Let μ_q be the group of roots of unity (over the prime field) of order dividing q. We first record the following consequence of local class field theory and of the multiplicative structure theory of p-adic fields.

THEOREM 27.1.1. *Let E be a finite extension of $\mathbb{Q}_p$. Suppose that $\mu_q \subseteq E$. Then $K_*^M(E)/(F^\times)^q$ is a q-Demuškin κ-structure with*

$$\dim_{\mathbb{Z}/q}(E^\times/q) = [E : \mathbb{Q}_p] + 2.$$

PROOF. Let u be the p-adic valuation on E. The field $\bar{E}_u$ is finite and has characteristic p. Hence $\bar{E}_u^\times/q = 1$. Since $\bar{E}_u^\times$ has no elements of order p, Lemma 3.2.4 shows that $G_u^q = (E^\times)^q \cap G_u$, so

$$G_u/((E^\times)^q \cap G_u) \cong G_u/q.$$

The sequences (3.2.5)–(3.2.6) (with $S = (E^\times)^q$) therefore imply that $E^\times/q \cong G_u/q \oplus \mathbb{Z}/q$. Now the group G_u is the direct sum of the (finite) group of roots of unity of p-power order in E and of a free $\mathbb{Z}_p$-module of rank $[E : \mathbb{Q}_p]$ [**Se2**, Ch. XIV, §4, Prop. 10]. Hence

$$G_u/q \cong \mu_q \oplus (\mathbb{Z}/q)^{[E:\mathbb{Q}_p]}.$$

We conclude that $E^\times/q$ is a free $\mathbb{Z}/q$-module of dimension $[E : \mathbb{Q}_p] + 2$.

By a theorem of Moore [**FV**, Ch. IX, Prop. 4.2], $K_2^M(E)/q$ is cyclic. The Hilbert symbol induces a non-degenerate surjective bilinear map

$$(E^\times/q) \times (E^\times/q) \to \mu_q$$

which factors through $K_2^M(E)/q$ ([**FV**, Ch. IV, §5], [**Se2**, Ch. XIV, §2]). It follows that $K_2^M(E)/q \cong \mathbb{Z}/q$ and the product

$$(E^\times/q) \times (E^\times/q) \to K_2^M(E)/q$$

is non-degenerate. □

PROPOSITION 27.1.2. *Let (F, v) be a valued field of characteristic* 0. *Suppose that v is discrete and that $\bar{F}_v$ is finite of characteristic p. Let $(\hat{F}, \hat{v})$ be the completion of (F, v). Then:*

(a) *$(\hat{F}, \hat{v})$ is a extension of $\mathbb{Q}_p$ with its p-adic valuation, and its degree is*

$$[\bar{F}_v : \mathbb{F}_p](v(F^\times) : \mathbb{Z}v(p)) \ < \infty;$$

(b) *$K_*^M(F)/(F^\times)^q(1 + q^2\mathfrak{m}_v) \cong K_*^M(\hat{F})/q$ in positive degrees via restriction.*

PROOF. (a) By Theorem 9.3.2(f), $(\hat{F}, \hat{v})$ is an immediate extension of (F, v). Therefore Theorem 12.4.2 shows that $O_{\hat{v}}$ is a free $W(\bar{F}_v)$-module of rank $(v(F^\times) : \mathbb{Z}v(p))$. Also, $W(\bar{F}_v)$ is a free $W(\mathbb{F}_p)$-module of rank $[\bar{F}_v : \mathbb{F}_p]$ (Theorem 12.3.8). It follows that $O_{\hat{v}}$ is a free $W(\mathbb{F}_p)$-module of rank

$$[\bar{F}_v : \mathbb{F}_p](v(F^\times) : \mathbb{Z}v(p)).$$

Consequently, the quotient field of $O_{\hat{v}}$ is an extension of the quotient field of $W(\mathbb{F}_p)$ of degree $\leq [\bar{F}_v : \mathbb{F}_p](v(F^\times) : \mathbb{Z}v(p))$. But the former field is $\hat{F}$, while the latter field is $\mathbb{Q}_p$ (by Example 12.3.7).

As noted in Theorem 12.4.3, the valuation ring $W(\bar{F}_v)$ extends $W(\mathbb{F}_p) = \mathbb{Z}_p$. Hence $O_{\hat{v}} \cap \mathbb{Q}_p$ is a valuation ring on $\mathbb{Q}_p$ containing $\mathbb{Z}_p$ and not containing $1/p$. Thus it coincides with $\mathbb{Z}_p$, i.e., this is an extension of valued fields. By the fundamental inequality (Theorem 14.2.1) and what we have already seen,

$$[\hat{F} : \mathbb{Q}_p] = [\bar{F}_v : \mathbb{F}_p](v(F^\times) : \mathbb{Z}v(p)).$$

(b) Since $(\hat{F}, \hat{v})$ is Henselian (Corollary 18.3.2), Proposition 18.2.1 implies that $1 + q^2\mathfrak{m}_{\hat{v}} \leq (\hat{F}^\times)^q$. Hence

$$(F^\times)^q(1 + q^2\mathfrak{m}_v) \leq (\hat{F}^\times)^q \cap F^\times.$$

Conversely, by Proposition 3.4.5 and the immediateness, $\hat{F}^\times = F^\times(1 + q\mathfrak{m}_{\hat{v}})$. The binomial formula therefore gives $(\hat{F}^\times)^q \leq (F^\times)^q(1 + q^2\mathfrak{m}_{\hat{v}})$. It follows that

$$(\hat{F}^\times)^q \cap F^\times = (F^\times)^q(1 + q^2\mathfrak{m}_v).$$

Since F is $\mathcal{T}_{\hat{v}}$-dense in its completion $\hat{F}$, Proposition 24.2.1 implies that the κ-structures

$$K_*^M(F)/(F^\times)^q(1 + q^2\mathfrak{m}_v), \quad K_*^M(\hat{F})/(\hat{F}^\times)^q$$

are isomorphic via restriction. Now recall the latter ring coincides with $K^M(\hat{F})/q$ in positive degrees (Proposition 24.1.3). □

27.2 A vanishing theorem

In the next section we will consider general rank 1 valuations with mixed characteristics. The analysis will be based on the following result.

THEOREM 27.2.1. *Let (F, v) be a valued field of rank* 1. *Let S be a $\mathcal{T}_v$-open subgroup of $F^\times$ such that $F^\times/S$ is finitely generated and $F^\times = SG_v$. Then $K_r^M(F)/S = 0$ for all $r \geq 2$.*

PROOF. For a positive $\alpha \in v(F^\times)$ let

$$W_\alpha = \{x \in F^\times \mid v(1-x) \geq \alpha\}.$$

Thus $W_\alpha \leq G_v$. Lemma 12.5.3 yields $\alpha > 0$ such that $F^\times = SG_v = SW_\alpha$. Since v has rank 1, the ordered abelian group $v(F^\times)$ embeds in $(\mathbb{R}, \leq_{\mathbb{R}})$ (Theorem 2.5.2). Hence the subsets

$$(1 - W_\alpha)^t, \quad t = 1, 2, 3, \dots$$

of F form a local system of $\mathcal{T}_v$-neighborhoods at 0.

It suffices to show that $aS \otimes bS \in \mathrm{St}_{F,2}(S)$ for all $a, b \in G_v$. Suppose that this is not the case. We define inductively a sequence $c_1, c_2, \dots \in G_v$ such that for each i,

$$1 - c_i \in (1-b)(1-W_\alpha)^{i-1}, \quad aS \otimes bc_i^{-1}S \in \mathrm{St}_{F,2}(S).$$

We take $c_1 = b$. Next suppose that c_i has already been constructed. Since $aS \otimes bS \notin \mathrm{St}_{F,2}(S)$ we have $c_i \neq 1$. We choose $y_i \in S$ such that

$$\frac{a}{1 - c_i^{-1}} \in y_i W_\alpha$$

and set

$$c_{i+1} = c_i(1 - y_i^{-1}a).$$

Since $c_i \in G_v$, also $c_i^{-1} \in G_v$, and we have $1 - c_i^{-1} \in \mathfrak{m}_v$. Therefore,

$$y_i^{-1}a \in (1 - c_i^{-1})W_\alpha \subseteq \mathfrak{m}_v.$$

Hence $c_{i+1} \in G_v$. Then

$$\frac{1 - c_{i+1}}{1 - c_i} = 1 - \frac{y_i^{-1}a}{1 - c_i^{-1}} \in 1 - W_\alpha,$$

so

$$1 - c_{i+1} \in (1-b)(1-W_\alpha)^i.$$

Furthermore,

$$\begin{aligned} aS \otimes bc_{i+1}^{-1}S &= aS \otimes bc_i^{-1}S - aS \otimes (1 - y_i^{-1}a)S \\ &= aS \otimes bc_i^{-1}S - y_i^{-1}aS \otimes (1 - y_i^{-1}a)S \in \mathrm{St}_{F,2}(S). \end{aligned}$$

This completes the inductive construction of the c_i.

Now we take a positive integer t such that $(1 - W_\alpha)^t$ is contained in the $\mathcal{T}_v$-open neighborhood $(1-b)^{-1}(1-S)$ of 0. Then

$$1 - c_{t+1} \in (1-b)(1-W_\alpha)^t \subseteq 1 - S,$$

so $c_{t+1} \in S$. We conclude that

$$aS \otimes bS = aS \otimes bc_{t+1}^{-1}S \in \mathrm{St}_{F,2}(S),$$

a contradiction. □

27.3 The general case

We first consider valuations of rank 1.

THEOREM 27.3.1. *Let v be a valuation of rank 1 on a field F such that* $\mathrm{char}\, F = 0$ *and* $\mathrm{char}\, \bar{F}_v = p$. *Let $S = (F^\times)^q(1+q^2\mathfrak{m}_v)$ and suppose that $(F^\times : S) < \infty$. Then either:*

(a) *v is discrete, $\bar{F}_v$ is finite, and $K_*^M(F)/S \cong K_*^M(\hat{F})/q$ in positive degrees, where $\hat{F}$ is the completion of F with respect to v; or*

(b) *$v(F^\times)$ is p-divisible and $K_r^M(F)/S = 0$ for all $r \geq 2$.*

PROOF. We have $v(S) = qv(F^\times)$ and $\bar{S}_v = (\bar{F}_v^\times)^q$. Since $S \leq (F^\times)^p(1+p\mathfrak{m}_v)$, the finiteness assumption implies that

$$(F^\times : (F^\times)^p(1+p\mathfrak{m}_v)) < \infty.$$

By Theorem 12.5.4, one of the following cases occurs:

CASE I: *v is discrete and $\bar{F}_v$ is finite.*
Then we apply Proposition 27.1.2(b).

CASE II: *$v(F^\times)$ is p-divisible and $\bar{F}_v$ is perfect.*
Then $v(S) = v(F^\times)$ and $\bar{S}_v = \bar{F}_v^\times$. The exact sequences (3.2.5)–(3.2.6) therefore show that $F^\times = SG_v$. Since S is $\mathcal{T}_v$-open in F, the vanishing theorem (Theorem 27.2.1) implies that $K_r^M(F)/S = 0$ for all $r \geq 2$. □

COROLLARY 27.3.2. *Let F, v and S be as in Theorem 27.3.1. Assume that $\mu_q \subseteq F^\times$. Then either $K_*^M(F)/S$ is a q-Demuškin κ-structure with* $\dim_{\mathbb{Z}/q} F^\times/S \geq 3$ *(in case (a)) or $K_r^M(F)/S = 0$ for all $r \geq 2$ (in case (b)).*

PROOF. In case (a), $\hat{F}$ is a finite extension of $\mathbb{Q}_p$, by Proposition 27.1.2(a). By Theorem 27.1.1, $K_*^M(\hat{F})/q$ is then a q-Demuškin κ-structure with

$$\dim_{\mathbb{Z}/q}(\hat{F}^\times/q) = [\hat{F} : \mathbb{Q}_p] + 2 \geq 3.$$ □

We note that if v is p-Henselian and F contains a root of unity of order p, then the subgroup S considered in theorem 27.3.1 is nothing but $(F^\times)^q$ (by Example 18.2.2(3)).

A more general type of valuations that give rise to Demuškin K-rings is obtained as follows. Let q be a p-power. Let v be a valuation on the field F and set again

$$S = (F^\times)^q(1+q^2\mathfrak{m}_v).$$

Following [Ef8], we call (F, v) an **arithmetically q-Demuškin** valued field if $\bar{F}_v$ is a finite field of characteristic p and there is an S-compatible coarsening u of v such that the quotient v/u is a discrete valuation, $u(F^\times) = u(S)$ and $\mathrm{char}\, \bar{F}_u = 0$ (in [Ef8, p. 334] the finiteness assumption on $\bar{F}_v$ is missing).

THEOREM 27.3.3. *Let (F, v) be an arithmetically q-Demuškin valued field and let S and u be as in the definition above. Then $K_*^M(F)/S \cong K_*^M(\hat{E})/q$ in positive degrees, where $\hat{E}$ is the completion of $(\bar{F}_u, v/u)$.*

In particular, if F contains a root of unity of order q, then $K_*^M(F)/S$ is q-Demuškin and $\dim_{\mathbb{Z}/q} F^\times/S \geq 3$.

PROOF. Since char $\bar{F}_u = 0$ we have

$$G_u = 1 + q^2\mathfrak{m}_u \le 1 + q^2\mathfrak{m}_v \le S,$$

i.e., u is S-compatible. As $u(F^\times) = u(S)$, Theorem 26.1.2 therefore gives

$$K_*^M(F)/S \cong K_*^M(\bar{F}_u)/\bar{S}_u.$$

To compute this κ-structure observe that

$$1 + q^2\mathfrak{m}_v \le 1 + q^2 O_v \le 1 + q^2 O_u \le O_u^\times.$$

Hence $\bar{S}_u = (\bar{F}_u^\times)^q(1 + q^2\mathfrak{m}_{v/u})$. Now by assumption, the valuation v/u on $\bar{F}_u$ is discrete with finite residue field $\bar{F}_v$ of characteristic p. Proposition 27.1.2(b) (for $\bar{F}_u$, v/u and $\bar{S}_u$) therefore implies that $K_*^M(\bar{F}_u)/\bar{S}_u \cong K_*^M(\hat{E})/q$.

The second assertion follows as before using Proposition 27.1.2(a) and Theorem 27.1.1. □

For the rest of this section we assume for simplicity that $q = p$ is prime.

It is presently unknown if all p-Demuškin generalized K-rings $K_*^M(F)/S$ with $\dim_{\mathbb{F}_p} F^\times/S \ge 3$ arise from arithmetically p-Demuškin valued fields as in Theorem 27.3.3. Still, we have the following partial result.

THEOREM 27.3.4. *Let v be a valuation on the field F and let again $S = (F^\times)^p(1+p^2\mathfrak{m}_v)$. Suppose that $v(F^\times) \ne v(S)$ and that $K_*^M(F)/S$ is a p-Demuškin κ-structure with $\dim_{\mathbb{F}_p} F^\times/S \ge 3$. Then (F, v) is arithmetically p-Demuškin.*

PROOF. We first observe that $v(S) = pv(F^\times)$, so the first assumption means that $v(F^\times)$ is not p-divisible.

We claim that char $\bar{F}_v = p$. Otherwise v would be S-compatible, so by Theorem 26.1.2,

$$K_*^M(F)/S \cong (K_*^M(\bar{F}_v)/\bar{S}_v)[v(F^\times)/v(S)].$$

However, as $v(F^\times) \ne v(S)$, this would contradict Proposition 23.4.6.

Next we claim that char $F = 0$. Indeed, assume that char $F = p$. Then $S = (F^\times)^p$. But by assumption, $(F^\times : S) < \infty$, and therefore Lemma 12.5.2 gives $F^\times = S$. This contradicts our assumption that $\dim_{\mathbb{F}_p} F^\times/S \ge 3$, and proves the claim.

Now since $F^\times/S$ is finite, its quotient $F^\times/(F^\times)^p(1+p\mathfrak{m}_v)$ is also finite. Corollary 12.5.5 therefore yields coarsenings $u^- \le u^+ \le v$ such that char $\bar{F}_{u^-} = 0$, char $\bar{F}_{u^+} = p$, the quotient valuation u^+/u^- has rank 1, and either:

(i) v/u^- is discrete, $\bar{F}_v$ is finite, and $v = u^+$; or
(ii) the value group of u^+/u^- is p-divisible.

One has

$$1 + p^2\mathfrak{m}_v \le 1 + p^2 O_v \le 1 + q^2 O_{u^-} \le O_{u^-}^\times.$$

Hence

$$u^-(S) = pu^-(F^\times) \quad , \quad \bar{S}_{u^-} = (\bar{F}_{u^-}^\times)^p(1 + p^2\mathfrak{m}_{v/u^-}).$$

Similarly,

$$u^+(S) = pu^+(F^\times) \quad , \quad \bar{S}_{u^+} = (\bar{F}_{u^+}^\times)^p(1 + p^2\mathfrak{m}_{v/u^+}) = (\bar{F}_{u^+}^\times)^p$$

(since $p = 0$ in $\bar{F}_{u^+}$). Proposition 5.2.1(h) gives rise to exact sequences

$$0 \to (u^+/u^-)(\bar{F}_{u^-}^\times)/(u^+/u^-)(\bar{S}_{u^-}) \to u^+(F^\times)/p \to u^-(F^\times)/p \to 0$$

and

$$0 \to (v/u^+)(\bar F_{u^+}^\times)/p \to v(F^\times)/p \to u^+(F^\times)/p \to 0.$$

The exact sequences (3.2.5)–(3.2.6) and the finiteness of $F^\times/S$ imply that $\bar F_{u^+}^\times/p = \bar F_{u^+}^\times/\bar S_{u^+}$ is also finite. By Lemma 12.5.2, $\bar F_{u^+}^\times/p = 1$. Hence $(v/u^+)(\bar F_{u^+}^\times)/p = 0$.

Furthermore

$$1 + \mathfrak{m}_{u^-} = 1 + p^2\mathfrak{m}_{u^-} \le 1 + p^2\mathfrak{m}_v \le S,$$

i.e., u^- is S-compatible. Therefore Theorem 26.1.2 gives

$$K_*^M(F)/S \cong (K_*^M(\bar F_{u^-})/\bar S_{u^-})[u^-(F^\times)/p].$$

Since $K_*^M(F)/S$ is p-Demuškin and $\dim_{\mathbb{F}_p}(F^\times/S) \ge 3$, Proposition 23.4.6 implies as before that $u^-(F^\times)/p = 0$.

Putting everything together, we conclude from the above exact sequences that

$$(u^+/u^-)(\bar F_{u^-}^\times)/(u^+/u^-)(\bar S_{u^-}) \cong u^+(F^\times)/p \cong v(F^\times)/p = v(F^\times)/v(S) \ne 0.$$

Since $(\bar F_{u^-}^\times)^p \le \bar S_{u^-}$, this rules out possibility (ii) above. We are left with possibility (i). Thus the conditions in the definition of an arithmetically p-Demuškin valued field are satisfied for (F, v) when we take $u = u^-$. □

It would also be desirable to deduce the existence of a valuation v as in Theorem 27.3.4 just from the assumption that $K_*^M(F)/S$ is p-Demuškin, where $(F^\times)^p \le S \le F^\times$ and $\dim_{\mathbb{F}_p} F^\times/S \ge 3$. For p odd, Proposition 27.3.5 below translates the existence of a valuation v with $v(F^\times) \ne v(S)$ into K-theoretic terms by means of the following notion:

DEFINITION. Let S be a subgroup of $F^\times$ and let $a \in F^\times$. We say that a is S-**primary** if for every $b \ne 0, 1$ in F at least one of the following symbols

$$\{a, b\}_S, \quad \{a, 1-b\}_S, \quad \{a, 1-b^{-1}\}_S$$

is zero in $K_2^M(F)/S$.

PROPOSITION 27.3.5. *Suppose that $K_*^M(F)/S$ is p-Demuškin with $p \ne 2$. Then the following conditions are equivalent:*

(a) *there exists a valuation v on F with $v(F^\times) \ne v(S)$;*
(b) *there exists an S-primary element $a \in F^\times \setminus S$.*

Moreover, for every S-primary element $a \in F^\times \setminus S$, the valuation v in (a) can be chosen so that $\{a, b\}_S = 0$ for all $b \in O_v^\times$.

PROOF. For $a \in F^\times$ let $\{a\}_S^\perp$ be the orthogonal complement of $\{a\}_S$ with respect to the perfect pairing

$$(F^\times/S) \times (F^\times/S) \to K_2^M(F)/S \cong \mathbb{Z}/p.$$

We note that the intermediate groups $S \le T < F^\times$ with $(F^\times : T) = p$ are precisely the complements $\{a\}_S^\perp$, where $a \in F^\times \setminus S$. Thus (b) means that there exists $S \le T < F^\times$ such that for every $b \in F^\times$, at least one of $b, 1-b, 1-b^{-1}$ is in T. By Theorem 11.4.5, this is equivalent to (a). Moreover, when these equivalent conditions hold, Theorem 11.4.5 also gives a valuation $v \in \mathrm{Val}(T, T)$. Then $O_v^\times \le T = \{a\}_S^\perp$, and

$$(v(F^\times) : v(S)) = (F^\times : SO_v^\times) \ge (F^\times : T) > 1.$$ □

CHAPTER 28

Decompositions of K-Rings

We now turn to study local–global properties for K-rings, in particular, their decomposition into direct products. The first main result of this chapter is Theorem 28.2.2, which gives such decompositions in a quite general topological situation. It will then be applied in §§28.3–28.4 to prove a decomposition theorem of a more arithmetical nature (Theorem 28.4.3).

28.1 The basic criterion

Our starting point is the following key lemma, which is inspired by the proof of [**JWd**, Th. 4.3].

LEMMA 28.1.1. *Let F be a field, let $S_1, \ldots, S_n$ be subgroups of $F^\times$. Let $S = S_1 \cap \cdots \cap S_n$, and suppose that the canonical homomorphism*

$$F^\times \to \prod_{i=1}^{n} F^\times/S_i$$

is surjective. Moreover, suppose that for every $a, b \in F^\times$ one has $aS \otimes bS \in \mathrm{St}_{F,2}(S)$, provided that either:

(1) *there exist distinct $1 \leq i, j \leq n$ such that $a \in \bigcap_{l \neq i} S_l$, and $b \in \bigcap_{l \neq j} S_l$; or*
(2) *there exists $1 \leq i \leq n$ such that $a, b \in \bigcap_{l \neq i} S_l$ and $1 \in aS_i + bS_i$.*

Then the product morphism

$$\prod \mathrm{Res}\colon K_*^M(F)/S \to \prod_{i=1}^{n} K_*^M(F)/S_i$$

is an isomorphism of κ-structures.

PROOF. Since the homomorphism $F^\times \to \prod_{i=1}^n F^\times/S_i$ is surjective, so is the canonical restriction homomorphism $K_r^M(F)/S \to \prod_{i=1}^n K_r^M(F)/S_i$ for every $r \geq 1$. It is clearly injective for $r = 1$. We show its injectivity for $r \geq 2$.

For each $1 \leq i \leq n$ let

$$\rho_i\colon F^\times/S \to F^\times/S_i, \quad \mathrm{Res}_i^r\colon K_r^M(F)/S \to K_r^M(F)/S_i$$

be the canonical projections. Let $\delta_i\colon F^\times/S_i \to F^\times/S$ be the monomorphism induced by the isomorphism $F^\times/S \xrightarrow{\sim} \prod_{i=1}^n F^\times/S_i$, i.e., $\delta_i(aS_i) = bS$, where $b \in \bigcap_{l \neq i} S_l$ and $bS_i = aS_i$. Then

$$\rho_j \circ \delta_i = \begin{cases} 1 & \text{if } i \neq j, \\ \mathrm{Id} & \text{if } i = j. \end{cases}$$

We have

$$F^\times/S = \prod_{i=1}^n \mathrm{Im}(\delta_i).$$

The assumptions then imply that:

(1') $\mathrm{Im}(\delta_{i_1}) \otimes \cdots \otimes \mathrm{Im}(\delta_{i_r}) \subseteq \mathrm{St}_{F,r}(S)$ for all $i_1, \dots, i_r$ which are not all equal;

(2') $\delta_i^{\otimes r}(\mathrm{St}_{F,r}(S_i)) \subseteq \mathrm{St}_{F,r}(S)$ for all i.

In light of (2'), there exist homomorphisms ϵ_i, $i = 1, \dots, n$, such that for each i, j the following diagram commutes and has exact rows:

$$\begin{array}{ccccccccc}
1 & \longrightarrow & \mathrm{St}_{F,r}(S_i) & \longrightarrow & (F^\times/S_i)^{\otimes r} & \longrightarrow & K_r^M(F)/S_i & \longrightarrow & 0 \\
 & & \Big\downarrow{\scriptstyle \delta_i^{\otimes r}} & & \Big\downarrow{\scriptstyle \delta_i^{\otimes r}} & & \Big\downarrow{\scriptstyle \epsilon_i} & & \\
1 & \longrightarrow & \mathrm{St}_{F,r}(S) & \longrightarrow & (F^\times/S)^{\otimes r} & \longrightarrow & K_r^M(F)/S & \longrightarrow & 0 \\
 & & & & \Big\downarrow{\scriptstyle \rho_j^{\otimes r}} & & \Big\downarrow{\scriptstyle \mathrm{Res}_j^r} & & \\
 & & & & (F^\times/S_j)^{\otimes r} & \longrightarrow & K_r^M(F)/S_j & \longrightarrow & 0.
\end{array}$$

Then

$$\mathrm{Res}_j^r \circ \epsilon_i = \begin{cases} 0 & \text{if } i \neq j, \\ \mathrm{Id} & \text{if } i = j. \end{cases}$$

Furthermore, $K_r^M(F)/S$ is generated by the images of $\mathrm{Im}(\delta_{i_1}) \otimes \cdots \otimes \mathrm{Im}(\delta_{i_r})$, where $1 \leq i_1, \dots, i_r \leq n$. By (1') these images are trivial unless $i_1 = \cdots = i_r$. Hence

$$K_r^M(F)/S = \prod_{i=1}^n \mathrm{Im}(\epsilon_i).$$

Now let $\alpha_i \in K_r^M(F)/S_i$, $i = 1, \dots, n$, and suppose that $\prod_{i=1}^n \epsilon_i(\alpha_i)$ is mapped by $\prod_{i=1}^n \mathrm{Res}_i^r$ trivially. Then for each j we have

$$\alpha_j = \prod_{i=1}^n (\mathrm{Res}_j^r \circ \epsilon_i)(\alpha_i) = \mathrm{Res}_j^r\Big(\prod_{i=1}^n \epsilon_i(\alpha_i)\Big) = 0.$$

This gives the desired injectivity. □

Corollary 28.1.2. *Let F be a field and let $S_1, \dots, S_n$ be subgroups of $F^\times$ containing a subgroup S. Consider the homomorphism*

$$\mathrm{Res}^r \colon K_r^M(F)/S \to \prod_{i=1}^n K_r^M(F)/S_i.$$

Suppose that:

(i) Res^1 *is bijective;*

(ii) *the restriction of* Res^2 *to the image of $F^\times \times F^\times$ in $K_2^M(F)/S$ is injective.*

Then

$$\prod \mathrm{Res} \colon K_*^M(F)/S \to \prod_{i=1}^n K_*^M(F)/S_i$$

is an isomorphism of κ-structures.

PROOF. By (i), $S = S_1 \cap \cdots \cap S_n$. Next suppose that $a, b \in F^\times$ and suppose that one of conditions (1) and (2) of Lemma 28.1.1 holds. We need to show that $aS \otimes bS \in \mathrm{St}_{F,2}(S)$. Now in both cases, for each i we have $\{a, b\}_{S_i} = 0$ in $K_2^M(F)/S_i$. By (ii), $\{a, b\}_S = 0$, so $aS \otimes bS \in \mathrm{St}_{F,2}(S)$. □

When condition (i) of Corollary 28.1.2 is weakened to surjectivity only, we have:

PROPOSITION 28.1.3. *Let p be a prime number, let F be a field, and let $S, S_1, \dots, S_n$ be subgroups of $F^\times$ such that $(F^\times)^p \le S \le S_i$ for each i. Consider the homomorphism*

$$\mathrm{Res}^r \colon K_r^M(F)/S \to \prod_{i=1}^{n} K_r^M(F)/S_i.$$

Suppose that:
(i) Res^1 *is surjective;*
(ii) Res^2 *is injective.*
Then there exists an intermediate subgroup $S \le S_0 \le F^\times$ such that the restriction morphism

$$K_*^M(F)/S \to \prod_{i=0}^{n} K_*^M(F)/S_i$$

is an isomorphism, and $K_r^M(F)/S_0 = 0$ for all $r \ge 2$.

PROOF. Since $F^\times/S$ is an $\mathbb{F}_p$-linear space, its subspace $(\bigcap_{i=1}^n S_i)/S$ has a complement. Thus there exists a subgroup S_0 of $F^\times$ with

$$S = \bigcap_{i=0}^{n} S_i, \quad F^\times = \Big(\bigcap_{i=1}^{n} S_i\Big) S_0.$$

The projection $F^\times/S \to \prod_{i=0}^n F^\times/S_i$ is clearly injective. We show that it is also surjective, i.e., that for every $a, a_1, \dots, a_n \in F^\times$ one has

$$aS_0 \cap \bigcap_{i=1}^{n} a_i S_i \ne \emptyset.$$

Indeed, assumption (i) yields $b \in \bigcap_{i=1}^n a_i S_i$. We can write $b^{-1}a = cd$ for some $c \in \bigcap_{i=1}^n S_i$ and $d \in S_0$. Then

$$ad^{-1} = bc \in aS_0 \cap \bigcap_{i=0}^{n} a_i S_i,$$

as desired.

We conclude that the restriction

$$K_r^M(F)/S \to \prod_{i=0}^{n} K_r^M(F)/S_i$$

is an isomorphism for $r = 1$. By assumption (ii) it is injective for $r = 2$. By Corollary 28.1.2, it is an isomorphism for all $r \ge 1$. This proves the first assertion.

Since

$$\mathrm{Res}^2 \colon K_2^M(F)/S \to \prod_{i=1}^{n} K_2^M(F)/S_i$$

is injective and

$$\mathrm{Res}^2\colon K_2^M(F)/S \to \prod_{i=0}^{n} K_2^M(F)/S_i$$

is bijective, necessarily $K_2^M(F)/S_0 = 0$. Hence $K_r^M(F)/S_0 = 0$ for all $r \geq 2$. □

28.2 Topological decompositions

We denote the closure of a subset S of the field F with respect to a topology $\mathcal{T}$ on F by $\mathrm{Cl}_{\mathcal{T}}(S)$.

LEMMA 28.2.1. *Let $\mathcal{T}$ be a ring topology on a field F and let S be a $\mathcal{T}$-open neighborhood of* 1. *Let $a, b \in F^\times$ satisfy*

$$1 \in a\mathrm{Cl}_{\mathcal{T}}(S) + b\mathrm{Cl}_{\mathcal{T}}(S).$$

Then there exist non-empty $\mathcal{T}$-open subsets V and W of S with $aV + bW \subseteq S$.

PROOF. Take $s_1, s_2 \in \mathrm{Cl}_{\mathcal{T}}(S)$ with $1 = as_1 + bs_2$. In particular, $as_1 + bs_2 \in S$. By continuity, there exist $\mathcal{T}$-open neighborhoods V', W' of s_1, s_2, respectively, such that $aV' + bW' \subseteq S$. Then $V = V' \cap S$ and $W = W' \cap S$ are non-empty $\mathcal{T}$-open sets as required. □

Topologies $\mathcal{T}_1, \ldots, \mathcal{T}_n$ on F are called **independent** if any intersection $U_1 \cap \cdots \cap U_n$ of non-empty $\mathcal{T}_i$-open sets U_i, $i = 1, \ldots, n$, is also non-empty. For example, if $\lambda_1, \ldots, \lambda_n$ are independent non-trivial localities on F, then the corresponding topologies $\mathcal{T}_{\lambda_1}, \ldots, \mathcal{T}_{\lambda_n}$ are independent, by the weak approximation theorem (Theorem 10.1.7).

THEOREM 28.2.2. *Let $\mathcal{T}_1, \ldots, \mathcal{T}_n$ be independent ring topologies on a field F. For each $1 \leq i \leq n$ let S_i be a $\mathcal{T}_i$-open subgroup of $F^\times$ with $0 \in \mathrm{Cl}_{\mathcal{T}_i}(S_i)$, and let $S = S_1 \cap \cdots \cap S_n$. Then*

$$K_*^M(F)/S \cong \prod_{i=1}^{n} K_*^M(F)/S_i$$

via restriction.

PROOF. The canonical homomorphism $F^\times/S \to \prod_{i=1}^n F^\times/S_i$ is clearly injective. By the independence assumption, it is surjective, whence an isomorphism.

Next let $a, b \in F^\times$, and suppose that at least one of conditions (1) and (2) of Lemma 28.1.1 holds. It suffices to show that $1 \in aS + bS$. Now for each $1 \leq l \leq n$ either $a \in S_l$, $b \in S_l$, or $1 \in aS_l + bS_l$. Since $0 \in \mathrm{Cl}_{\mathcal{T}_l}(S_l)$ and $S_l \subseteq \mathrm{Cl}_{\mathcal{T}_l}(S_l)$ we therefore have $1 \in a\mathrm{Cl}_{\mathcal{T}_l}(S_l) + b\mathrm{Cl}_{\mathcal{T}_l}(S_l)$ for all l. Consequently Lemma 28.2.1 yields non-empty $\mathcal{T}_l$-open subsets V_l, W_l of S_l such that $aV_l + bW_l \subseteq S_l$, $l = 1, \ldots, n$. Then

$$a\Big(\bigcap_{l=1}^{n} V_l\Big) + b\Big(\bigcap_{l=1}^{n} W_l\Big) \subseteq \bigcap_{l=1}^{n} S_l = S.$$

By the independence again, the subsets $\bigcap_{l=1}^n V_l$ and $\bigcap_{l=1}^n W_l$ of S are non-empty. Hence $(aS + bS) \cap S \neq \emptyset$, so $1 \in aS + bS$. □

The topological assumption on 0 in Theorem 28.2.2 is satisfied in the following important situation.

LEMMA 28.2.3. *Let m be a positive integer and let λ be a locality on the field F. Let S be a $\mathcal{T}_\lambda$-open subgroup of $F^\times$ containing $(F^\times)^m$. Then $0 \in \mathrm{Cl}_{\mathcal{T}_\lambda}(S)$.*

PROOF. If λ is the trivial valuation, then $\mathcal{T}_\lambda$ is the indiscrete topology, so $\mathrm{Cl}_{\mathcal{T}_\lambda}(S) = F$, and we are done.

Suppose that $\lambda = v$ is a non-trivial valuation. Then the sets $W_a = aO_v$, where $0 \neq a \in \mathfrak{m}_v$, form a local system of $\mathcal{T}_\lambda$-neighborhoods at 0. But $a^m \in S \cap W_a$, so this intersection is non-empty.

Next suppose that $\lambda = P$ is an ordering. Then the sets $W'_a = (-a, a)_P$, where $a \in (0,1)_P$, form a local system of $\mathcal{T}_\lambda$-neighborhoods at 0. Again, $a^m \in S \cap W'_a$, showing that this intersection is non-empty. □

From Theorem 28.2.2 and Lemma 28.2.3 we obtain the following topological decomposition theorem for K-rings.

COROLLARY 28.2.4. *Let $\lambda_1, \dots, \lambda_n$ be independent localities on a field F and let m be a positive integer. For each i let S_i be a $\mathcal{T}_{\lambda_i}$-open subgroup of $F^\times$ containing $(F^\times)^m$ and let $S = S_1 \cap \cdots \cap S_n$. Then*

$$K_*^M(F)/S \cong \prod_{i=1}^n K_*^M(F)/S_i$$

via restriction.

28.3 Local pairs

This section introduces some terminology which will be needed for the proof of the arithmetical decomposition Theorem 28.4.3 below. We will refine the previous analysis of the tree of localities on a field, as studied in §7.

Define a **local pair** on the field F to be a pair (λ, S), where λ is a locality on F and S is a subgroup of $F^\times$. The local pair is **compatible** if λ is S-compatible, i.e., if $G_\lambda \leq S$. We partially order the collection of all local pairs on F by setting $(\lambda_1, S_1) \leq (\lambda_2, S_2)$ if and only if λ_1 is coarser than λ_2 and $S_1 \leq S_2$.

LEMMA 28.3.1.
(a) *Any collection (λ_i, S_i), $i \in I$, of local pairs on F has a largest lower bound $(\mu, S) = \inf_i(\lambda_i, S_i)$.*
(b) *If each pair (λ_i, S_i) is compatible, then so is (μ, S).*

PROOF. For (a) take μ to be the finest common coarsening of the λ_i (Corollary 7.3.2), and take $S = \bigcap_{i \in I} S_i$.

For (b) suppose that each (λ_i, S) is compatible. Then

$$G_\mu \leq \bigcap_{i \in I} G_{\lambda_i} \leq \bigcap_{i \in I} S_i = S. \qquad \square$$

PROPOSITION 28.3.2. *Let u be a valuation on the field F and let (λ, S) be a local pair on F with λ finer than u. Then:*
(a) *if (λ, S) is compatible, then so are (u, S) and $(\lambda/u, \bar{S}_u)$;*
(b) *if λ is a valuation and (u, S) and $(\lambda/u, \bar{S}_u)$ are both compatible, then so is (λ, S).*

PROOF. (a) As $G_u \le G_\lambda \le S$, the pair (u,S) is compatible. By Proposition 7.2.1(a),

$$G_{\lambda/u} = \pi_u(G_\lambda \cap O_u^\times) \le \pi_u(S \cap O_u^\times) = \bar{S}_u,$$

so $(\lambda/u, \bar{S}_u)$ is compatible as well.

(b) By assumption, the kernel G_u of $\pi_u\colon O_u^\times \to \bar{F}_u^\times$ is contained in $S \cap O_u^\times$. Further, $G_\lambda \le O_\lambda^\times \le O_u^\times$, so

$$\pi_u(G_\lambda) = G_{\lambda/u} \le \bar{S}_u = \pi_u(S \cap O_u^\times).$$

Hence $G_\lambda \le S \cap O_u^\times \le S$, as desired. □

We say that a local pair (λ, S) is **degenerate** if either:

(i) λ is a valuation, $G_\lambda \le S$, and $F^\times = SO_\lambda^\times$; or
(ii) λ is an ordering and $S = F^\times$.

By definition, a degenerate local pair is compatible. If v is a valuation on F and (v,S) is degenerate, then $v(F^\times) = v(S)$, so Theorem 26.1.2 gives a *canonical* κ-structure isomorphism $K_*^M(F)/S \cong K_*^M(\bar{F}_v)/\bar{S}_v$. In particular, $F^\times/S \cong \bar{F}_v^\times/\bar{S}_v$ canonically. If P is an ordering and (P,S) is degenerate, then $K_*^M(F)/S = K_*^M(F)/F^\times \cong \mathbf{0}$. We further observe:

REMARK 28.3.3. If (λ,S) is degenerate and $S \le S' \le F^\times$, then (λ, S') is also degenerate.

In analogy with Proposition 28.3.2 we have:

PROPOSITION 28.3.4. *Let u be a valuation on the field F and let (λ,S) be a local pair on F with λ finer than u. Then (λ,S) is degenerate if and only if (u,S) and $(\lambda/u, \bar{S}_u)$ are both degenerate.*

PROOF. First, suppose that λ is an ordering. If (λ,S) is degenerate, then $S = F^\times$, whence $\bar{S}_u = \bar{F}_u^\times$, so both (u,S), $(\lambda/u,\bar{S}_u)$ are degenerate. Conversely, suppose that (u,S), $(\lambda/u,\bar{S}_u)$ are degenerate. Then $F^\times = SO_u^\times$, and $\bar{S}_u = \bar{F}_u^\times$. The latter equality means that $S \ge O_u^\times$, so by the former equality, $S = F^\times$. Therefore (λ,S) is degenerate, as desired. Next, if $G_u \not\le S$, then $G_\lambda \not\le S$. Therefore none of the pairs (λ,S), (u,S) is degenerate, and we are done again.

It remains to consider the case where λ is a valuation and $G_u \le S$. Then $G_u \le S \cap O_u^\times$, as well as $\bar{F}_u^\times = \pi_u(O_u^\times)$ and $\bar{S}_u O_{\lambda/u}^\times = \pi_u((S \cap O_u^\times)O_\lambda^\times)$. Therefore $\bar{F}_u^\times = \bar{S}_u O_{\lambda/u}^\times$ if and only if $O_u^\times = (S \cap O_u^\times)O_\lambda^\times$. Also, as $O_\lambda^\times \le O_u^\times$, we have $F^\times = SO_\lambda^\times$ if and only if both $F^\times = SO_u^\times$ and $O_u^\times = (S \cap O_u^\times)O_\lambda^\times$. The assertion follows from these two equivalences. □

PROPOSITION 28.3.5. *Let (u,S) be a degenerate local pair on the field F, with u a valuation. There is an order-preserving bijection*

$$(\mu, U) \mapsto (\mu/u, \bar{U}_u)$$

between the non-degenerate compatible local pairs (μ,U) on F with $(u,S) \le (\mu,U)$ and the non-degenerate compatible local pairs (μ^0,U^0) on $\bar{F}_u$ with $(0,\bar{S}_u) \le (\mu^0,U^0)$ (i.e., with $\bar{S}_u \le U^0$).

PROOF. We have already observed that $F^\times/S \cong \bar{F}_u^\times/\bar{S}_u$ canonically. Therefore the map $U \mapsto \bar{U}_u$ is an order-preserving bijection between the intermediate groups $S \le U \le F^\times$ and the intermediate groups $\bar{S}_u \le U^0 \le \bar{F}_u^\times$.

By Proposition 7.2.1, the map

$$\mu \mapsto \mu^0 = \mu/u$$

from the collection of all localities on F finer than u to the collection of all localities on $\bar{F}_u$ is order-preserving, and is a bijection on the (equivalence classes of) valuations. Since (u, S) is degenerate, Corollary 26.2.2 shows that this map is also a bijection between the orderings P containing S and the orderings $\bar{P}$ containing $\bar{S}_u$; in fact, $P = S\pi_u^{-1}(\bar{P})$.

Since (u, S) is degenerate (hence compatible), Proposition 28.3.2 and Proposition 28.3.4 imply that if (μ, U) is a non-degenerate compatible pair, then so is $(\mu^0, \bar{U}_u)$, and the converse is true when μ^0 is a valuation.

Finally, suppose that (μ^0, U^0) is a non-degenerate compatible pair, where $\bar{S}_u \le U^0$ and where $\mu^0 = \bar{P}$ is an ordering. Then $\bar{P} = G_{\mu^0} \le U^0 < \bar{F}_u^\times$, so necessarily $\bar{P} = U^0$. For the corresponding pair (μ, U) under the bijection above, both μ and U are the ordering $S\pi_u^{-1}(\bar{P})$. Observe that this pair is also non-degenerate and compatible. □

28.4 Arithmetical decompositions

A **minimal non-degenerate compatible system** on the field F consists of finitely many non-degenerate compatible local pairs

$$(\lambda_1, U_1), \dots, (\lambda_n, U_n)$$

on F such that there is no list of non-degenerate compatible local pairs

$$(\mu_1, W_1), \dots, (\mu_m, W_k)$$

on F with $\inf_i(\lambda_i, U_i) = \inf_j(\mu_j, W_j)$ and with $k < n$.

REMARKS 28.4.1.

(1) Any sublist of a minimal compatible non-degenerate system is also a minimal compatible non-degenerate system.

(2) Let $(\lambda_1, U_1), \dots, (\lambda_n, U_n)$ be non-degenerate compatible local pairs on F and let $S = U_1 \cap \dots \cap U_n$. For each $1 \le i \le n$ the pair $(\lambda_i, SG_{\lambda_i})$ is also a non-degenerate and compatible. Moreover,

$$S = (SG_{\lambda_1}) \cap \dots \cap (SG_{\lambda_n}),$$

so

$$\inf_i(\lambda_i, U_i) = \inf_i(\lambda_i, SG_{\lambda_i}).$$

Consequently, if $(\lambda_1, U_1), \dots, (\lambda_n, U_n)$ is a minimal non-degenerate compatible system on F then so is

$$(\lambda_1, SG_{\lambda_1}), \dots, (\lambda_n, SG_{\lambda_n}).$$

From Proposition 28.3.5 we deduce the following "push-down" property:

PROPOSITION 28.4.2. *Let (u, S) be a degenerate local pair on F, with u a valuation. Suppose that $(\lambda_1, U_1), \dots, (\lambda_n, U_n)$ is a minimal non-degenerate compatible system on F such that $(u, S) \leq (\lambda_i, U_i)$ for every i. Then*

$$(\lambda_1/u, (\overline{U_1})_u), \dots, (\lambda_n/u, (\overline{U_n})_u)$$

is a minimal non-degenerate compatible system on $\bar{F}_u$.

We can now state our main result. Its proof is modelled after [**Ef4**] and is similar in spirit to that of Theorem 10.2.1.

THEOREM 28.4.3. *Let F be a field, let*

$$(\lambda_1, U_1), \dots, (\lambda_n, U_n)$$

be a minimal non-degenerate compatible system on F, let $S = U_1 \cap \dots \cap U_n$, and suppose that $(F^\times)^m \leq S$ for some positive integer m. Then:

(a) *Res: $K_*^M(F)/S \to \prod_{i=1}^n K_*^M(F)/U_i$ is an isomorphism of κ-structures;*
(b) *$U_i = SG_{\lambda_i}$, $i = 1, \dots, n$.*

PROOF. (a) We argue by induction on n. When $n = 1$, the assertion is trivial. So suppose that $n \geq 2$. We have $(\rho, S) = \inf_i(\lambda_i, U_i)$ for some locality ρ on F. Since each local pair (λ_i, U_i) is compatible, so is (ρ, S) (Lemma 28.3.1(b)). The minimality of the system implies that (ρ, S) must be degenerate, and therefore (ρ, U_i) is degenerate for every i (Remark 28.3.3). But as (λ_i, U_i) is non-degenerate, $U_i \neq F^\times$, which implies that ρ is a valuation; we will henceforth denote it by u. We further set

$$\lambda_i^0 = \lambda_i/u, \quad S^0 = \bar{S}_u, \quad U_i^0 = (\overline{U_i})_u.$$

Note that $(\bar{F}_u^\times)^m = \pi_u((O_u^\times)^m) \leq S^0$. Since the local pairs $(u, U_1), \dots, (u, U_n)$ are degenerate, $\lambda_i \neq u$ for each i, so λ_i^0 is non-trivial. Furthermore, there is a canonical isomorphism $F^\times/S \xrightarrow{\sim} \bar{F}_u^\times/S^0$; it maps U_i/S onto U_i^0/S^0, $i = 1, \dots, n$. As $S = U_1 \cap \dots \cap U_n$ we have $S^0 = U_1^0 \cap \dots \cap U_n^0$.

Next let $\mathcal{T}_1, \dots, \mathcal{T}_k$ be the distinct topologies on $\bar{F}_u$ induced by $\lambda_1^0, \dots, \lambda_n^0$. For each $1 \leq j \leq k$ let $I(j)$ be the set of all $1 \leq i \leq n$ such that λ_i^0 induces $\mathcal{T}_j$. Let

$$(\mu_j^0, W_j^0) = \inf_{i \in I(j)} (\lambda_i^0, U_i^0).$$

By Proposition 10.1.5, μ_j^0 is non-trivial. By Lemma 8.1.3, $\mu_1^0, \dots, \mu_k^0$ induce on $\bar{F}_u$ the distinct topologies $\mathcal{T}_1, \dots, \mathcal{T}_k$, respectively. Also,

$$S^0 = U_1^0 \cap \dots \cap U_n^0 = W_1^0 \cap \dots \cap W_k^0.$$

By Proposition 28.3.2(a), each local pair (λ_i^0, U_i^0) is compatible. It follows from Lemma 28.3.1(b) that the local pairs (μ_j^0, W_j^0), $j = 1, \dots, k$, are also compatible. Therefore W_j^0 is $\mathcal{T}_{\mu_j^0}$-open. Corollary 28.2.4 now shows that

$$K_*^M(\bar{F}_u)/S^0 \cong \prod_{j=1}^k K_*^M(\bar{F}_u)/W_j^0 \tag{28.4.1}$$

via restriction.

We claim that $|I(j)| < n$ for every j. Otherwise $k = 1$ and μ_1^0 is a non-trivial valuation which is coarser than $\lambda_1^0, \dots, \lambda_n^0$. Proposition 7.2.1(d) then yields a valuation μ on F which is strictly finer than u such that $\mu_1^0 = \mu/u$, and $\lambda_1, \dots, \lambda_n$

are finer than μ. This contradicts the fact that u is the finest common coarsening of $\lambda_1, \dots, \lambda_n$, and thus proves the claim.

For every j, the local pairs (λ_i, U_i), $i \in I(j)$, form a minimal non-degenerate compatible system on F (Remark 28.4.1(1)). By Proposition 28.4.2, the pairs (λ_i^0, U_i^0), $i \in I(j)$, therefore form a minimal non-degenerate compatible system on $\bar{F}_u$. The induction hypothesis now implies that

$$K_*^M(\bar{F}_u)/W_j^0 \cong \prod_{i \in I(j)} K_*^M(\bar{F}_u)/U_i^0 \tag{28.4.2}$$

via restriction. Combining (28.4.1) and (28.4.2) we obtain that

$$K_*^M(\bar{F}_u)/S^0 \cong \prod_{j=1}^{k} \prod_{i \in I(j)} K_*^M(\bar{F}_u)/U_i^0 = \prod_{i=1}^{n} K_*^M(\bar{F}_u)/U_i^0 \tag{28.4.3}$$

via restriction.

Finally, since (u, S) is degenerate, so is each pair (u, U_i) (by Remark 28.3.3 again). Thus $u(F^\times) = u(S) = u(U_i)$. By Remark 26.1.4, there is a commutative square of canonical morphisms

$$\begin{array}{ccc} K_*^M(F)/S & \xrightarrow{\sim} & K_*^M(\bar{F}_u)/S^0 \\ {\scriptstyle \mathrm{Res}}\downarrow & & \downarrow{\scriptstyle \mathrm{Res}} \\ K_*^M(F)/U_i & \xrightarrow{\sim} & K_*^M(\bar{F}_u)/U_i^0. \end{array}$$

Consequently, the isomorphism (28.4.3) lifts to an isomorphism as required in (a).

(b) By Remark 28.4.1(2),

$$(\lambda_1, SG_{\lambda_1}), \dots, (\lambda_n, SG_{\lambda_n})$$

is also a minimal non-degenerate compatible system of local pairs. We may therefore apply to it part (a) to obtain a canonical isomorphism

$$K_*^M(F)/S \cong \prod_{i=1}^{n} K_*^M(F)/SG_{\lambda_i}.$$

When combined with the isomorphism of (a), this shows that

$$\prod \mathrm{Res}\colon \prod_{i=1}^{n} K_*^M(F)/SG_{\lambda_i} \to \prod_{i=1}^{n} K_*^M(F)/U_i$$

is also an isomorphism. In particular, the projection

$$\prod_{i=1}^{n} F^\times/SG_{\lambda_i} \to \prod_{i=1}^{n} F^\times/U_i$$

is an isomorphism, and the assertion follows. □

As a special case, we deduce the following result, which is essentially due to Jacob [**J1**] (for classical Milnor K-rings modulo 2). His proof is based on a formal analysis, due to Marshall [**Mar2**], of the so-called "finite spaces of orderings". It is then transferred to the arithmetical context using a predecessor of the explicit construction of the valuation ring $O(S, H)$, discussed in §11. Analogous results in the context of Witt rings of quadratic forms were obtained by Craven ([**Cr1**], [**Cr2**]). See also [**Min**] for a related result in the context of maximal pro-2 Galois groups.

COROLLARY 28.4.4. *Let S be a preordering on the field F with $(F^\times : S) < \infty$. Then there exist localities $\lambda_1, \ldots, \lambda_n$ on F such that*

$$K_*^M(F)/S \cong \prod_{i=1}^n K_*^M(F)/SG_{\lambda_i}$$

via restriction, and the local pairs $(\lambda_i, SG_{\lambda_i})$, $i = 1, \ldots, n$, are non-degenerate.

PROOF. Being a preordering, S is an intersection of orderings on F (Proposition 6.1.8). Since $(F^\times : S) < \infty$, it is in fact the intersection of finitely many orderings $P_1, \ldots, P_k$. By definition, the local pairs

$$(P_1, P_1), \ldots, (P_k, P_k)$$

are compatible and non-degenerate. Then $\inf_j(P_j, P_j) = (\mu, S)$ for some locality μ on F. We can choose a minimal non-degenerate compatible system

$$(\lambda_1, S_1), \ldots, (\lambda_n, S_n)$$

on F with $(\mu, S) = \inf_i(\lambda_i, S_i)$. By Theorem 28.4.3,

$$\mathrm{Res}\colon K_*^M(F)/S \to \prod_{i=1}^n K_*^M(F)/S_i$$

is an isomorphism of κ-structures, and $S_i = SG_{\lambda_i}$, $i = 1, \ldots, n$. □

CHAPTER 29

Realization of κ-Structures

In this final chapter we ask what κ-structures occur as generalized Milnor K-rings. While even a hypothetical answer to this question still seems to be beyond reach, there are some special cases where either the answer is known, or at least one has a clear conjecture. After developing in §29.1 some techniques for realizing basic κ-structure constructions, we give in §29.2 one of the main results of the reduced theory of ordered fields (in a K-theoretic language): the structure of fields F modulo a preordering S with $F^\times/S$ finite. The elementary type conjecture, which extends this result, is then briefly explained in §29.3.

29.1 Basic constructions

We say that a κ-structure A is **realizable** if there exists a field F and a subgroup S of $F^\times$ such that $A \cong K_*^M(F)/S$ as κ-structures.

29.1.1 EXAMPLES.

(1) The trivial κ-structure $\mathbf{0}$ is realizable as $K_*^M(F)/F^\times$ for any field F.

(2) κ is realizable as $K_*^M(\mathbb{R})/2^s$ for any $s \geq 1$, and also as $K_*^M(\mathbb{Q})/P$, where P is the unique ordering on $\mathbb{Q}$ (Theorem 25.1.2).

(3) Let $A = \bigoplus_{r=0}^\infty A_r$, where

$$A_0 = \mathbb{Z}, \quad A_1 = \{0, \epsilon_A\} \cong \mathbb{Z}/2, \quad A_2 = \{0, \epsilon_A^2\} \cong \mathbb{Z}/2,$$

and $A_r = 0$ for all $r \geq 3$. It is clearly a κ-structure. However, A is *not* realizable. Indeed, if $A \cong K_*^M(F)/S$ for some field F and a subgroup S of $F^\times$, then Theorem 25.1.2 would imply that S is an ordering. But then we would get that $A \cong \kappa$, contrary to $A_3 = 0$.

(4) More generally, suppose that F is a field, S is a subgroup of $F^\times$, and $K_r^M(F)/S \neq 0$ for some $r \geq 3$. Define a κ-structure $A = \bigoplus_{r=0}^\infty A_r$ by

$$A_0 = \mathbb{Z}, \quad A_1 = F^\times/S, \quad A_2 = K_2^M(F)/S,$$

$A_r = 0$ for $r \geq 3$, and with the same multiplication map $A_1 \times A_1 \to A_2$ as in $K_*^M(F)/S$. Then there are no field E and a subgroup T of $E^\times$ such that $K_*^M(E)/T \cong A$. Indeed, for such E and T the structures $K_*^M(F)/S$ and $K_*^M(E)/T$ would be isomorphic up to degree 2. But by Proposition 24.2.2, they would then be isomorphic as κ-structures – a contradiction.

Next we show that the class of realizable κ-structures is closed under some of the fundamental algebraic constructions studied in §23.2, starting with extensions.

PROPOSITION 29.1.2. *Let $\bar{F}$ be a field, $\bar{S}$ a subgroup of $\bar{F}^\times$, and Δ an abelian group which is generated by $\leq \alpha$ elements, where α is a cardinal number. Then there* *exists a purely transcendental field extension F of $\bar{F}$ such that* $\mathrm{tr.deg}(F/\bar{F}) = \alpha$, *and a subgroup S of $F^\times$ containing $\bar{S}$ such that*

$$K_*^M(F)/S \cong (K_*^M(\bar{F})/\bar{S})[\Delta]$$

as κ-structures.

PROOF. Considering α as an ordinal number, we may equip $\Gamma = \mathbb{Z}^\alpha$ with the lexicographic order with respect to the standard order on $\mathbb{Z}$. We fix a group epimorphism $\Gamma \to \Delta$ and denote its kernel by Γ_0.

Now let $F = \bar{F}(\Gamma)$ be the field of generalized rational functions over $\bar{F}$ with exponents in Γ (§2.9) and let v be its canonical valuation as in Example 4.2.4. Recall that $v(F^\times) = \Gamma$ and $\bar{F}_v = \bar{F}$. By the usual abuse of notation, t^γ will stand for the map in $\bar{F}(\Gamma)$ with value 1 at γ and value 0 elsewhere. Likewise, we identify the elements of $\bar{F}$ as the maps in F with support in $\{0\}$. By Proposition 2.9.1 and Remark 2.9.2, F is purely transcendental over $\bar{F}$ and

$$\mathrm{tr.deg}(F/\bar{F}) = \mathrm{rr}(\Gamma) = \dim_{\mathbb{Q}}(\mathbb{Q}^\alpha) = \alpha.$$

Next let

$$S = \{t^\gamma \mid \gamma \in \Gamma_0\} \cdot G_v \cdot \bar{S}.$$

Then $\bar{S}_v = \bar{S}$ and $v(S) = \Gamma_0$. Therefore Theorem 26.1.2 and Example 26.1.1(5) give a κ-structure isomorphism

$$K_*^M(F)/S \cong (K_*^M(\bar{F}_v)/\bar{S}_v)[v(F^\times)/v(S)] \cong (K_*^M(\bar{F})/\bar{S})[\Delta]. \qquad \square$$

Taking $\Delta = 0$ in the previous construction we obtain:

COROLLARY 29.1.3. *Let $\bar{F}$ be a field, let $\bar{S}$ be a subgroup of $\bar{F}^\times$, and let α be a cardinal number. There exists a purely transcendental field extension F of $\bar{F}$ with* $\mathrm{tr.deg}(F/\bar{F}) = \alpha$, *as well as a subgroup S of $F^\times$ containing $\bar{S}$, such that*

$$K_*^M(F)/S \cong K_*^M(\bar{F})/\bar{S}.$$

LEMMA 29.1.4. *Let F_1 be a finite separable extension of a field F_0 and let $S_1 \leq F_1^\times$. Then there exists a subgroup S of $F_0(t)^\times$ such that*

$$K_*^M(F_0(t))/S \cong K_*^M(F_1)/S_1.$$

PROOF. Suppose that F_1 is obtained from F_0 by adjoining a root of the nonconstant polynomial $g(X) \in F_0[X]$. Let v be the $g(t)$-adic valuation on $F_0(t)$ (Example 4.1.3). Recall that v is a discrete valuation with uniformizer $g(t)$ and with residue field F_1. For

$$S = \langle g(t) \rangle \cdot \pi_v^{-1}(S_1)$$

we therefore have $v(S) = v(F(t)^\times)$ and $\bar{S}_v = S_1$. Moreover, $G_v \leq S$, so Theorem 26.1.2 gives the desired isomorphism. $\square$

Direct products of κ-structures can now be realized as follows:

THEOREM 29.1.5. *Let $F_1, \dots, F_n$ be purely transcendental extensions of a common infinite subfield F_0. Let S_0 be a subgroup of $F_0^\times$, and for each $1 \le i \le n$ let S_i be a subgroup of $F_i^\times$ containing S_0 and $(F_i^\times)^m$, where m is a fixed positive integer. Then there exist a purely transcendental extension F of F_0 and a subgroup S of $F^\times$ containing S_0 and $(F^\times)^m$ such that*

$$K_*^M(F)/S \cong \prod_{i=1}^n K_*^M(F_i)/S_i$$

and

$$\mathrm{tr.deg}(F/F_0) = \max_{1 \le i \le n} \mathrm{tr.deg}(F_i/F_0) + 1.$$

PROOF. In light of Corollary 29.1.3 we may replace $F_1, \dots, F_n$ by some purely transcendental extensions to assume that the transcendence degrees $\mathrm{tr.deg}(F_i/F_0)$, $i = 1, \dots, n$, are all equal. We may therefore assume, without loss of generality, that $F_1 = \cdots = F_n$.

Now let t be a transcendental element over F_1, let $F = F_1(t)$, and choose distinct elements $c_1, \dots, c_n$ of F_1. For $1 \le i \le n$ let v_i be the $(t - c_i)$-adic valuation on F (Example 4.1.3). It is discrete with uniformizer $t - c_i$ and has residue field F_1. As before, Theorem 26.1.2 gives

$$K_*^M(F)/\langle t - c_i \rangle \pi_{v_i}^{-1}(S_i) \cong K_*^M(F_1)/S_i = K_*^M(F_i)/S_i.$$

We take

$$S = \bigcap_{i=1}^n \Big(\langle t - c_i \rangle \pi_{v_i}^{-1}(S_i) \Big).$$

Note that $S_i \le S$ for each i, whence $S_0 \le S$. Furthermore, for each i one has $(O_{v_i}^\times)^m \le \pi_{v_i}^{-1}(S_i)$. Since $F^\times = \langle t - c_i \rangle O_{v_i}^\times$, this implies that $(F^\times)^m \le S$.

Since $v_1, \dots, v_n$ are distinct and discrete, they are independent. Corollary 28.2.4 therefore gives a κ-structure isomorphism

$$K_*^M(F)/S \cong \prod_{i=1}^n K_*^M(F)/\langle t - c_i \rangle \pi_{v_i}^{-1}(S_i) \cong \prod_{i=1}^n K_*^M(F_i)/S_i. \qquad \square$$

A similar construction, in the category of group schemes, was given by M. Kula [**Kul**].

29.2 K-rings modulo preorderings of finite index

The next result originates from results of Craven [**Cr1**] in quadratic form theory, and of Marshall in the more formal category of spaces of orderings ([**Mar2**], [**Mar4**]). It is based on Jacob's theorem (Corollary 28.4.4) and the constructions of the previous section.

Let $\mathcal{A}$ be the minimal class of (isomorphism types of) κ-structures such that:

(i) $\kappa \in \mathcal{A}$;
(ii) if $A, B \in \mathcal{A}$, then $A \times B \in \mathcal{A}$;
(iii) if $A \in \mathcal{A}$ and Δ is a finite abelian 2-group, then also $A[\Delta] \in \mathcal{A}$.

For every $A \in \mathcal{A}$ the group A_1 is non-trivial of exponent 2 (in fact $\epsilon_A \neq 0$), and the Bockstein map

$$\mathrm{Bock}_A \colon A_1 \to A_2, \quad a \mapsto a^2,$$

is injective. Indeed, this is obvious for $A = \kappa$, and follows inductively for direct products and for extensions as above using Lemma 23.3.4.

THEOREM 29.2.1. *The following conditions on a non-trivial κ-structure A are equivalent:*

(a) $A \in \mathcal{A}$;

(b) *there exists a purely transcendental extension F of $\mathbb{Q}$ of finite transcendence degree, as well as a preordering S on F such that $(F^\times : S) < \infty$ and $A \cong K_*^M(F)/S$;*

(c) *there is a field F and a preordering S on F such that $(F^\times : S) < \infty$ and $A \cong K_*^M(F)/S$.*

PROOF. (a)$\Rightarrow$(b): Given $A \in \mathcal{A}$, it is enough to find a field F as above and a subgroup S of $F^\times$ with $A \cong K_*^M(F)/S$; indeed, then necessarily $(F^\times : S) = |A_1| < \infty$, and the previous remark and Proposition 25.1.4 imply that S is a preordering on F.

Now to construct such a subgroup we argue by induction on the construction of A. For $A = \kappa$ we take $F = \mathbb{Q}$ and take S to be the unique ordering on $\mathbb{Q}$ (see Example 29.1.1(2)). For direct products we use Theorem 29.1.5 with $m = 2$. For extensions by finite 2-groups we apply Proposition 29.1.2.

(b)$\Rightarrow$(c): Trivial.

(c)$\Rightarrow$(a): We argue by induction on $(F^\times : S)$.

Corollary 28.4.4 yields localities $\lambda_1, \dots, \lambda_m$ on F such that

$$K_*^M(F)/S \cong \prod_{i=1}^{n} K_*^M(F)/SG_{\lambda_i} \tag{29.2.1}$$

via restriction, and the local pairs $(\lambda_i, SG_{\lambda_i})$, $i = 1, \dots, n$, are non-degenerate. It now suffices to show that $K_*^M(F)/SG_{\lambda_i} \in \mathcal{A}$ for every i.

If $\lambda_i = P$ is an ordering, then $SG_{\lambda_i} = P$, so by Theorem 25.1.2,

$$K_*^M(F)/SG_{\lambda_i} \cong \kappa$$

and we are done.

Next suppose that $\lambda_i = u$ is valuation. As (u, SG_u) is non-degenerate, $u(F^\times) \neq u(S)$. Since $F^\times/S \cong A_1$ has exponent 2, Theorem 26.1.2 and Example 26.1.1(2) give

$$K_*^M(F)/SG_u \cong (K_*^M(\bar{F}_u)/\bar{S}_u)[u(F^\times)/u(S)].$$

In particular,

$$(F^\times : S) \geq (F^\times : SG_u) = (\bar{F}_u^\times : \bar{S}_u)(u(F^\times) : u(S)) > (\bar{F}_u^\times : \bar{S}_u).$$

As $\{-1\}_S \neq 0$, it follows from the κ-structure isomorphism (29.2.1) that $\{-1\}_{SG_u} \neq 0$, i.e., $-1 \notin SG_u$. Therefore Lemma 7.1.3 shows that $\bar{S}_u$ is a preordering. The induction hypothesis implies that $K_*^M(\bar{F}_u)/\bar{S}_u \in \mathcal{A}$. Since $u(F^\times)/u(S)$ is a finite 2-group, we conclude that the extension $K_*^M(F)/SG_u$ is also in $\mathcal{A}$. □

As an application, we now recover a result of Bröcker [**Br2**], which counts the number of orderings on a field F containing a given preordering S of finite index in $F^\times$. To this end we denote the number of morphisms from a κ-structure A to κ by $\nu(A)$ (possibly $\nu(A) = \infty$). Recall that when $A = K_*^M(F)/S$ these epimorphisms correspond bijectively to the orderings on F containing S (Corollary 25.1.3). Thus, in view of Theorem 29.2.1, the problem is to compute $\nu(A)$ for $A \in \mathcal{A}$. This value is given by the following proposition.

PROPOSITION 29.2.2.
(a) $\nu(\kappa) = 1$;
(b) *for κ-structures A, B one has $\nu(A \times B) = \nu(A) + \nu(B)$;*
(c) *for a κ-structure A one has $\nu(A[\mathbb{Z}/2]) = 2\nu(A)$.*

PROOF. (a) is immediate.

(b) We have structural epimorphisms

$$\pi_A : A \times B \to A, \quad \pi_B : A \times B \to B.$$

From every κ-structure morphism $A \to \kappa$ (resp., $B \to \kappa$) we therefore obtain by composition with π_A (resp., π_B) a morphism $A \times B \to \kappa$. This gives $\nu(A) + \nu(B)$ such morphisms.

Conversely, let $\varphi : A \times B \to \kappa$ be a morphism. If there were $a \in A_1$ and $b \in B_1$ with $\varphi_1((a, 0)), \varphi_1((0, b)) \neq 0$, then $\varphi_1((a, 0)), \varphi_1((0, b))$ were necessarily both ϵ, whence

$$0 \neq \epsilon^2 = \varphi_1((a, 0)) \cdot \varphi_1((0, b)) = \varphi_2((a, 0) \cdot (0, b)) = \varphi_2((0, 0)) = 0,$$

a contradiction. Consequently, φ is trivial on either A_1 or B_1, so it breaks via π_A or π_B, i.e., it is one of the $\nu(A) + \nu(B)$ morphisms constructed above.

(c) This follows from the universal property of extensions (Lemma 23.2.2); see also the proof of the Baer–Krull correspondence (Theorem 26.2.1). □

29.3 κ-structures of elementary type

We now relax the assumption that S is a preordering, and require only that it contains $(F^\times)^q$ for some fixed prime power $q = p^s$, where $s \geq 1$. Then many more κ-structures can be realized. In fact, consider the minimal class $\mathcal{E}_q$ of (isomorphism types of) κ-structures such that:

(i) $K_*^M(F)/(F^\times)^q \in \mathcal{E}_q$ whenever F is $\mathbb{C}$, $\mathbb{R}$, or is a finite extension of $\mathbb{Q}_p$ containing a root of unity of order q;
(ii) if $A, B \in \mathcal{E}_q$, then also $A \times B \in \mathcal{E}_q$;
(iii) if $A \in \mathcal{E}_q$ and Δ is a finite abelian group with $q\Delta = 0$, then also $A[\Delta] \in \mathcal{E}_q$.

Of course, $K_*^M(\mathbb{C})/(\mathbb{C}^\times)^q$ is trivial, and $K_*^M(\mathbb{R})/(\mathbb{R}^\times)^q$ is κ if $p = 2$ and is trivial otherwise. We have also seen that the structures $K_*^M(F)/(F^\times)^q$, for a finite extension F of $\mathbb{Q}_p$ containing a root of unity of order q are q-Demuškin (Theorem 27.1.1). Note that every $A \in \mathcal{E}_q$ satisfies $qA_1 = 0$. We call $\mathcal{E}_q$ the class of κ-structures of **q-elementary type**. One has the following partial generalization of Theorem 29.2.1.

THEOREM 29.3.1. *Each of the following conditions on a κ-structure A implies the next one:*

(a) $A \in \mathcal{E}_q$;

(b) *there exists a purely transcendental extension F of $\mathbb{Q}$ of finite transcendence degree, as well as a subgroup S of $F^\times$ containing $(F^\times)^q$, such that $(F^\times : S) < \infty$ and*
$$A \cong K_*^M(F)/S;$$

(c) *there is a field F and a subgroup S of $F^\times$ containing $(F^\times)^q$ such that $(F^\times : S) < \infty$ and*
$$A \cong K_*^M(F)/S.$$

PROOF. (a)$\Rightarrow$(b): We have already seen in Example 29.1.1 that $\mathbf{0}$ and κ are realizable over the field $\mathbb{Q}$. Hence $K_*^M(\mathbb{C})/(\mathbb{C}^\times)^q$ and $K_*^M(\mathbb{R})/(\mathbb{R}^\times)^q$ are realizable as required in (b).

Next let F be a finite extension of $\mathbb{Q}_p$ containing a root of unity of order q. It follows from the Krasner-Ostrowski lemma that F is the compositum $F_1\mathbb{Q}_p$ for some finite extension F_1 of $\mathbb{Q}$ (see Example 18.5.4(1)). Hence F is the completion of F_1 with respect to the canonical p-adic valuation. Recall that $(F^\times)^q$ is open in $F^\times$ (Proposition 18.2.1). Let $S_0 = F_0^\times \cap (F^\times)^q$. By Proposition 24.2.1,
$$\mathrm{Res}\colon K_*^M(F_0)/S_0 \to K_*^M(F)/(F^\times)^q$$
is an isomorphism. Moreover, Lemma 29.1.4 yields a subgroup S of $\mathbb{Q}(t)^\times$ such that
$$K_*^M(\mathbb{Q}(t))/S \cong K_*^M(F_0)/S_0.$$
In particular, $\mathbb{Q}(t)^\times/S \cong F^\times/q$, so $(\mathbb{Q}(t)^\times : S) < \infty$ and $(\mathbb{Q}_p(t)^\times)^q \leq S$.

This shows that all the structures in (i) above are realizable as is required in (b).

Next we realize direct products. To this end assume that $A, B \in \mathcal{E}_q$ are realizable as in (b). Then $qA_1 = qB_1 = 0$. By Theorem 29.1.5, $A \times B$ is also realizable as $K_*^M(F)/S$ for some purely transcendental extension F of $\mathbb{Q}$ of finite transcendence degree and some subgroup S of $F^\times$. Furthermore, since $A_1 \oplus B_1$ is finite and $q(A_1 \oplus B_1) = 0$, we must have $(F^\times : S) < \infty$ and $(F^\times)^q \leq S$.

Similarly, the class of structures A satisfying (b) is closed under extensions by finite abelian groups Δ with $q\Delta = 0$, by Proposition 29.1.2.

(b)$\Rightarrow$(c): Trivial. □

In view of Theorem 29.2.1, it would be interesting to know if conditions (a)–(c) of Theorem 29.3.1 are in fact equivalent, say, for $q = p$ prime. This would mean in particular that every structure of the form $K_*^M(F)/(F^\times)^p$ with $(F^\times : (F^\times)^p) < \infty$ is of p-elementary type. A variant of this question for $p = 2$ was posed by Marshall [**Mar1**] in the context of "abstract Witt rings". The conjecture that this is indeed so has been studied by many authors under the name "*the elementary type conjecture*". Jacob and Ware transformed it to the context of maximal pro-2 Galois groups of fields ([**JWr1**], [**JWr2**]). A stronger pro-p arithmetical version of the conjecture has been stated in [**Ef2**] (see also [**Ef5**, Question 4.8]), and has been verified for many important families of fields. See [**Mar5**] for further details.

Bibliography

[Ab] S. Abhyankar, *Two notes on formal power series*, Proc. Amer. Math. Soc. **7** (1956), 903–905.

[AEJ] J.K. Arason, R. Elman and B. Jacob, *Rigid elements, valuations, and realization of Witt rings*, J. Algebra **110**, 449–467.

[Ar] E. Artin, *Über die Zerlegung definiter Funktionen in Quadrate*, Abh. Math. Sem. Univ. Hamburg **5** (1927), 100–115.

[AS1] E. Artin and O. Schreier, *Algebraische Konstruktion reeller Körper*, Abh. Math. Sem. Univ. Hamburg **5** (1927), 85–99.

[AS2] E. Artin and O. Schreier, *Eine Kennzeichnung der reell abgeschlosenen Körper*, Abh. Math. Sem. Univ. Hamburg **5**, 225–231.

[Ax] J. Ax, *A metamathematical approach to some problems in number theory*, In: Proc. Sympos. Pure Math. **20** (Institute on Number Theory, Stony Brook, N. Y., 1969), Amer. Math. Soc., Providence, R.I., 1971, pp. 161–190.

[Ba] R. Baer, *Über nicht archimedisch geordnete Körper*, Sitz. Ber. der Heidelberger Akad. Abh. (1927), 3–13.

[BaT] H. Bass and J. Tate, *The Milnor ring of a global field*, In: Algebraic K-Theory (Battelle Institute Conference 1972), H. Bass (ed.), Vol. II, Springer Lect. Notes Math. 342, 1973, pp. 349–446.

[Be1] E. Becker, *Euklidische Körper und euklidische Hüllen von Körpern*, J. reine angew. Math. **268–269** (1974), 41–52.

[Be2] E. Becker, *Partial orders on a field and valuation rings*, Comm. Algebra **18** (1976), 1932–1976.

[Be3] E. Becker, *Hereditarily-pythagorean Fields and Orderings of higher Level*, IMPA Lect. Notes **29**, IMPA, Rio de Janeiro, 1978.

[BeBG] E. Becker, R. Berr, and D. Gondard, *Valuation fans and residually real closed Henselian fields*, J. Algebra **215** (1999), 574–602.

[BeK] E. Becker and E. Köpping, *Reduzierte quadratische Formen und Semiordnungen reeller Körper*, Abh. Math. Sem. Univ. Hamburg **46** (1977), 143–177.

[BeS] E. Becker and K.-J. Spitzlay, *Zum Satz von Artin–Schreier über die Eindeutigkeit des reellen Abschlusses eines angeordneten Körpers*, Comment. Math. Helv. **50** (1975), 81–87.

[BoT] F. Bogomolov and Y. Tschinkel, *Reconstruction of function fields*, a preprint, 2003.

[Bou1] N. Bourbaki, *Elements of Mathematics, Commutative Algebra*, Chapters 1–7, Springer-Verlag, Berlin, 1989.

[Bou2] N. Bourbaki, *Elements of Mathematics, Algebra II*, Springer-Verlag, Berlin, 1990.

[Br1] L. Bröcker, *Characterization of fans and hereditarily pythagorean fields*, Math. Z. **151** (1976), 149–163.

[Br2] L. Bröcker, *Über die Anzahl der Anordnungen eines kommutativen Körpers*, Arch. Math. **31** (1978), 133–136.

[Ch] C. Chevalley, *Algebraic Functions of one Variable*, Math. Surveys VI, Amer. Math. Soc., New York, 1951.

[Cr1] T.C. Craven, *Characterizing reduced Witt rings of fields*, J. Algebra **53** (1978), 68–77.

[Cr2] T.C. Craven, *Characterizing reduced Witt rings II*, Pac. J. Math. **80** (1979), 341–349.

[D] M. Deuring, *Verzweigungstheorie bewerteter Körper*, Math. Ann. **105** (1931), 277–307.

[Di] V. Diekert, *Über die absolute Galoisgruppe dyadischer Zahlkörper*, J. reine angew. Math. **350** (1984), 152–172.

[Ef1] I. Efrat, *A Galois-theoretic characterization of p-adically closed fields*, Israel J. Math. **91** (1995), 273–284.

[Ef2] I. Efrat, *Orderings, valuations and free products of Galois groups*, In: Séminaire de Structures Algébriques Ordonnées, Lecture Notes No. **54**, University of Paris VII, 1995.

[Ef3] I. Efrat, *Finitely generated pro-p Galois groups of p-Henselian fields*, J. Pure Appl. Algebra **138** (1999), 215–228.

[Ef4] I. Efrat, *Free pro-p product decompositions of Galois groups*, Math. Z. **225** (1997), 245–261.

[Ef5] I. Efrat, *Pro-p Galois groups of algebraic extensions of* $\mathbb{Q}$, J. Number Theory **64** (1997), 84–99.

[Ef6] I. Efrat, *Construction of valuations from K-theory*, Math. Res. Letters **6** (1999), 335–344.

[Ef7] I. Efrat, *The local correspondence over absolute fields – an algebraic approach*, Intern. Math. Res. Notices **2000:23** (2000), 1213–1223.

[Ef8] I. Efrat, *Demuškin fields with valuations*, Math. Z. **243** (2003), 333–353.

[Ef9] I. Efrat, *Quotients of Milnor K-rings, orderings and valuations*, Pac. J. Math. (to appear).

[Ef10] I. Efrat, *Compatible valuations and generalized Milnor K-theory*, Trans. Amer. Math. Soc. (to appear).

[EfF] I. Efrat and I. Fesenko, *Fields Galois-equivalent to a local field of positive characteristic*, Math. Res. Letters **6** (1999), 345–356.

[En] O. Endler, *Valuation Theory*, Springer-Verlag, Berlin, 1972.

[EnEg] O. Endler and A.J. Engler, *Fields with Henselian valuation rings*, Math. Z. **152** (1977), 191–193.

[Eg1] A.J. Engler, *Fields with two incomparable Henselian valuation rings*, manuscripta math. **23** (1978), 373–385.

[Eg2] A.J. Engler, *Totally real rigid elements and* F_π*-henselian valuation rings*, Comm. Algebra **25** (1997), 3673–3697.

[FV] I. Fesenko and S.V. Vostokov, *Local Fields and their Extensions – A Constructive Approach*, Amer. Math. Soc., Providence, RI, 2002.

[FK] I. Fesenko and M. Kurihara (eds.), *Invitation to higher local fields* (Conference Proceedings, Münster, August–September 1999), Geometry & Topology Publications, Coventry, 2000.

[FJr] M. Fried and M. Jarden, *Field Arithmetic*, Springer-Verlag, Heidelberg, 2005.

[Fu] L. Fuchs, *Abelian groups*, Pergamon Press, New York–Oxford–London–Paris, 1960.

[G1] A. Grothendieck, *Esquisse d'un program*, In: Geometric Galois Actions: 1. Around Grothendieck's esquisse d'un programme, L. Schneps et al. (eds.), Lect. Note Ser. **242**, Cambridge University Press, Lond. Math. Soc., 1997, pp. 5–48.

[G2] A. Grothendieck, *A letter to G. Faltings*, In: Geometric Galois Actions: 1. Around Grothendieck's esquisse d'un programme, L. Schneps et al. (eds.), Lect. Note Ser. **242**, Cambridge University Press, Lond. Math. Soc., 1997, pp. 49–58.

[Ha] H. Hahn, *Über die nichtarchimedischen Grössensysteme*, Sitz.-Ber. d. Wiener Akad., Math.-Nat. Klasse, Abt. IIa **116** (1907), 601–653.

[Has] H. Hasse, *Zahlentheorie*, Akademie-Verlag, Berlin, 1949.

[HasS] H. Hasse and F.K. Schmidt, *Die Struktur discret bewerteter Körper*, J. reine angew. Math. **170** (1933), 4–63.

[He] K. Hensel, *Theorie der algebraischer Zahlen*, Teubner, Leipzig, 1908.

[Hi] D. Hilbert, *Grundlagen der Geometrie*, B.G. Teubner, Stuttgart, 1987.

[Hö] O. Hölder, *Die Axiome der Quantität und die Lehre vom Maß*, Ber. Verh. Sächs. Ges. Wiss. Leipzig, Math.-Phys. Cl. **53** (1901), 1–64; J. Math. Psych. **40** (1996), 235–252 (English translation).

[HwJ] Y.S. Hwang and B. Jacob, *Brauer group analogues of results relating the Witt ring to valuations and Galois theory*, Canad. J. Math. **47** (1995), 527–543.

[I] K. Iwasawa, *On Galois groups of local fields*, Trans. Amer. Math. Soc. **80** (1955), 448–469.

[J1] B. Jacob, *On the structure of pythagorean fields*, J. Algebra **68** (1981), 247–267.

[J2] B. Jacob, *Fans, real valuations, and hereditarily-Pythagorean fields*, Pac. J. Math **93** (1981), 95–105.

[JWd] B. Jacob and A. Wadsworth, *A new construction of noncrossed product algebras*, Trans. Amer. Math. Soc. **293** (1986), 693–722.

[JWr1] B. Jacob and R. Ware, *A recursive description of the maximal pro-2 Galois group via Witt rings*, Math. Z. **200** (1989), 379–396.

[JWr2] B. Jacob and R. Ware, *Realizing dyadic factors of elementary type Witt rings and pro-2 Galois groups*, Math. Z. **208** (1991), 193–208.

[Jak1] A.V. Jakovlev, *The Galois group of the algebraic closure of a local field*, Izv. Akad. Nauk SSSR, Ser. Mat. **32** (1968), 1283–1322 (Russian); Math. USSR, Izv. **2** (1968), 1231–1269 (English Translation).

[Jak2] A.V. Jakovlev, *Remarks on my paper: "The Galois group of the algebraic closure of a local field"*, Izv. Akad. Nauk SSSR, Ser. Mat. **42** (1978), 212–213 (Russian); Math. USSR Izv. **12** (1978), 205–206 (English Translation).

[Jn] U. Jannsen, *Über Galoisgruppen lokaler Körper*, Invent. Math. **70** (1982/83), 53–69.

[JnW] U. Jannsen and K. Wingberg, *Die Struktur der absoluten Galoisgruppe $\mathfrak{p}$-adischer Zahlkörper*, Invent. Math. **70** (1982/83), 71–98.

[Jr] M. Jarden, *Intersections of local algebraic extensions of a Hilbertian field*, In: Generators and Relations in Groups and Geometries (Lucca 1990), A. Barlotti et al. (eds.), NATO Adv. Sci. Inst. Ser. C Math. Phys. Sci. **333**, Kluwer, Dordrecht, 1991, pp. 343–405.

[Kah] B. Kahn, *La conjecture de Milnor (d'aprés V. Voevodsky)*, Sém. Bourbaki 1996/97, Astérisque **245** (1997), 379–418.

[Kap1] I. Kaplansky, *Maximal fields with valuations I*, Duke J. Math. **9** (1942), 303–321.

[Kap2] I. Kaplansky, *Infinite abelian groups*, The University of Michigan Press, Ann Arbor, Michigan, 1969.

[Ked] K.S. Kedlaya, *The algebraic closure of the power series field in positive characteristic*, Proc. Amer. Math. Soc. **129** (2001), 3461–3470.

[Ker] I. Kersten, *Brauergruppen von Körpern*, Vieweg, Braunschweig, 1990.

[Kn] M. Knebusch, *On the extension of real places*, Comment. Math. Helv. **48** (1973), 354–369.

[KnS] M. Knebusch and C. Scheiderer, *Einführung in die reelle Algebra*, Vieweg, Braunschweig, 1989.

[Ko1] H. Koch, *Über die Galoissche Gruppe der algebraischen Abschliessung eines Potenzreihenkörpers mit endlichem Konstantenkörper*, Math. Nachr. **35** (1967), 323–327.

[Ko2] H. Koch, *The Galois group of a p-closed extension of a local field*, Dokl. Akad. Nauk SSSR **238** (1978), 19–22 (Russian); Soviet Math. Dokl. **19** (1978), 10–13 (English translation).

[Ko3] H. Koch, *Galois theory of p-extensions*, Springer-Verlag, Berlin, 2002.

[Koe1] J. Koenigsmann, *From p-rigid elements to valuations (with a Galois-characterisation of p-adic fields)* (with an appendix by F. Pop), J. reine angew. Math. **465** (1995), 165–182.

[Koe2] J. Koenigsmann, *Encoding valuations in absolute Galois groups*, In: Valuation theory and its applications, Vol. II (Saskatoon 1999), F.-V. Kuhlmann et al. (eds.), Fields Inst. Commun. **33**, Amer. Math. Soc., Providence, RI, 2003, pp. 107–132.

[Kra] M. Krasner, *Sém. d'Algébre*, Inst. Henri Poincaré, Paris, 1964.

[KreN] M.G. Krein and M.A. Naimark, *The method of symmetric and Hermitian forms in the theory of separation of the roots of algebraic equations*, Kharkov, 1936 (Russian); Lin. Multilin. Algebra **10** (1981), 265–308 (English translation).

[Kru1] W. Krull, *Galoissche Theorie bewerteter Körper*, Sitz.-Ber. d. Bayerischen Akademie, Math.-nat. Abteilung (1930), 225–238.

[Kru2] W. Krull, *Allgemeine Bewertungstheorie*, J. reine angew. Math. **167** (1931), 160–196.

[Ku] F.-V. Kuhlmann, *Henselian function fields and tame fields*, a manuscript, Heidelberg, 1990.

[KuPR] F.-V. Kuhlmann, M. Pank and P. Roquette, *Immediate and purely wild extensions of valued fields*, manuscripta math. **55** (1986), 39–67.

[KuKM] F.-V. Kuhlmann, S. Kuhlmann and M. Marshall (eds.), *Valuation theory and its applications* (Saskatoon 1999), Fields Inst. Commun. **32–33**, Amer. Math. Soc., Providence, RI, 2002/3.

[Kul] M. Kula, *Fields with prescribed quadratic form schemes*, Math. Z. **167** (1979), 201–212.

[Kür] J. Kürschák, *Über Limesbildung und allgemeine Körpertheorie*, J. reine angew. Math. **142** (1913), 211-253.

[Lam1] T.Y. Lam, *The theory of ordered fields*, In: Ring Theory and Algebra III (Norman, Oklahoma, 1979), B.R. MacDonald (ed.), Lecture Notes in Pure and Applied Math. **55**, Marcel Dekker, New York/Basel, 1980, pp. 1–152.

[Lam2] T.Y. Lam, *Orderings, valuations and quadratic forms*, Conf. Board of the Mathematical Sciences **52**, Amer. Math. Soc., Providence, RI, 1983.

[Lam3] T.Y. Lam, *Introduction to Quadratic Forms over Fields*, Amer. Math. Soc., Providence, RI, 2005.

[Lg] S. Lang, *Algebra*, Addison–Wesley Publishing Company, Reading, Massachusetts, 1984.

[Len] H. Lenstra, *Construction of the ring of Witt vectors*, a manuscript, 2002.

[Lz1] M. Lazard, *Détermination des anneaux p-adiques et π-adiques dont les anneaux des restes sont parfaits*, Sém. Krasner **9**, Fac. Sci. Paris, Paris, 1953/4.

[Lz2] M. Lazard, *Bemerkungen zur Theorie der bewerteten Körper und Ringe*, Math. Nach. **12** (1954), 67–73.

[Mac1] S. MacLane, *Subfields and automorphisms of p-adic fields*, Ann. Math. **40** (1939), 423–442.

[Mac2] S. MacLane, *The universality of formal power series*, Bull. Amer. Math. Soc. **45** (1939), 888–890.

[Mar1] M. Marshall, *Abstract Witt Rings*, Queen's Pap. Pure Appl. Math. **57**, Kingston, 1980.

[Mar2] M. Marshall, *Spaces of orderings IV*, Canad. J. Math. **32** (1980), 603–627.

[Mar3] M. Marshall, *An approximation theorem for coarse V-topologies on rings*, Canad. Math. Bull. **37** (1994), 527–533.

[Mar4] M. Marshall, *Spaces of Orderings and Abstract Real Spectra*, Springer Lect. Notes Math. **1636**, Springer-Verlag, Berlin–Heidelberg, 1996.

[Mar5] M. Marshall, *The elementary type conjecture in quadratic form theory*, Cont. Math. **344** (2004), 275–293.

[Mat] H. Matsumura, *Commutative algebra*, W.A. Benjamin, Inc., New York, 1970.

[MelS] O.V. Mel'nikov and A.A. Sharomet, *The Galois group of a multidimensional local field of positive characteristic*, Matem. Sbornik **180** (1989), 1132–1146 (Russian); Math. USSR-Sb. **67** (1990), 595–610 (English translation).

[MerS1] A.S. Merkur'ev and A.A. Suslin, *K-cohomology of Brauer-Severi varieties and the norm residue homomorphism*, Izv. Akad. Nauk SSSR, Ser. Mat. **46** (1982), 1011–1046 (Russian); Math. USSR Izv. **21** (1983), 307–340 (English translation).

[MerS2] A.S. Merkurjev and A.A. Suslin, *The Norm residue symbol of degree* 3, Izv. Akad. Nauk SSSR, Ser. Mat. **54** (1990), 339–356 (Russian); Math. USSR Izv. **36** (1991), 346–368 (English translation).

[Mi] J. Milnor, *Algebraic K-theory and quadratic forms*, Invent. math. **9** (1970), 318–344.

[Min] J. Mináč, *Galois groups of some 2-extensions of ordered fields*, C.R. Math. Rep. Acad. Sci. Canada **8** (1986), 103–108.

[N1] J. Neukirch, *Eine algebraische Kennzeichnung der Henselkörper*, J. reine angew. Math. **231** (1968), 75–81.

[N2] J. Neukirch, *Kennzeichnung der p-adischen und der endlichen algebraischen Zahlkörper*, Invent. Math. **6** (1969), 296–314.

[N3] J. Neukirch, *Algebraic Number Theory*, Springer-Verlag, Berlin, 1999.

[NSW] J. Neukirch, A. Schmidt and K. Wingberg, *Cohomology of Number Fields*, Springer-Verlag, Berlin–Heidelberg, 2000.

[O1] A. Ostrowski, *Über sogennate perfekte Körper*, J. reine angew. Math. **147** (1917), 191–204.

[O2] A. Ostrowski, *Über einige Lösungen der Funktionalgleichung* $\varphi(x)\cdot\varphi(y) = \varphi(x\cdot y)$, Acta Math. **41** (1918), 271–284.

[O3] A. Ostrowski, *Untersuchungen zur arithmetischen Theorie der Körper*, Math. Z. **39** (1934), 269–404.

[P1] F. Pop, *Galoissche Kennzeichnung p-adisch abgeschlossener Körper*, J. reine angew. Math. **392** (1988), 145–175.

[P2] F. Pop, *On Grothendieck's conjecture of birational anabelian geometry*, Ann. Math. **139** (1994), 145–182.

[P3] F. Pop, *Glimpses of Grothendieck's anabelian geometry*, In: Geometric Galois Actions: 1. Around Grothendieck's esquisse d'un programme, L. Schneps et al. (eds.), Lect. Note Ser. **242**, Cambridge University Press, Lond. Math. Soc., 1997, pp. 113–126.

[Pr] A. Prestel, *Lectures on Formally real Fields*, Lect. Notes Math. **1093**, Springer-Verlag, Berlin, 1984.

[Ra1] F.J. Rayner, *Relatively complete fields*, Proc. Edinburgh Math. Soc. **11** (1958), 131–133.

[Ra2] F.J. Rayner, *An algebraically closed field*, Glasgow Math. J. **9** (1968), 146–151.

[Ri1] P. Ribenboim, *Théorie des Valuations*, Les Presses de l'Université de Montréal, Montréal, 1968.

[Ri2] P. Ribenboim, *Equivalent forms of Hensel's lemma*, Expo. Math. **3** (1985), 3–24.

[Ri3] P. Ribenboim, *The Theory of Classical Valuations*, Springer-Verlag, New York, 1999.

[Ris] L. Ribes, *Introduction to profinite Groups and Galois Cohomology*, Queen's Pap. Pure Appl. Math. **24**, Kingston, 1970.

[Ro] P. Roquette, *History of valuation theory I*, In: Valuation theory and its applications, Vol. I (Saskatoon 1999), F.-V. Kuhlmann et al. (eds), Fields Inst. Commun. **32**, Amer. Math. Soc., Providence, RI, 2002, pp. 291–355.

[Ry] K. Rychlik, *Zur Bewertungstheorie der algebraischen Körper*, J. reine angew. Math. **153** (1923), 94–107.

[Sa] I.R. Šafarevič, *On p-extensions*, Amer. Math. Soc. Transl. Ser. 2 **4** (1956), 59–72.

[Sch1] W. Scharlau, *Über die Brauer-Gruppe eines Hensel-Körpers*, Abh. Math. Sem. Univ. Hamburg **33** (1969), 243–249.

[Sch2] W. Scharlau, *Quadratic and Hermitian Forms*, Springer-Verlag, Berlin, 1985.

[Schi] O.F.G. Schilling, *The Theory of Valuations*, Math. Surveys **4**, Amer. Math. Soc., New York, 1950.

[Schm] F.K. Schmidt, *Mehrfach perfekte Körper*, Math. Ann. **108** (1933), 1–25.

[Se1] J.-P. Serre, *Extensions de corps ordonnés*, C.R. Acad. Sci. Paris **229** (1949), 576–577.

[Se2] J.-P. Serre, *Local Fields*, Springer-Verlag, Berlin, 1979.

[Se3] J.-P. Serre, *Galois Cohomology*, Springer-Verlag, Berlin 2002.

[Se4] J.-P. Serre, *Topics in Galois Theory*, Jones and Barlett, Boston, 1992.

[Sp] M. Spiess, *An arithmetic proof of Pop's Theorem concerning Galois groups of function fields over number fields*, J. reine angew. Math. **478** (1996), 107–126.

[St] A.L. Stone, *Nonstandard analysis in topological algebra*, In: Applications of Model-Theory to Algebra, Analysis and Probability Theory (Pasadena, Calif., 1967), W.A.J. Luxemburg (ed.), Holt, Rinehart and Winston, New York, 1969, pp. 285–299.

[Ste] D. Ştefănescu, *A method to obtain algebraic elements over $K((t))$ in positive characteristic*, Bull. Math. Soc. Sci. Math. R.S. Roumanie (N.S.) **26 (74)** (1982), 77–91.

[Sz] T. Szamuely, *Groupes de Galois de corps de type fini (d'aprés Pop)*, Sém. Bourbaki 2002/3, Astérisque **294** (2004), 403–431.

[Szy] K. Szymiczek, *Quadratic forms over fields*, Dissertationes Math. (Rosprawy Mat.) **152** (1977).

[T] O. Teichmüller, *Über die Struktur diskret bewerteter perfekte Körper*, Math. Nachr. Ges. Wissensch. Göttingen, Kl. I (1936), 151–161.

[U1] K. Uchida, *Isomorphisms of Galois groups of algebraic function fields*, Ann. Math. **106** (1977), 589–598.

[U2] K. Uchida, *Isomorphisms of Galois groups of solvably closed Galois extensions*, Tohôku Math. J. **31** (1979), 359–362.

[U3] K. Uchida, *Homomorphisms of Galois groups of solvably closed Galois extensions*, J. Math. Soc. Japan **33** (1981), 595–604.

[V1] V. Voevodsky, *Motivic cohomology with $\mathbb{Z}/2$-coefficients*, Publ. Math. IHES **98** (2003), 59–104.

[V2] V. Voevodsky, *On motivic cohomology with $\mathbb{Z}/l$-coefficients*, a preprint.

[Wad] A.R. Wadsworth, *p-Henselian fields: K-theory, Galois cohomology, and graded Witt rings*, Pac. J. Math. **105** (1983), 473–496.

[War1] R. Ware, *When are Witt rings groups rings? II*, Pac. J. Math. **76** (1978), 541–564.

[War2] R. Ware, *Valuation rings and rigid elements in fields*, Canad. J. Math. **33** (1981), 1338–1355.

[War3] R. Ware, *Galois groups of maximal p-extensions*, Trans. Amer. Math. Soc. **333** (1992), 721–728.

[We] H. Weber, *Zu einem Problem von H.J. Kowalsky*, Abh. Braunschweig. Wiss. Ges. **29** (1978), 127–134.

[Win] K. Wingberg, *Der Eindeutigkeitssatz für Demuškinformationen*, Invent. Math. **70** (1982-83), 99–113.

[Wit] E. Witt, *Zyklische Körper und Algebren der Charakteristik p vom Grad p^n. Struktur diskret bewerteter Körper mit vollkommenem Restklassenkörper der Charakteristik p*, J. reine angew. Math. **176** (1937), 126–140.

[ZS] O. Zariski and P. Samuel, *Commutative Algebra II*, Springer-Verlag, New York–Heidelberg, 1975.

Glossary of Notation

$\mathbb{N}$	the non-negative integers
$\mathbb{Z}$	the rational integers
$\mathbb{Q}$	the rational numbers
$\mathbb{R}$	the real numbers
$\mathbb{C}$	the complex numbers
$\mathbb{F}_q$	the field of q elements
$\mathbb{Z}_p$	the p-adic integers
$\mathbb{Q}_p$	the field of p-adic numbers
$\hat{\mathbb{Z}}$	the p-adic completion of $\mathbb{Z}$
$\lvert\cdot\rvert_\infty$	Archimedian absolute value on $\mathbb{Q}$, ix
$\lvert\cdot\rvert_p$	p-adic absolute value on $\mathbb{Q}$, ix
$F^\times$	multiplicative group of F, ix
$\mathrm{Im}(f)$	image of the map f, xiii
$\mathrm{Ker}(f)$	kernel of the homomorphism f, xiii
$\mathrm{Coker}(f)$	cokernel of the homomorphism f, xiii
${}_nA$	n-torsion subgroup of an abelian group A, xiii
$R^\times$	group of invertible elements in a ring R, xiii
E^G	fixed field of E under the automorphism group G, xiii
$\mathrm{tr.deg}(E/F)$	relative transcendence degree of an extension E/F, xiii
Γ_{tor}	torsion subgroup of Γ, 5
Γ_p	p-primary subgroup of Γ, 5
Γ_{div}	divisible hull of Γ, 7
ι_Γ	canonical map $\Gamma \to \Gamma_{\mathrm{div}}$, 8
$\frac{1}{n}\Gamma$	group of all α in Γ_{div} with $n\alpha \in \Gamma$, 10
$\frac{1}{l^\infty}\Gamma$	l-divisible hull of Γ, 10
$\mathrm{rr}(\Gamma)$	rational rank of Γ, 11
$\chi_{\bar{\mu}}(\Gamma)$	group of characters of Γ into $\bar{\mu}$, 12
$\leq_{\mathrm{div}}$	extension of $\leq$ to the divisible hull, 16
$\mathrm{rank}(\Gamma)$	rank of Γ, 17
Γ_Σ	Σ-core of Γ, 19
Γ_{inf}	group of infinitesimal elements in Γ, 20
$\leq_{\mathbb{R}}$	standard order on $\mathbb{R}$, 20

K^Γ	set of maps $f: \Gamma \to K$, 27
$\mathrm{Supp}(f)$	support of the map f, 27
Γ_f	cut of $f \in K^\Gamma$, 27
$f \preceq f'$	$f, f' \in K^\Gamma$ coincide on Γ_f, 27
$K((\Gamma))$	formal power series field, 28
$K[[\Gamma]]$	series in $K((\Gamma))$ with non-negative support, 29
$K_{\mathrm{Puis}}((\Gamma))$	generalized Puiseux series field, 30
$K(\Gamma)$	generalized rational function field, 30
$\mathrm{rank}(v)$	rank of a valuation v, 39
v_O	valuation associated with a valuation ring O, 39
O_v	valuation ring of v, 39
$\bar{F}_v$	residue field of (F, v), 40
G_v	principal unit group of v, 40
$\mathfrak{m}_v$	valuation ideal of v, 40
$O_v^\times$	group of v-units, 40
π_v	place associated with v, 40
$\bar{S}_v$	push-down of S under v, 41
v_π	valuation corresponding to a prime π in a UFD, 47
v_∞	degree valuation on $K(t)$, 48
$v_{\pi,\mathrm{cont}}$	π-content valuation, 53
$\mathrm{cont}(f)$	content of a polynomial f, 53
$v^\#$	mirror valuation of v, 53
$u \leq v$	u is coarser than v, v finer than u, 56
v/u	quotient valuation, 56
$<_P$	strict ordering relation corresponding to an ordering P, 63
$\leq_P$	coarse ordering relation corresponding to an ordering P, 63
$(a, b)_P$	open interval relative to P, 63
$(a, \infty)_P$	infinite open interval relative to P, 63
$(-\infty, b)_P$	infinite open interval relative to P, 63
$[a, b]_P$	closed interval relative to P, 64
$\lvert \cdot \rvert_P$	absolute value corresponding to P, 64
ΣS	non-zero sums of elements of S, 64
ΣF^2	non-zero sums of squares, 65
G_λ	69
$\lambda_1 \leq \lambda_2$	coarsening relation for localities, 69
$O_R(P)$	valuation ring associated with P, 71
$\mathcal{T}_\lambda$	topology of locality λ, 75
$\mathcal{T}^0$	$\mathcal{T}$-neighborhoods of 0, 79
d_P	metric induced by an ordering P, 81
$d_{v,c}$	metric induced by a valuation v, 81

d_v	metric induced by a valuation v, 82
$\mathrm{Val}(S)$	set of all S-compatible valuations, 95
$\mathrm{Val}(S,H)$	subset of $\mathrm{Val}(S)$, 95
$v_{(1)}(S,H)$	supremum of $\mathrm{Val}(S,H)$, 96
$v_{(2)}(S,H)$	infimum of $\mathrm{Val}(S)\setminus\mathrm{Val}(S,H)$, 96
v_S	S-core of v, 99
$v^*(S,H)$	99
$A(S)$	100
$O^-(S,H)$	$F\setminus H$-part of $O(S,H)$, 100
$O^+(S,H)$	H-part of $O(S,H)$, 100
$O(S,H)$	explicit construction of a valuation ring, 100
H_S	105
B^{p^∞}	107
ρ_v	Teichmüller character, 108
$\sum_{i=0}^\infty \rho(\bar{a}_i)\lambda^i$	λ-adic expansion, 110
$A[X_i^{p^{-n}} \mid i\in I, n\in\mathbb{N}]$	112
f_i^*	generic coefficients in p-perfect structures, 112–113
$W(\bar{A})$	115
$W(\bar{F})$	Witt vector ring over $\bar{F}$, 116
W_α	118
$(G:H)$	(supernatural) index of a subgroup H of a profinite group G, 125
$\lvert G\rvert$	(supernatural) order of a profinite group G, 125
$[E:F]$	(supernatural) degree of an algebraic field extension E/F, 125
$\mathrm{Aut}(E/F)$	automorphism group of a normal extension E/F, 126
$\mathrm{Gal}(E/F)$	Galois group of a Galois extension E/F, 126
$e(u/v)$	ramification index, 128
$f(u/v)$	inertia degree, 128
$N_{E/F}$	the norm map, 131
$Z(u/v)$	decomposition group, 133
E_Z	decomposition field, 133
u_Z	induced valuation on decomposition field, 133
$\bar{f}(X)$	residue polynomial, 136
F_{sep}	separable closure of F, 139
F_{sol}	solvable closure of F, 139
$F(p)$	maximal pro-p Galois extension of F, 139
$T(u/v)$	inertia group, 141
E_T	inertia field, 142
u_T	valuation induced on inertia field, 142
$\chi(u/v)$	group of characters of valued field extension, 144
$V(u/v)$	ramification group, 145

E_V	ramification field, 145
u_V	valuation induced on ramification field, 145
$\mathrm{Tr}_{E/F}$	trace map, 145
$H\backslash G/U$	set of double cosets of G, 151
$d(u/v)$	defect, 154
$\wp(X)$	Artin–Schreier polynomial $X^p - X$, 165
$\mathbb{Q}_{p,\mathrm{alg}}$	field of algebraic p-adic numbers, 167
$\mathbb{C}_p$	completion of algebraic closure of $\mathbb{Q}_p$, 173
sgn_P	sign map associated with P, 182
$V_P(a_0,\dots,a_m)$	number of sign changes in $a_0,\dots,a_m$ relative to P, 182
$W_P(c)$	182
G_F	absolute Galois group of F, 201
Frob_q	Frobenius automorphism, 201
$G_F(p)$	maximal pro-p Galois group of F, 202
$\chi_{\bar{F},p}$	pro-p cyclotomic character of $\bar{F}$, 202
$\mathrm{Tens}(\Gamma)$	tensor algebra over Γ, 209
κ	tensor algebra over $\{\pm 1\}$, 209
ϵ	unique non-zero element of κ_1, 209
ϵ_A	image of ϵ in a κ-structure A, 209
$\mathbf{0}$	trivial κ-structure, 209
$\prod_{i\in I} A_i$	direct product of κ-structures, 210
$\otimes_\kappa$	tensor product of κ-structures, 210
$A[\Gamma]$	extension of A by Γ, 211
$\bigwedge_*(\Gamma)$	alternating algebra over Γ, 213
Bock_A	Bockstein operator of A, 214
$\mathrm{St}_{F,r}(S)$	group of Steinberg elements in $(F^\times/S)^{\otimes r}$, 217
$K_r^M(F)/S$	Milnor K-group of F modulo S of degree r, 217
$K_*^M(F)/S$	Milnor K-ring of F modulo S, 217
$K_r^M(F)$	Milnor K-group of F of degree r, 218
$K_*^M(F)$	Milnor K-ring of F, 218
$\{a_1,\dots,a_r\}_S$	symbol in $K_r^M(F)/S$, 218
$\{a_1,\dots,a_r\}$	symbol in $K_r^M(F)$, 219
Res	restriction morphism, 219
$\cup$	cup product, 221
$\hat{F}^{(n,p)}$	descending sequence of a pro-p group $\hat{F}$, 222
trg	transgression map, 222
$\mathrm{Bock}_{F,S}$	Bockstein operator in $K_*^M(F)/S$, 225
T_M	236
N_S	subgroup generated by the non-p-rigid elements, 237

$\mathrm{Cl}_{\mathcal{T}}$	$\mathcal{T}$-closure, 256
$\mathcal{A}$	set of κ-structures realizable by preorderings of finite index, 265
$\nu(A)$	number of morphisms $A \to \kappa$, 267
$\mathcal{E}_q$	set of q-elementary type κ-structures, 267

Index

Titles in This Series

124 **Ido Efrat,** Valuations, orderings, and Milnor K-theory, 2006

123 **Barbara Fantechi, Lothar Göttsche, Luc Illusie, Steven L. Kleiman, Nitin Nitsure, and Angelo Vistoli,** Fundamental algebraic geometry: Grothendieck's FGA explained, 2005

122 **Antonio Giambruno and Mikhail Zaicev, Editors,** Polynomial identities and asymptotic methods, 2005

121 **Anton Zettl,** Sturm-Liouville theory, 2005

120 **Barry Simon,** Trace ideals and their applications, 2005

119 **Tian Ma and Shouhong Wang,** Geometric theory of incompressible flows with applications to fluid dynamics, 2005

118 **Alexandru Buium,** Arithmetic differential equations, 2005

117 **Volodymyr Nekrashevych,** Self-similar groups, 2005

116 **Alexander Koldobsky,** Fourier analysis in convex geometry, 2005

115 **Carlos Julio Moreno,** Advanced analytic number theory: L-functions, 2005

114 **Gregory F. Lawler,** Conformally invariant processes in the plane, 2005

113 **William G. Dwyer, Philip S. Hirschhorn, Daniel M. Kan, and Jeffrey H. Smith,** Homotopy limit functors on model categories and homotopical categories, 2004

112 **Michael Aschbacher and Stephen D. Smith,** The classification of quasithin groups II. Main theorems: The classification of simple QTKE-groups, 2004

111 **Michael Aschbacher and Stephen D. Smith,** The classification of quasithin groups I. Structure of strongly quasithin K-groups, 2004

110 **Bennett Chow and Dan Knopf,** The Ricci flow: An introduction, 2004

109 **Goro Shimura,** Arithmetic and analytic theories of quadratic forms and Clifford groups, 2004

108 **Michael Farber,** Topology of closed one-forms, 2004

107 **Jens Carsten Jantzen,** Representations of algebraic groups, 2003

106 **Hiroyuki Yoshida,** Absolute CM-periods, 2003

105 **Charalambos D. Aliprantis and Owen Burkinshaw,** Locally solid Riesz spaces with applications to economics, second edition, 2003

104 **Graham Everest, Alf van der Poorten, Igor Shparlinski, and Thomas Ward,** Recurrence sequences, 2003

103 **Octav Cornea, Gregory Lupton, John Oprea, and Daniel Tanré,** Lusternik-Schnirelmann category, 2003

102 **Linda Rass and John Radcliffe,** Spatial deterministic epidemics, 2003

101 **Eli Glasner,** Ergodic theory via joinings, 2003

100 **Peter Duren and Alexander Schuster,** Bergman spaces, 2004

99 **Philip S. Hirschhorn,** Model categories and their localizations, 2003

98 **Victor Guillemin, Viktor Ginzburg, and Yael Karshon,** Moment maps, cobordisms, and Hamiltonian group actions, 2002

97 **V. A. Vassiliev,** Applied Picard-Lefschetz theory, 2002

96 **Martin Markl, Steve Shnider, and Jim Stasheff,** Operads in algebra, topology and physics, 2002

95 **Seiichi Kamada,** Braid and knot theory in dimension four, 2002

94 **Mara D. Neusel and Larry Smith,** Invariant theory of finite groups, 2002

93 **Nikolai K. Nikolski,** Operators, functions, and systems: An easy reading. Volume 2: Model operators and systems, 2002

92 **Nikolai K. Nikolski,** Operators, functions, and systems: An easy reading. Volume 1: Hardy, Hankel, and Toeplitz, 2002

91 **Richard Montgomery,** A tour of subriemannian geometries, their geodesics and applications, 2002
90 **Christian Gérard and Izabella Łaba,** Multiparticle quantum scattering in constant magnetic fields, 2002
89 **Michel Ledoux,** The concentration of measure phenomenon, 2001
88 **Edward Frenkel and David Ben-Zvi,** Vertex algebras and algebraic curves, second edition, 2004
87 **Bruno Poizat,** Stable groups, 2001
86 **Stanley N. Burris,** Number theoretic density and logical limit laws, 2001
85 **V. A. Kozlov, V. G. Maz'ya, and J. Rossmann,** Spectral problems associated with corner singularities of solutions to elliptic equations, 2001
84 **László Fuchs and Luigi Salce,** Modules over non-Noetherian domains, 2001
83 **Sigurdur Helgason,** Groups and geometric analysis: Integral geometry, invariant differential operators, and spherical functions, 2000
82 **Goro Shimura,** Arithmeticity in the theory of automorphic forms, 2000
81 **Michael E. Taylor,** Tools for PDE: Pseudodifferential operators, paradifferential operators, and layer potentials, 2000
80 **Lindsay N. Childs,** Taming wild extensions: Hopf algebras and local Galois module theory, 2000
79 **Joseph A. Cima and William T. Ross,** The backward shift on the Hardy space, 2000
78 **Boris A. Kupershmidt,** KP or mKP: Noncommutative mathematics of Lagrangian, Hamiltonian, and integrable systems, 2000
77 **Fumio Hiai and Dénes Petz,** The semicircle law, free random variables and entropy, 2000
76 **Frederick P. Gardiner and Nikola Lakic,** Quasiconformal Teichmüller theory, 2000
75 **Greg Hjorth,** Classification and orbit equivalence relations, 2000
74 **Daniel W. Stroock,** An introduction to the analysis of paths on a Riemannian manifold, 2000
73 **John Locker,** Spectral theory of non-self-adjoint two-point differential operators, 2000
72 **Gerald Teschl,** Jacobi operators and completely integrable nonlinear lattices, 1999
71 **Lajos Pukánszky,** Characters of connected Lie groups, 1999
70 **Carmen Chicone and Yuri Latushkin,** Evolution semigroups in dynamical systems and differential equations, 1999
69 **C. T. C. Wall (A. A. Ranicki, Editor),** Surgery on compact manifolds, second edition, 1999
68 **David A. Cox and Sheldon Katz,** Mirror symmetry and algebraic geometry, 1999
67 **A. Borel and N. Wallach,** Continuous cohomology, discrete subgroups, and representations of reductive groups, second edition, 2000
66 **Yu. Ilyashenko and Weigu Li,** Nonlocal bifurcations, 1999
65 **Carl Faith,** Rings and things and a fine array of twentieth century associative algebra, 1999
64 **Rene A. Carmona and Boris Rozovskii, Editors,** Stochastic partial differential equations: Six perspectives, 1999
63 **Mark Hovey,** Model categories, 1999
62 **Vladimir I. Bogachev,** Gaussian measures, 1998

For a complete list of titles in this series, visit the AMS Bookstore at **www.ams.org/bookstore/**.